Nanotechnology-Enhanced Food Packaging

Nanotechnology-Enhanced Food Packaging

Edited by

Jyotishkumar Parameswaranpillai
Radhakrishnan Edayileveettil Krishnankutty
Aswathy Jayakumar
Sanjay Mavinkere Rangappa
Suchart Siengchin

WILEY-VCH

Editors

Dr. Jyotishkumar Parameswaranpillai
School of Biosciences, Mar Athanasios
College for Advanced Studies
Tiruvalla (MACFAST)
Pathanamthitta
689101 Kerala
India

Dr. Radhakrishnan Edayileveettil Krishnankutty
Mahatma Gandhi University
School of Biosciences
P.D. Hills
686560 Kottayam, Kerala
India

Dr. Aswathy Jayakumar
King Mongkut's University of Technology
North Bangkok
Innovation in Design and Engineering
1518 Pracharaj 1
Wongsawang Road, Bangsue
10800 Bangkok
Thailand

Dr. Sanjay Mavinkere Rangappa
King Mongkut's University of Technology
North Bangkok
Department of Materials & Production Engineering
1518 Pracharat 1
Wongsawang Road, Bangsue
10800 Bangkok
Thailand

Prof. Dr. -Ing. habil. Suchart Siengchin
King Mongkut's University of Technology
North Bangkok
Department of Materials & Production Engineering
1518 Pracharat 1
Wongsawang Road, Bangsue
10800 Bangkok
Thailand

Cover Image: © Liyao Xie/Getty Images

Library of Congress Card No.: applied for

British Library Cataloguing-in-Publication Data
A catalogue record for this book is available from the British Library.

Bibliographic information published by the Deutsche Nationalbibliothek
The Deutsche Nationalbibliothek lists this publication in the Deutsche Nationalbibliografie; detailed bibliographic data are available on the Internet at <http://dnb.d-nb.de>.

Print ISBN: 978-3-527-34773-5
ePDF ISBN: 978-3-527-82770-1
ePub ISBN: 978-3-527-82772-5
oBook ISBN: 978-3-527-82771-8

Typesetting Straive, Chennai, India
Printing and Binding CPI Group (UK) Ltd, Croydon, CR0 4YY

Printed on acid-free paper

C106082_081121

Contents

Preface

The introduction of nanotechnology in food packaging has led to enormous growth in the food packaging industry. Studies have shown that nanotechnology in food packaging can improve the quality and shelf life of food products. The incorporation of organic and inorganic nanomaterials in traditional packaging material has gained much interest due to the enhanced thermomechanical, barrier, and antimicrobial properties. The active packaging and intelligent/smart packaging system offers enhanced features that can extend the quality and shelf life of food products. The active system encompasses the incorporation of antimicrobial agents, preservatives, and gas absorbers, whereas the intelligent/smart system uses chemical sensors, time–temperature indicators, gas sensors, and freshness indicators. The concerns regarding the toxicity, risk, and migration aspects of nanoparticles are also rising; however, these assessments are not still conclusive.

Recently, there is a tremendous increase in the number of research articles and patents in the field of nanocomposites. However, only a few books have been published in the area of nanotechnology-enhanced food packaging. Due to this, we believe it is befitting to judiciously edit a book on *Nanotechnology-Enhanced Food Packaging*. This book is a collection of chapters with the recent developments in nanotechnological innovations in food packaging. The chapters provide a comprehensive review on types of nanoparticles, their processing, characterization, development of nanoparticle-based packaging systems, and applications. In addition to this, the book discusses the use of nanoparticles in the development of active, smart, and intelligent packaging. We are confident that the present book will benefit researchers working in both academia and industry.

The book includes 15 chapters that cover the recent advantages of *Nanotechnology-Enhanced Food Packaging*. Chapter 1, "Introduction to Nanotechnology-Enhanced Food Packaging Industry," discusses an overview of nanotechnology in food packaging. Chapter 2, "An Overview of Biopolymers in Food Packaging Systems," is designed to give a deep insight into the various biopolymers used in food packaging. Chapter 3, "Nanostructures Based on Starch, Their Preparation, Processing, and Application in Packaging," deals with starch-based nanostructures, synthesis, and their potential application in food packaging. Chapter 4, "Cellulose Nanostructures and Its Application as Effective Food Packaging Systems," gives an overview of cellulose, its source, structure, properties, synthesis, and application as food packaging materials. Chapter 5, "Chitosan-Based Nanoparticles and Their Applications in Food Industry," summarizes an update about chitosan-based nanoparticles and their application in the food industry and packaging. Chapter 6, "Nutrients-Based Nanocarriers and Its Application in Packaging Systems," summarizes the recent developments in nanocarriers and their application in active food packaging. Chapter 7, "Active Packaging Systems Based on Metal and Metal Oxide Nanoparticles," emphasizes the recent progress in the application of metal and metal oxide nanoparticles in packaging application. Chapter 8, "Fabrication of Intelligent Packaging Systems Using Nano-Indicators and Sensors," discusses various nanomaterials-based sensors and their application in intelligent packaging. Chapter 9, "Nanostructure-Based Edible Coatings as a Function of Food Preservation," discusses the recent developments in nanotechnology-enabled edible coatings. Chapter 10, "An Overview of Higher Barrier Packaging Using Nanoadditives," highlights the recent developments in higher barrier packaging films with a special focus on gas and moisture barrier properties. Chapter 11, "Nanostructure-Based Multilayer Food Packaging Films," gives an overview of the various multilayer packaging and their applications in food protection and preservation. Chapter 12, "Characterization Techniques for Nanostructures in Food Packaging," discusses the various methods used for the development of nanocomposites and techniques used for the characterization of nanostructures in food packaging. Chapter 13, "Biodegradability Assessment of Biopolymer-Based Films," gives an overview of the biodegradation of biopolymers. Chapter 14, "Nanobiotechnology in Food Preservation and Molecular Perspective," gives an overview of the use of nanomaterials on food preservation, postharvest storage, and regulation and the safety aspects. Chapter 15, "Environmental and Toxicological Aspects of Nanostructures in Food Packaging," discusses the various nanoparticles in food packaging, toxicity measurements, migration issues, and environmental impact of nanoparticles.

Thanks to the authors for their contribution.

13 April 2021

Dr. Jyotishkumar Parameswaranpillai (Thailand)
Dr. Radhakrishnan Edayileveettil Krishnankutty (India)
Dr. Aswathy Jayakumar (India)
Dr. Sanjay Mavinkere Rangappa (Thailand)
Prof. Dr. -Ing. habil. Suchart Siengchin (Thailand)

1

Introduction to Nanotechnology-Enhanced Food Packaging Industry

Kunal Singha[1], Baburaj Regubalan[2], Pintu Pandit[1], Subhankar Maity[3], and Shakeel Ahmed[4]

[1] *National Institute of Fashion Technology, Department of Textile Design, Mithapur Farms, Patna 800001, India*
[2] *Kalasalingam Academy of Research and Education, Department of Food Processing Technology, Krishnankoil, Tamilnadu 626128, India*
[3] *Uttar Pradesh Textile Technology Institute, Department of Textile Technology, Kanpur, Uttar Pradesh 208001, India*
[4] *Government Degree College Mendhar, Department of Chemistry, Mendhar, Jammu and Kashmir 185211, India*

1.1 Introduction

Today creating biodegradable and natural materials based on biodegradable food packaging materials is a major global challenge for the environment. However, the use of bio-based packaging products such as food grade or biodegradable films from recycled sources could address the problem of waste in at least some way. The correct choice of products and packaging technology thus enables consistency and freshness of products to be maintained over the time required to be promoted and used. Even though, the availability of bio-based food packaging is limited in the market due to its low gas barrier and mechanical properties. As a result, these natural polymers were frequently mixed or chemically modified with other synthetic polymers to expand their packaging applications. Bio-based packaging has many essential features, including traditional packaging, such as the preservation and securing of products, ensuring nutritional integrity and health, and providing awareness to the consumers. Another nanotechnology that may help to minimize waste from the processing of packaged food is the use of nanocomposites in the processing. The use of nanocomposites that seek to facilitate the use of biologically degradable films protects fresh food and enhances the durability of it.

Nanotechnology involves the manufacturing, manipulating, and characterizing of nanosized objects, particles, and materials with a dimension of approximately 1–100 nm. Although the size of the material is reduced to the nanoscale range, its physical and chemical properties are magnified greatly from those of the macroscale structures made of the same material. Consequently, the nanoscale (1–100 nm) systems may have some implications, but successful implementations of the same for

Nanotechnology-Enhanced Food Packaging, First Edition. Edited by Jyotishkumar Parameswaranpillai, Radhakrishnan Edayileveettil Krishnankutty, Aswathy Jayakumar, Sanjay Mavinkere Rangappa, and Suchart Siengchin.

serving mankind are invaluable. Nanotechnology provides a range of significant improvements to enhance health, stability, and quality of life and to create assertive impacts on the environment [1, 2].

The packaging systems are protective shields that secure, manage, transports, store, retains, and marks any entity in the supply chain from raw materials to end users. These functions are required to accurately define any type of packaging, and how a variety of requirements, such as mechanical, thermal, and barrier characteristics, are preserved depending on the type of products to be packaged. Nanomaterials are gradually being used in the food packaging industry; therefore a variety of advanced nanomaterial technology is being researched and developed for packaging materials. There have been studies of approximately 500 nano-packaging materials for industrial use, while nanotechnology is expected to manufacture 25% of all food packings. Nano-packaging can also conceive antimicrobials, minerals, enzymes, flavors, and nutraceuticals to enhance shelf life and performance. In this line, antimicrobial films are used as packaging material to improve the shelf life of perishable foods like fruits and dairy products [3–6].

Nanotechnology's potential role in the food technology sector is probably the most exciting in the immediate future, and it is emerging as one of the fastest developing areas of nano-research of agriculture and food. New developments in food processing, labeling, nutraceutical delivery, quality assurance, and healthy food have also been seen. Many organizations, scientists, inventors, and industries are developing new technologies, protocols, and products, which directly apply nanotechnology to farming and food products. Companies are now designing packaging materials that prolong food and beverage life and boost consumer safety through the use of nanotechnology in daily-based consumer packaging [6]. Food processing and control are the main subjects of food industry-related nanotechnology research and development. Effective and intelligent packaging is the leading advancement of food packaging that aims to enhance product quality and consistency as well as to maximize product longevity. Most businesses and industrial design nano-packaging like time–temperature indicators (TTIs) can react to undesirable or damaging changes under the worse effects of climatic pollution. They can self-repair themselves, thus making this nano-packaging as "active and smart packaging."

Nanotechnology enhances the delivery of nutraceuticals, vitamins, or fragile micronutrients to everyday foods by creating small, edible capsules based on released nanoparticles to targeted locations in the body. Relevant health consequences are reduced frequency of cardiac disease, stroke, neurodegenerative diseases, and cancer [7, 8]. Nanoparticles are also used to introduce multiple functionalities such as color and odors as well as to be used as storage tanks for drug releases or fungicides. Despite considerable development in this area, nanotechnology remains a rare topic for food packaging, nanotechnology, and food science and technology. This chapter explores this knowledge gap by closely analyzing current developments in nano-package technology for food and drug systems and particular applications that gain immediate customer adoption and regulatory attention. This article examines this knowledge gap on the topics covered, which include bio-based packaging for environmentally safe biodegradable packaging; improved packages to enhance barrier properties, mechanical durability, and flexibility; active packaging of antimicrobials, flavor absorbers, and oxygen scavenging; and intelligent package features like freshness indicator, ripeness indicator, radio-frequency identification (RFID) , and TTI. This chapter concludes with a concise overview of future nano-packaging technologies possibilities.

1.2 Nanotechnology Applications in Food Processing

Nanostructured food ingredients are developed to facilitate sensory attributes like appearance, taste, texture, and flavor. Nanotechnology increases the durability of different foods and reduces food waste caused by microbial infestation. Nanocarriers are presently used as a supply system without interfering with their basic morphology to transport food additives into food products. The particle size can directly affect the delivery of bioactive compounds to different sites since some cell lines have noticed that it is efficient to absorb only submicron nanoparticles but not larger microparticles [9–13].

Nanotechnology provides effective distribution systems with all the functionality mentioned earlier for encapsulation formulation, emulsions, biopolymer matrices, clear solutions, and colloids. Nano-polymers are intended to replace traditional products for food packaging. Nanosensors may show the existence of pathogenic microbes, toxins, and adulterants in food [14]. Nanoparticles have greater characteristics of encapsulation and release performance than traditional embossing methods. By nanoencapsulation of the masks scent or taste, the interactions between active ingredients and the food network that govern the release of active agents can be monitored. This method guarantees the supply of desired food ingredients at the desired level of production, storage, and usage. This nano-packing method is consistent with other ingredients in the device against moisture, fire, chemicals, and biological degradation. Moreover, these nanotechnologies-based food nutrition delivery systems can reach deep into the tissues and effectively distribute active agents to the target sites in the body because of their smaller scale [15–18].

1.2.1 Nanotechnology Applications in Preserving Meat Density, Taste, and Presentation

Nanotechnology offers several options to improve meat quality and taste. Nano-encapsulation techniques have been widely applied to enhance flavor release and retention and maintain the balance of food. These bioactive molecules nanocarriers are popular for their safety and supply-based feature worldwide. Rutin is a popular dietary flavonoid, but its use is limited in the food industry [19]. Its low solubility and ferritin nanoencapsulation have improved the solubility and the thermal and UV stability of the ferritin-trapped routines compared with the free routine. Thus, nanoemulsions are widely used in producing lipid-soluble biological compounds that can be generated with easy processing methods utilizing natural foods and can also be engineered to increase water dispersion and bioavailability. Nano-packagings are important ways to boost the bioavailability of nutraceutical compounds because of their subcellular size, which contributes to higher bioavailability than large particles and produces faster and longer releases of encapsulating food nutrient compounds. Many metal oxides such as titanium dioxide and silicon (SiO_2) are widely used as colorants or flow agents in foodstuffs. SiO_2 nanomaterial is also one of the nanomaterials widely common in foodstuffs with flavors [20].

These nanotechnology-based applications of modern nanocarriers can help the food industry preserve the meat color and taste and make the food colorful, visually attractive, and presentable for the consumers.

1.2.2 Nanotechnology Applications for Maintaining the Food Nutrient Value

Many bioactive compounds like lipids, short-chain fatty acids, functional chelates, probiotics, and antioxidants are vulnerable to acids and enzymes in the stomach and duodenum. The encapsulation of these bioactive compounds allows them not just to avoid such adverse conditions but also to readily assimilate easily. In addition to that, small edible nanoparticles are developed for significant health gain to increase the regular supply of pharmaceutical items, probiotics, vitamins, and fragile micronutrients. The different strategies for encapsulating miniatures to include nutrients such as protein and antioxidants more useful for specific nutritional and health benefits include nanocomposite, nano-emulsification, and nanostructure. Polymeric nanoparticles are sufficient to secure and transmit bioactive compounds to specific bioactive compound encapsulation functions (flavonoids and vitamins) [21–23].

1.3 Nanotechnology Functions for Preserving or Shelf Life

In functional foods, the bioactive components are sensitive to external factors and eventually deteriorated during transport and storage; nanoencapsulation of these bioactive nanocomponents improves its shelf life. Especially, nanoemulsion-based edible coating controls the fruit ripening and extends the shelf life of perishable commodity. Also, consumable nanocoats could be a barrier to the moisture on various foods and could provide colors, flavors, nutrients, enzymes, and anti-brews [24–26].

1.4 Nanotechnology in Food Packaging

In food packaging, monolayer films cannot satisfy all the requirement because different food commodities require different barrier and mechanical properties. Polymer nanocomposites are the latest materials aimed at solving this problem. Polymer nanocomposites are prepared by dispersing an inert, nanoscale filler in a polymeric matrix. The widely used filler materials are silica (SiO_2), clay, cellulose-based nanofibers, graphene, silicate nanoplatelets, starch nanocrystals, carbon nanotubes, chitin or chitosan nanoparticles, etc. The nanocomposite can enhance barrier properties, flame resistance, better thermal properties, and alterations in surface wettability and hydrophobicity. European Food Safety Authority approved that the nano-TiN to use in package material can contact with food material. It is widely used in processing aid and to improve mechanical strength of polyester (PET).

Intelligent and successful food nano-based packaging offers many advantages, including improved mechanical strength packaging products, barrier properties, antimicrobial film for nano-sensing (Figure 1.1) pathogen identification, and food safety warning over traditional packaging techniques. Nanocomposites can also be used as active ingredients in packaging and coating material to improve food packaging. Several researchers were involved in investigating the antimicrobial effects of organic compounds in polymeric matrices, such as organic acids, essential oils, and nisin. However, these compounds do not comply with the many processing stages that require high temperatures and pressures because they are highly sensitive to these physical conditions. The use of inorganic nanoparticles allows for good antibacterial activity at low concentrations and increased stability under intense conditions [27]. The use of these nanoparticles in antimicrobial food packaging was therefore very important in recent years.

Nanocor® supplies specially designed plastic nanocomposites (nanoclays) owned by AMCOL International. Durethan® is used in food packaging and medical fields. It provides excellent gas and moisture barrier properties, strength, toughness, and

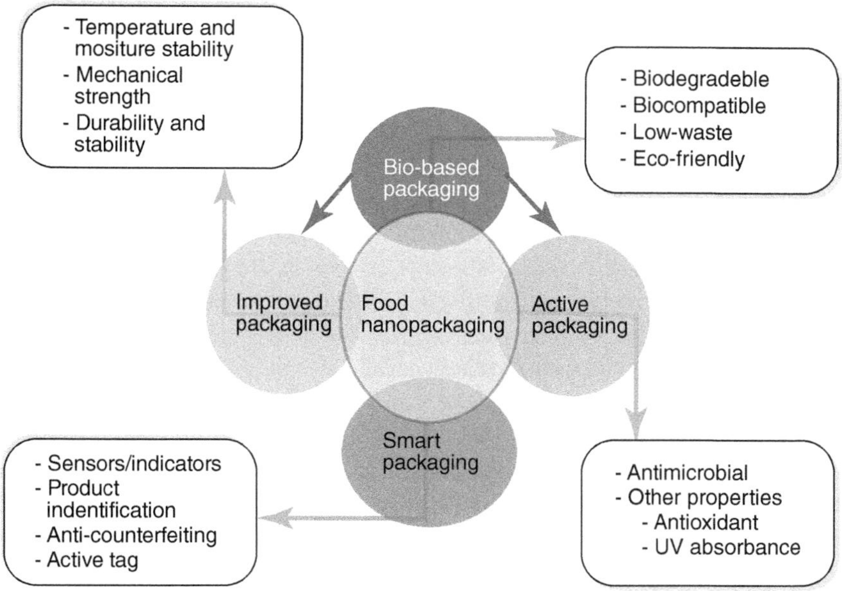

Figure 1.1 Features of food nano-packaging applied in the food industry [2].

chemical resistance. In South Korea, Hite Pitcher beer bottles were made out of Aegis™ OXCE (nylon 6 nanoclay composite) developed by Honeywell Polymer. It has high-oxygen barrier properties designed for alcoholic beverage and beer. A milk bottle and baby mug incorporated with silver nanoparticles have been developed by Baby Dream Co., Ltd., an infant product company in South Korea.

Antimicrobial packaging is not limited to antimicrobial products, but nano-compositions and nanolaminates are widely used in product packaging to resist intense mechanical and thermal shocks, which increase food shelf life. The incorporation of nanoparticles into packaging materials provides quality foods with longer durability. Moreover, to ensure the highest food-grading quality and standard, polymer composites are designed to supply both thermostable and usable packaging materials. Numerous inorganic or organic fillers are used to produce better polymer composites. The addition of nanoparticles in polymers has made it possible to develop robust, cost-effective packaging material.

1.4.1 Usages of Nanosensors in Pathogen and Adulterant Detection in the Food Industry

Nanomaterials for use in the development of biosensors include high responsiveness and other modern features. In dietary microbiology, nanosensors or nanobiosensors are used to detect pathogens in processing plants and foodstuffs, to measure accessible foodstuffs, and to alert customers and suppliers to food health. The nanosensor serves as an indicator of changes in environmental conditions, such as humidity or storage temperature, microbial contamination, or product degradation. To achieve potential biosensor applications, former researchers have studied specific nanostructures such as thin films, nanoparticles, nanorods, and nanofibers. These thin, film-based optical immunosensors have contributed to efficient and highly responsive detection systems for microbial or cell detection. These immunosensors are used to immobilize specific anticorps, antigens, or protein molecules on thin nanofilms or sensor chips that transmit signals for the detection of target molecules. Dimethyl siloxane combined with carbohydrate biosensors has been very carefully identified and used for microorganisms, contaminants, and other food and beverage-related items due to their quick identification, usability and cost-effectiveness. The contaminants connected to such nanotubes induce observable shifts of conductivity of waterborne contaminants in the identification of waterborne toxins. Therefore, the use of an electronic nose or tongue consisting of several nanosensors tracks food by communicating scents of foodstuffs or gas signals [28–30].

Adulteration is one of the key challenges faced in the food sector. Nanosensors have better sensitivity and accuracy than other sensors, for example, gold nanoparticles functionalized with cyanuric acid groups selectively bind to melamine, a common adulterant used to inflate the protein content in pet foods and infant formulas. Similarly, melamine adulteration in raw milk can also be detected up to 2.5 ppb using nanosensors.

1.4.2 Nanotechnology Applications in Food Safety Issues

In addition to all of the benefits for the food sector in nanotechnology, the protection in nanomaterials cannot be overlooked. Many researchers have tackled nanomaterial protection issues, with an emphasis on the potential transfer of nanoparticles from packaging to food and their impact on the health of customers. Nano-packaged food products must be acquired in more studies to determine the danger of its nanocomponents because its physicochemical properties in nanostats are completely different from those of macrostats. Furthermore, the small size of such nanomaterials will raise the likelihood of body and tissue bioaccumulation. Dissolution is caused by several influences including the composition of the soil, concentration, soil strength, aggregation, and adsorption of particles [3, 31, 32].

The value of the application of nanometer-scale structures in the food industry has also increased in recent years, and research efforts in this field have been strongly oriented. As nanobiotechnology advances, devices or materials dependent on this technology become less and more responsive. Its applicability in the fields of food packaging and food safety is well known. Promising findings were also obtained for food safety nanomaterials that can protect food against heat, contaminants, and harsh environmental conditions (Figure 1.2). They deliver excellent logistics systems for the delivery of bioactive compounds to the targeted body tissue sites. In the use of nanotechnology in the food system, consistency in the safety risks and environmental impacts should be the priority, and compulsory testing of the relevant nano-foodstuffs until they can be used in the market is necessary [2, 33].

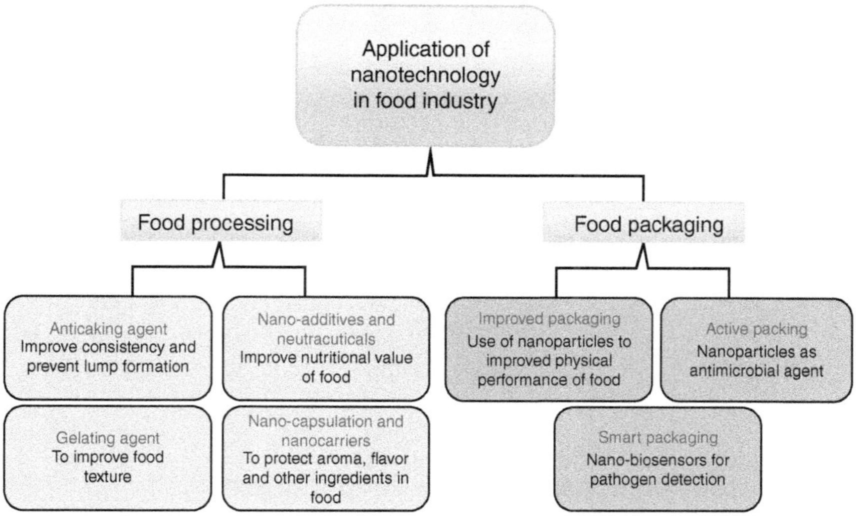

Figure 1.2 Diverse applications of nanotechnology in the food industry [3].

1.4.3 Bio-Based Nano-packaging in Food Industry

Bio-based nano-wrapping paper is a highly recyclable film used for food items for controlling moisture transfer and exchange of gas like CO_2 and O_2 for improved protection and for ensuring nutritional and sensory uniformity and reliability. Moreover, such type of materials in packaging are more socially friendly than conventional packaging films. As another type of packaging, biological packaging provides a barrier between consumer goods and their environment, thereby shielding them from harmful effects of microbes, relative humidity, and sunlight. The fundamental feature that separates biodegradable films from other packaging approaches is that the behavior of living things degrades these biodegradable sheets. This kind of body is the most common, as it is environmentally sustainable since all decomposition materials, e.g. carbon dioxide, biomass, and water, are fully reusable. Chemical packaging does not (or less) use fossil fuels to be used for consumer processing, and instead it uses green energy to recycle incineration power [34]. Therefore, nano-based bio-packaging sheets are very much popular these days due to their higher biodegradability than nonbiodegradable chemical sheets.

The promising design of bio-based nanocomposites for synthetic polymers drives new work on nanocomposites for use in food packaging. The classification of biopolymers is available in the literature. The use of nanotechnology on such polymers can offer new ways to improve both their properties and their cost-effectiveness. The most well-studied bionanocomposites are starch and derivatives such as polylactic acid (butylene succinate), polyhydroxybutyrate, and aliphatic polyester [35–38].

The three main applications of nanomaterial in food packaging were further described as the research trends in food packaging using nanotechnology, where nanoreinforcement, active nanocomposite packaging, and smart nanocomposites are mainly involved in food packaging. In active packaging, the well-known ingredients of antimicrobial packaging are silver, gold, and metal oxide nanoparticles. Silver nanoparticles found in several commercial applications are most widely studied. It may also be suitable for other active packaging fields, such as ethylene removers [39–43]. Nanosensors can respond to external changes in stimuli. The latest developments in smart food packaging polymer nanomaterials include indicators of oxygen, freshness, and pathogens [44, 45]. EcoSphere Biolatex® is a novel technology developed by the EcoSynthetix company. It is a biopolymer nanosphere with a granule size of 50–150 nm and a higher surface area than native starch. It possesses high solid dispersion in water. Native starch is replaced by EcoSphere Biolatex in adhesives. Instead of traditional adhesive, nanosphere starch is being used in McDonald's hamburger clamshell in the United States.

1.5 Nanocoating Applications in Food Industry

Food coating can be described as a thin film of an edible composite material to prevent mass transfer. Such coatings can serve as barriers to moisture and gases. Coatings are added directly and formed by applying a fluid film-forming solution or liquid compounds on the food component. Edible surface materials may be classified into two categories: soluble polysaccharides and lipids. Alginates, cellulose, pectin, starch, chitosan, and other polysaccharides are suitable polysaccharides. Many lipid compounds have been used for producing culinary films and clothes such as animal or vegetable fats, waxes, acylglycerols, and fatty acids, which can act as an appropriate lipid. Lipid films have outstanding moisture trapping properties or as binding agents for gloss applying to chocolate. Waxes are widely used to cover fruit and vegetables to slow ventilation and to prevent moisture loss [46].

Edible coatings are used in a broad variety of products from nuts, herbs, poultry, sweets, cheese, candies, bakery, and fried food. Few research work has documented nanoparticles being incorporated into coating films to enhance their physical properties by enhancing the release of oxygen; montmorillonite clay was applied to pectins. Similarly, gelatin and montmorillonite-derived nanocomposites have been used to greatly change physical properties. There was also a considerable increase in chitosan/layered nanocomposites stability. Nanoparticles can be used as antimicrobial and additive carriers. It can also be used to stabilize additives and control their diffusion effectively in food and the various regions, e.g. surface vs. bulk of the food system. This control may be useful for long-term food storage or for conveying certain desired characteristics like flavor to a food system. In this way, the United States has also produced an edible antibacterial nanocoating that can be applied directly to baking products, released by the Sono-Tec Corporation [46–49].

The three main applications of nanomaterials in food packaging were also described as research trends in food packaging using nanotechnology, where nanoreinforcement, active nanocomposite packaging, and intelligent nanocomposite packaging are the main issues for food packaging. The presence of nanoparticles in the polymer matrix products increases the properties of the commodity in better packaging. In addition to barrier properties, strength, rigidity, dimensional stability, and material heat tolerance may be strengthened by inserting nanoclays or SiO_2 nanoparticles. Nanoparticles are specially developed for applications of antimicrobial packaging in active packaging as active agents, silver, gold, and metal–oxygen nanoparticles with the antimicrobial function being the most studied nanoparticles with silver nanoparticles already present in several commercial applications. This is suitable for other fields of active packaging such as ethylene layer removers. Nanoparticles may be used as reactive particles in packaging materials for smart packaging to warn about the quality of the drug packed. To interact and to classify the drug, the so-called nanosensors can respond to external stimulation adjustments to ensure its consistency and health. Recent developments in polymer nanomaterials are oxygen indicators, freshness indicators, and pathogenic in smart food packaging.

1.6 Nanocoats Used in Food Manufacturing

Edible coatings are either applied to or formed directly on foods, whereas edible films are self-supporting structures used to wrap food products and also located in between two food components. It is a fact that such coatings may act as a barrier to heat, moisture, and gas. Coatings are added and formulated either by applying a liquid film-forming solution or directly on the food component with molten compounds. Edible coatings may be divided into two categories: polysaccharides and lipids and both of these categories are hydro-soluble. The best lipids are waxes, acylglycerols, and fatty acids, and among them, lipid films have excellent moisture capture properties and are used to color gloss on candy products as coloring agents. On the other side, waxes are commonly used to cover fruit and vegetables to delay breathing and rising moisture shortages [50].

Today, an edible coating is used for a large range of foodstuffs including nuts, potatoes, poultry, candy, dairy, cookies, pastry, and French fried products. Nonetheless, few research studies have identified nanoparticles in coating films to enhance their physical properties. Montmorillonite clay has been applied to pectins to reduce oxygen diffusion. Also, nanocomposites prepared for gelatin and montmorillonite have been used to improve physical properties. The efficiency of chitosan-layered nanocomposites was also greatly improved [51].

1.7 Importance of Nanolamine in Food Business

Nanolamines offer food scientists with various directions to manufacture modern food industry nanolaminate films. A nanoclay consists of two or more layers of physically or chemically connected nanometer-like content. One of the most efficient nanolamine methods is based on a layer deposition technique that covers loaded surfaces with interfacial films composed of several nanolayers of different materials. Nanolaminates have some benefits over conventional manufacturing methods for edible coatings and films and can also be used for several essential applications in the food and milk field.

Specific layers of adsorbing substances can be created, such as proteins, polysaccharides, lipids, and colloidal particles. Films are prepared with some active functional agents, such as antioxidants, antibrowning agents, enzymes, flavorings, and colors. These nanolaminated lacquerings could be entirely produced with the same nanocoating by simple processes such as dipping or washing from edible ingredients. The composition, thickness, structure, and properties of the laminate formed around the object depends on the dipping and coating process. These include changes in the adsorbed substances type in dipping solutions, the total number of dipping steps used, the order in which the item is introduced through the various dipping solutions, the solution, and the environmental conditions used [51].

1.8 Antimicrobial Films Used in Food Industry

There is currently substantial focus given to the use in packaging products of antimicrobial substances (such as silver nanoparticles and silver coatings). Antimicrobial films may help monitor the production and spoilage of pathogenic microorganisms.

According to the accepted structural stability and barrier characteristics, the nanomaterials and the antimicrobial characteristics of the antimicrobials impregnated in the image; it is highly beneficial to create an antimicrobial photo. This film allows nanomaterials to add more powerful copies of biological molecules.

A layer-by-layer incorporation of antimicrobial peptides such as nisin may also contribute to the creation of antimicrobial films. Nisin acts as a depolarizing agent in bacterial membranes and creates pores in lipid bilayers. Nanofilm multilayer peptides intercalated different peptides charged at neutral pH, which was much more stable than when peptide film only stabilized electrostatic interactions.

There have also been records of nanoscale chitosan antibacterial action. A potential antimicrobial pathway includes interactions between the positive and the negative chitosan cell membranes, raising the membrane permeability and eventually contributing to the breakdown and leakage of intracellular content. The ineffectiveness of both rough chitosan and engineered nanoparticles at pH levels above 6 is consistent with observation given the lack of protonated amino groups [52].

1.9 Nano-scavenging Oxygen Film Used in Food or Eating Substances

Oxygen (O_2) is responsible either directly or indirectly for the deterioration of many foods. For example, direct oxidation reactions lead to fruit browning and vegetable oils rancidity. Degradation of food by indirect action of O_2 includes aerobic microorganism food spoilage. The inclusion of O_2 scavengers in the food kit will also hold O_2 rates very small, which are beneficial for many purposes because they will increase the food's life.

Successful production of oxygen scavenger films was achieved by applying titanium nanoparticles (TiO_2) to different polymers, which are used to pack a wide range of oxygen-sensitive products. In particular, the emphasis was on the photocatalytic behavior of ultraviolet nanocrystalline titania. Since TiO_2 acts by a photocatalytic mechanism, the requirement for ultraviolet absorption (UVA) light is its major drawback [53].

1.10 UV-Proof Processing of Foods Using Nanometal Oxides

The film based on nanocrystalline titanium (TiO_2) is the commonly used material for UV absorption. During the exposure to sunlight, the effectiveness of TiO_2-coated film exposure to sunlight inactivates TiO_2 visible photo- catalytic absorption in the context of UV irradiation. Doping TiO_2 with silver has been reported to have greatly improved photocatalytic bacterial inactivation. The resulting combination was good antibacterial properties of nanoparticles $TiO_2/Ag+$ in a polyvinyl chloride (PVC) nanocomposite.

1.11 Nano-intelligent Food Labeling

In smart/natural, nanomaterials are used to monitor biochemical or microbial modifications in products, such as the identification of particular food contaminants or unique food spoilage markers. Nanoparticles may be used as reactive particles in packaging materials as regards smart packaging to notify the state of the packed product. To interact, warn, and classify the drug, the so-called nanosensors can respond to external stimulation adjustments to ensure its consistency and health. The latest innovations for smart food packaging polymer nanomaterials include spoilage triggers, oxygen markers, detection of items, and traceability [54].

1.12 Nanotechnology-Aided Freshness and Spoilage Indicators

The chemical interactions of nanosensors with spoilage components produced during the deterioration of food resulted in color change and state the level of freshness. The electrical, electronic, magnetic, and optical properties of polymers or electrically active conjugated polymers play an important role in chemical or electrical oxidation. Particularly electrochemical-polymerized conducting polymers may switch from oxidized (doped) to reduced (undoped) isolating state, which is the basis for many applications. The product indicator includes polyaniline film, which responds to several fundamental volatile amines released by noticeable colors during fish spoilage. Color variations were well linked in terms of overall volatile amine concentrations and microbial fish sample development rates in terms of the gross polyaniline (Milkfish) color variation [55].

Intelligent package has the potential to improve food safety and reduce food bone illness. Food spoilage is induced by microorganisms whose metabolism creates volatile compounds that can be identified by the conduction and/or recognition of micro-orientations dependent on gas emissions and food-freshness detections through performing polymer nanocomposites or metal oxides. Polymer nanocomposite-based sensors are used to conduct particles that are integrated into the polymer insulation matrix. The sensor resistance changes establish a pattern that adapts to the studied material. Conducting polymer nanocomposite sensors in black and polyaniline carbon were designed for the detection and identification of foodborne pathogens by producing a specific response pattern for each microorganism (for example, Salmonella sp., Bacillus parahemolyticus). For example, chicken freshness was analyzed based on the fragrance using a neural network to analyze metallic performance results such as tin and indium oxide gas sensors. In food packaging, a device that has several nanosensors, which are extremely susceptible to spoilage markers, creates a color change that indicates when the food is harmed.

1.13 Nanotechnology-Aided Oxygen Indicators in Food Industry

Metal nanoparticles can be easily used to generate oxygen and to cultivate aerobic microorganisms during the storage of food. There has been growing interest in developing nontoxic and irreversible sensors of oxygen in food-free, oxygen-free systems such as vacuum or nitrogen packaging. A UV colorimetric oxygen indicator was developed with UVA light that uses titania nanoparticles (TiO_2) to photosensitize the reduction in polymer encapsulation of methylene blue by triethanolamine. The sensor bleaches through UV irradiation and stays colorless before oxygen is added to the initial blue light. The survival time is relative to the amount of access to oxygen [4, 56].

1.14 Application of Nanotechnology in Product Identification and Anti-counterfeiting

Nanoparticles can be used as some smart food packaging as a food safety tracking device or to avoid falsification. BioMerieux has developed the Food Expert ID® multi detection test for nano-monitoring responses to food scares. Nanobarcodes for individual objects or pellets were produced by the US Oxonica Inc., which must be interpreted using a modified microscope for anti-counterfeiting purposes. Commercially available nanobars are made of inert metals, such as nickel, platinum gold, and silver, by electroplating into templates that define the particle diameter, which then releases stripped nanorods from templates [1–4].

1.15 Usages of Nanotechnology in Traceability and Active Tags in Food and Drug Industry

Radiofrequency recognition usually involves package stickers in food and drug or pharmaceutical industries. The brands are electronic radio-frequency sensor-based mechanisms used for transferring data from a tag connected to an object and automated recognition of the object. RFID is an improvement on previous manual tracking systems or bar codes. It is extremely robust and can work at extreme temperatures and pressures and can be detected over 100 m, and many tags can be played at the same time. Nanotechnology also allows for cost-effective RFID tags in sensor packaging. Smaller, more compact nano-enabling tags may be placed on thin labels [1–3, 55]. It is a fact that when concerning public health, an evaluation of the possible migration into food of packaging components and an evaluation of their potential danger are critical for a thorough risk assessment. However, very little research has been conducted so far on the impact of nanomaterials on absorption or possible association of food contact materials with nanomaterial-dependent food components [56]. Thin film transistor is the key part of RFID tags, and it can be embedded in food packages; a researcher came upon with cheaper printable thin film transistor made up of carbon nanotube-filled inks. It can be easily printed on papers and plastics [57].

1.16 Conclusions

In the last ten years, nanotechnology offers enormous opportunities for creative food packaging technologies that favor customers and businesses alike. Even at an early stage of improvement of the material properties of packaging, nanotechnology will have significant advantages and will require continuing investments of nanotechnological applications in packaging materials. Nanotechnology may offer a range of advantages in the context of advanced functional properties to render food packaging. Nanotechnology can improve food safety, making it convenient and creating modern product processing and innovative product and storage functions. However, all technologies are still at an intermediate level, and most of them, at least in the short term, are looking for good quality products. Also, nanomaterials can be used for the production of packaging, which keeps the product fresher indoors for longer food life and improves food safety. Smart packaging with nanosensors can also provide customers with inside knowledge of the food condition. Sensors can warn or inform consumers of the exact nutritional status contained therein before the food is spoiled. Nanotechnology is transforming the production of the entire packaging company.

References

1 Kuswandi, B. (2016). Nanotechnology in food packaging. In: *Nanoscience in Food and Agriculture*, Sustainable Agriculture Reviews 20, vol. 1 (eds. S. Ranjan et al.), 151–183. Cham: Springer https://doi.org/10.1007/978-3-319-39303-2_6.

2 Kuswandi, B. (2017). Environmental friendly food nano-packaging. *Environ. Chem. Lett.* 15 (2): 205–221.

3 Singh, T., Shukla, S., Kumar, P. et al. (2017). Application of nanotechnology in food science: perception and overview. *Front. Microbiol.* 8: 1501.

4 Dasgupta, N., Ranjan, S., Mundekkad, D. et al. (2015). Nanotechnology in agro-food: from field to plate. *Food Res. Int.* 69: 381–400.

5 Aguero, A., Quiles-Carrillo, L., Jorda-Vilaplana, A. et al. (2019). Effect of different compatibilizers on environmentally friendly composites from poly (lactic acid) and diatomaceous earth. *Polym. Int.* 68 (5): 893–903.

6 Adeyeye, S.A.O. (2019). Food packaging and nanotechnology: safeguarding consumer health and safety. *Nutr. Food Sci.*

7 Thekkethil, A.J., Nair, R., and Madhavan, A. (2019). The Role of Nanotechnology In Food Safety: A Review. In: *2019 International Conference on Computational Intelligence and Knowledge Economy (ICCIKE)* (pp. 405–409). IEEE, Dubai, UAE, DOI:10.1109/ICCIKE47802.2019.9004412.

8 Estevez-Areco, S., Guz, L., Famá, L. et al. (2019). Bioactive starch nanocomposite films with antioxidant activity and enhanced mechanical properties obtained by extrusion followed by thermo-compression. *Food Hydrocolloids* 96: 518–528.

9 Campohermoso, M.A., Contreras-Esquivel, J.C., and Artés-Hernández, F. (2019). Potassium permanganate-based ethylene scavengers for fresh horticultural produce as an active packaging. *Food Eng. Rev.* 11 (3): 159–183.

10 Dar, A.H., Rashid, N., Majid, I. et al. (2020). Nanotechnology interventions in aquaculture and seafood preservation. *Crit. Rev. Food Sci. Nutr.* 60 (11): 1912–1921.

11 Kim, I., Viswanathan, K., Kasi, G. et al. (2020). ZnO nanostructures in active antibacterial food packaging: preparation methods, antimicrobial mechanisms, safety issues, future prospects, and challenges. *Food Rev. Int.*: 1–29.

12 da Cruz, R.M. (ed.) (2019). *Food Packaging: Innovations and Shelf-Life*. CRC Press.

13 Azeredo, H.M., Otoni, C.G., Corrêa, D.S. et al. (2019). Nanostructured antimicrobials in food packaging—recent advances. *Biotechnol. J.* (12): 14, 1900068.

14 Liao, C., Li, Y., and Tjong, S.C. (2019). Bactericidal and cytotoxic properties of silver nanoparticles. *Int. J. Mol. Sci.* 20 (2): 449.

15 Huang, Y., Mei, L., Chen, X., and Wang, Q. (2018). Recent developments in food packaging based on nanomaterials. *Nanomaterials* 8 (10): 830.

16 Wang, Y., Yang, Y., Shi, Y. et al. (2020). Antibiotic-free antibacterial strategies enabled by nanomaterials: progress and perspectives. *Adv. Mater.* 32 (18): 1904106.

17 Stavitskaya, A., Batasheva, S., Vinokurov, V. et al. (2019). Antimicrobial applications of clay nanotube-based composites. *Nanomaterials* 9 (5): 708.

18 López-Rubio, A., Fabra, M.J., and Martínez-Sanz, M. (2019). Food packaging based on nanomaterials. *Nanomaterials* 9 (9): 1224.

19 Álvarez-Hernández, M.H., Martínez-Hernández, G.B., Avalos-Belmontes, F. et al. (2019). Potassium permanganate-based ethylene scavengers for fresh horticultural produce as an active packaging. *Food Eng. Rev.* 11 (3): 159–183.

20 Verma, M.L., Dhanya, B.S., Rani, V. et al. (2020). Carbohydrate and protein based biopolymeric nanoparticles: current status and biotechnological applications. *Int. J. Biol. Macromol.*

21 Wang, Y.L., Lee, Y.H., Chiu, I.J. et al. (2020). Potent impact of plastic nanomaterials and micromaterials on the food chain and human health. *Int. J. Mol. Sci.* 21 (5): 1727.

22 Liao, C., Li, Y., and Tjong, S.C. (2020). Visible-light active titanium dioxide nanomaterials with bactericidal properties. *Nanomaterials* 10 (1): 124.

23 Araújo, D.J.C., Vilarinho, M.C.L.G., and Machado, A.V. (2019). Agroindustrial residues as cellulose source for food packaging applications. In: *Wastes: Solutions, Treatments and Opportunities III*, 217. CRC Press.

24 Alvarez-Pérez, O.B., Torres-León, C., Diaz-Herrera, R. et al. (2019). Packaging biodegradability. In: *Food Packaging: Innovations and Shelf-Life* (ed. R.M. da Cruz), 32. CRC Press.

25 Fahmy, H.M., Eldin, R.E.S., Serea, E.S.A. et al. (2020). Advances in nanotechnology and antibacterial properties of biodegradable food packaging materials. *RSC Adv.* 10 (35): 20467–20484.

26 Saravanakumar, K., Sathiyaseelan, A., Mariadoss, A.V.A. et al. (2020). Physical and bioactivities of biopolymeric films incorporated with cellulose, sodium alginate and copper oxide nanoparticles for food packaging application. *Int. J. Biol. Macromol.* 153: 207–214.

27 Sozer, N. and Kokini, J.L. (2009). Nanotechnology and its applications in the food sector. *Trends Biotechnol.* 27 (2): 82–89.

28 Rashidi, L. and Khosravi-Darani, K. (2011). The applications of nanotechnology in food industry. *Crit. Rev. Food Sci. Nutr.* 51 (8): 723–730.

29 Reza Mozafari, M., Johnson, C., Hatziantoniou, S., and Demetzos, C. (2008). Nanoliposomes and their applications in food nanotechnology. *J. Liposome Res.* 18 (4): 309–327.

30 García, M., Forbe, T., and Gonzalez, E. (2010). Potential applications of nanotechnology in the agro-food sector. *Food Sci. Technol.* 30 (3): 573–581.

31 Kim, D.M. and Lee, G.D. (2006). Introduction to the technology, applications, products, markets, R&D, and perspectives of nanofoods in the food industry. *J. Food Sci. Nutr.* 11 (4): 348–357.

32 Biswal, S.K., Nayak, A.K., Parida, U.K., and Nayak, P.L. (2012). Applications of nanotechnology in agriculture and food sciences. *Int. J. Sci. Innov. Discovery* 2 (1): 21–36.

33 Monge, M. and Moreno-Arribas, M.V. (2016). Applications of nanotechnology in wine production and quality and safety control. In: *Wine Safety, Consumer Preference, and Human Health*, 51–69. Cham: Springer.

34 Sozer, N. and Kokini, J.L. (2012). The applications of nanotechnology. In: *Chemical Analysis of Food—Techniques and Applications*, 145–176. *Academic, Elsevier.*

35 Shewan, H.M. and Stokes, J.R. (2013). Review of techniques to manufacture micro-hydrogel particles for the food industry and their applications. *J. Food Eng.* 119 (4): 781–792.

36 Thiruvengadam, M., Rajakumar, G., and Chung, I.M. (2018). Nanotechnology: current uses and future applications in the food industry. *3 Biotech* 8 (1): 74.

37 Wesley, S.J., Raja, P., Raj, A.A., and Tiroutchelvamae, D. (2014). Review on-nanotechnology applications in food packaging and safety. *Int. J. Engine Res.* 3 (11): 645–651.

38 Ponce, A.G., Ayala-Zavala, J.F., Marcovich, N.E. et al. (2018). Nanotechnology trends in the food industry: Recent developments, risks, and regulation. In: *Impact of Nanoscience in the Food Industry*, 113–141. Academic Press.

39 Kour, H., Malik, A.A., Ahmad, N. et al. (2014). Nanotechnology–new lifeline for food industry. *Crit. Rev. Food Sci. Nutr.*, (just-accepted).

40 Mikiciuk, J., Mikiciuk, E., and Szterk, A. (2017). Physico-chemical properties and inhibitory effects of commercial colloidal silver nanoparticles as potential antimicrobial agent in the food industry. *J. Food Process. Preserv.* 41 (2): e12793.

41 Trujillo, L.E., Ávalos, R., Granda, S. et al. (2016). Nanotechnology applications for food and bioprocessing industries. *Biol. Med.* 8 (3): 1.

42 Jampílek, J. and Králová, K. (2018). Benefits and potential risks of nanotechnology applications in crop protection. In: *Nanobiotechnology Applications in Plant Protection*, 189–246. Cham: Springer.

43 dos Santos Pires, A.C., Soares, N.D.F.F., da Silva, L.H.M. et al. (2010). Polydiacetylene as a biosensor: fundamentals and applications in the food industry. *Food Bioprocess Technol.* 3 (2): 172–181.

44 Augustin, M.A. and Oliver, C.M. (2012). An overview of the development and applications of nanoscale materials in the food industry. In: *Nanotechnology in the Food, Beverage and Nutraceutical Industries*, 3–39. Woodhead Publishing.

45 Rai, V.R. and Bai, J.A. (2018). *Nanotechnology Applications in the Food Industry*. CRC Press.

46 Pillai, S.D. and Shayanfar, S. (2017). Electron beam technology and other irradiation technology applications in the food industry. In: *Applications of Radiation Chemistry in the Fields of Industry, Biotechnology and Environment*, 249–268. Cham: Springer.

47 Prakash, A., Sen, S., and Dixit, R. (2013). The emerging usage and applications of nanotechnology in food processing industries: the new age of nanofood. *Int. J. Pharm. Sci. Rev. Res* 22 (1): 107–111.

48 Mahfoudhi, N., Ksouri, R., and Hamdi, S. (2016). Nanoemulsions as potential delivery systems for bioactive compounds in food systems: preparation, characterization, and applications in food industry. In: *Emulsions*, 365–403. Academic Press.

49 Wang, K., Sun, D.W., and Pu, H. (2017). Emerging non-destructive terahertz spectroscopic imaging technique: principle and applications in the agri-food industry. *Trends Food Sci. Technol.* 67: 93–105.

50 Echiegu, E.A. (2017). Nanotechnology applications in the food industry. In: *Nanotechnology*, 153–171. Singapore: Springer.

51 Davarcioglu, B. (2017). Nanotechnology applications in food packaging industry. In: *Nanotechnology*, 87–113. Singapore: Springer.

52 Khan, A.S. (2017). Nanotechnology applications in food and agriculture. In: *2017 IEEE 17th International Conference on Nanotechnology (IEEE-NANO)*, July, 877–878. IEEE.

53 Baltić, M.Ž., Bošković, M., Ivanović, J. et al. (2013). Nanotechnology and its potential applications in meat industry. *Tehnol. Mesa* 54 (2): 168–175.

54 Vélez, M.A., Perotti, M.C., Santiago, L. et al. (2017). Bioactive compounds delivery using nanotechnology: design and applications in dairy food. In: *Nutrient Delivery*, 221–250. Academic Press.

55 Var, I. and Sağlam, S. (2015). Nanotechnology applications in the food industry. *GIDA-J. Food* 40 (2): 101–108.

56 Rouf, A., Kanojia, V., and Naik, H.R. (2017). Cell immobilization: an overview on techniques and its applications in food industry. *Int. J. Commun. Syst.* 5 (6): 1817–1824.

57 Jung, M., Kim, J., Noh, J. et al. (2010). All-printed and roll-to-roll-printable 13.56-MHz-operated 1-bit RF tag on plastic foils. *IEEE Trans. Electron Devices* 57 (3): 571–580.

2

An Overview of Biopolymers in Food Packaging Systems

Jéssica de Matos Fonseca☆*, Betina L. Koop*☆*, Thalles C. Trevisol, Cristiane Capello, Alcilene R. Monteiro, and Germán A. Valencia*

Federal University of Santa Catarina, Department of Chemical and Food Engineering, Florianópolis 88040-900, Brazil

2.1 Introduction

In the last years, polymers isolated from renewable raw materials have been studied as alternative to synthetic polymers for food packaging applications [1]. Synthetic polymers-based films are thin layers, between 50 and 80 μm. These materials have excellent barrier properties against several gases (e.g. H_2O, O_2, and CO_2), as well as against biological, chemical, and physical contaminants [1, 2]. However, synthetic polymers are derived from petroleum, a nonrenewable resource, and they can be classified as nondegradable materials, negatively impacting the ecosystems [3]. In this sense, several researchers have focused to replace synthetic polymers by polymers from natural sources, also called as biopolymers [4–9]. Several biopolymers have been used to manufacture films and coatings due to the good film-forming properties of these macromolecules. Furthermore, these macromolecules can be isolated directly from biomass or they can be synthetized using microbial routes, as well as by means of chemical synthesis using monomers from agro-resources [3, 10, 11].

Food packaging based on biopolymers are sensitive to the relative humidity and temperature. In recent years, the development of advanced films and coatings based on biopolymers has been explored and applied to extend the shelf life of foods, as well as to improve food safety, quality, and convenience to consumers [3]. The objective of this chapter is to review the new strategies to manufacture biopolymers-based films and coatings, aiming their applications on foods, as well as the prospects and limitations of these materials for food packaging sector.

☆ Jéssica de Matos Fonseca and Betina Luiza Koop were contributed equally.

Nanotechnology-Enhanced Food Packaging, First Edition. Edited by Jyotishkumar Parameswaranpillai, Radhakrishnan Edayileveettil Krishnankutty, Aswathy Jayakumar, Sanjay Mavinkere Rangappa, and Suchart Siengchin.

2.2 Main Polymers Isolated from Biomass

2.2.1 Casein and Whey

Milk proteins are conformed by casein and whey; casein is a milk-specific protein (~80%) having four different proteins (αS1-, αS1-, β-, and κ-caseins). In milk, this protein form colloidal micelles with particle size between 50 and 600 nm. Caseins can be coagulated using acid compounds or enzymes (rennet), given that the coagulation method impacts the physicochemical properties of this macromolecule [12]. Acid casein is obtained after its precipitation by the exposure of the milk below pH 4.6 (isoelectric point of casein). In the enzymatic method, caseins are coagulated using chymosin (rennet) enzymes where several chemical bonds are cleaved in κ-casein. Rennet caseins are insoluble in water, and they have a pH around 7.5 [13].

Acid caseins are also insoluble in water, so they are neutralized to pH 6.7 using potassium, sodium, calcium, or ammonium to obtain caseinates, which are soluble in water. In another way, casein can be neutralized using sodium hydroxide to produce a caseinate called as sodium caseinate. Similarly, sodium, ammonium, and potassium caseinates are obtained using others neutralizing agents. Some caseinates such as sodium, ammonium, and potassium caseinates have similar physical properties, and their solutions are viscous and translucent. In contrast, solutions based on calcium caseinate are turbid [14]. Sodium and calcium caseinates are the most utilized biopolymers due to their excellent solubility, gelation, viscosity, and emulsifying and foaming properties [15].

Casein is a macromolecule having a high degree of polar groups such as carboxyl and amino, which provide good film-forming properties. Furthermore, films based on casein have acceptable gas barrier against the oxygen and other nonpolar molecules, being used as film or coating to reduce the lipid oxidation of foods [16]. In addition, films and coatings based on casein can be used as active material by blending with vitamins, minerals, dyes, and bioactive compounds (e.g. antioxidant and antimicrobial), preserving the food quality, and increasing the food shelf life [17]. Other characteristics in casein such as high nutritional value, water solubility, and emulsification capability turn this macromolecule an important material to manufacture edible packaging [18]. Table 2.1 shows some recent studies that applied casein derivatives as food packaging materials with antimicrobial and antioxidant properties.

Whey is a protein obtained from cheese or coagulated dairy products; this macromolecule is considered as by-product and represents approximately 20% of the total milk protein. Whey has a water content oscillating between 85% and 90% and a solid fraction composed of lactose, proteins, lipids, and minerals [28, 29]. In the food sector, whey can be used in liquid form, as powdered cheese whey, as lactose, and as proteins isolate or concentrate [30]. The proteins from whey are separated (concentration) and applied in the food industry due to their excellent nutritional and functional properties. Whey protein concentrate (WPC) and whey protein isolate (WPI) are obtained from the removal of nonprotein fractions to attain a protein content higher than 25% and 90%, in the final product, respectively [28].

Table 2.1 Films and coatings based on casein and whey for food packaging applications.

Components	Production approach	Main results	References
CAS[a]/tannin from white peel grape, red peel grape, and oak bark	Casting	Films with improved water solubility and water vapor permeability. Bilayer films containing tannin have antioxidant property and antimicrobial activity against *E. Coli* and *L. innocua*	[19]
CAS[a]/LMP[b]/glycerol/natamycin	Casting	Films with antimicrobial activity against *P. chrysogenum* inoculated on cured meat	[20]
CAS[a]/arabic gum/essential oil from cinnamon and lemon	Dip coating	Films with antioxidant activity were used to extend the shelf life of guava up to 40 days	[21]
WPC[c]/glycerol/rosemary extract	Casting	The films delayed the salami's lipid oxidation for 30 days	[22]
WPC[c]/glycerol/extracts from Shirazi thyme, sage leaves, and cumin seeds	Coating	The coating inhibited the *A. flavus* growth in pistachio kernels contaminated	[23]
Whey protein/glycerol/green tea active extract	Casting	The films delayed the lipid oxidation of fresh salmon samples until the 14th day of storage	[24]
WPI[d]/pectin/alginate /glycerol/tween 20	Casting	Coatings were efficient to retain the humidity of bread	[25]
[e] WPNF/glycerol/trehalose	Casting	Coatings with hydrophobic surface and antioxidant activities	[26]
WPI[d]/PLA[f] resin/glycerol	Extrusion	Films reduced the lipid oxidation in baby foods	[27]

a) CAS: sodium caseinate.
b) LMP: low methoxyl pectin.
c) WPC: whey protein concentrate.
d) WPI: whey protein isolate.
e) WPNF: whey protein nanofibrils.
f) PLA: poly(lactic acid).

Whey proteins are composed of different globular proteins such as β-lactoglobulin, α-lactalbumin, bovine serum albumin, immunoglobulins, and the polypeptides proteose-peptone. Particularly, β-lactoglobulin is the major whey protein responsible for gelation and aggregation behavior in whey [31].

Films and coatings based on whey are transparent materials, and they have better mechanical and barrier properties when compared with those manufactured with polysaccharides. Protein network in whey can be modified by means of thermal treatments, and the resulting modified whey can be used to manufacture materials with improved tensile and barrier properties [29, 31].

Finally, films and coatings based on casein and whey have limited elasticity and water sensitivity, limiting their applications. The films elasticity can be improved using plasticizers such as glycerol or sorbitol into the film-forming solution [18, 29]. Table 2.1 shows some studies using WPC and WPI with addition of plasticizers for antimicrobial and antioxidant food packaging.

2.2.2 Cellulose and Derivatives

Cellulose is considered the most abundant biopolymer in the world, being a component of the plant tissues and vegetal cell wall. This biopolymer can be obtained from several plants such as cotton, sugarcane bagasse, eucalyptus, and wood, among others, which it become an environmental and economically feasible by-product [32]. In addition, derivatives of cellulose have also been used for food packaging applications, especially the cellulose ethers as carboxymethyl cellulose (CMC), ethylcellulose (EC), hydroxyethyl cellulose (HEC), hydroxypropyl methylcellulose (HPMC), and methylcellulose (MC) [33].

Cellulose is conformed of β-D-glucose monomers, which are 6-membered rings called pyranoses. The β-D-glucose units are bonded by acetal linkages (oxygen atoms) produced from dehydration reactions between the C-1 of the pyranose and C-4 of a neighbor pyranose [34]. The equatorial position of β-D-glucose hydroxyl functional groups causes a considerable linearization of cellulose chains that facilitates the formation of long chains (fibers) of the biopolymer [34]. The hydrogen bonds between hydroxyl groups (–OH) from different cellulose chains promote the formation of microcrystalline regions, while other micro-regions are characterized by a disordered structure (amorphous) due to low amount of hydrogen bonds [35].

Strong hydrogen bonds of the cellulose crystalline regions caused it to become mechanical and thermally resistant and insoluble in several solvents including water [36]. On the other hand, hydroxyl groups of cellulose amorphous regions are more distant between them than hydroxyl groups of crystalline regions, which allows physical interactions between cellulose and other molecules. In the case of water, cellulose is able to absorb a large number of molecules without dissolving them. This characterizes a high swelling ability of this biopolymer [33].

Cellulose ethers (MC, EC, CMC, HEC, HPMC) are biopolymers produced from substitution reactions of cellulose hydroxyl groups or alkylation and differ as to the substituting group and number of hydroxyl groups substituted (substituting degree) [33]. Generally, the cellulose etherification reactions are performed in alkaline medium using halides and alkyl sulfates as etherifying agents. The biggest advantage of the cellulose ethers in comparison with cellulose is their higher water solubility [36]. These materials are used to modify the rheology of solutions by changing of viscosity, increasing of water swelling ability, stabilizing of suspensions, gelling, and emulsifying and to form films and coatings that are more flexible than cellulose [35, 37].

Applications of films and coatings based on cellulose and its derivatives are presented in Table 2.2. It is observed that cellulose and its derivatives have the ability to form composites with a variety of other biopolymers for both active and smart packaging applications.

2.2.3 Chitin and Chitosan

Chitin is considered the second most abundant biopolymer in the world. Usually, this biopolymer is isolated from the exoskeletons of crustaceans such as crabs, lobsters, shrimps, squid, among others [51–53]. The molecular formula of chitin consists of the 2-acetamido-2-deoxy-β-D-glucose through a β (1,4) linkage, considered a polysaccharide (cellulose) with hydroxyl groups at position C-2 replaced by an acetamido group [51].

Chitosan is a derivative obtained by the deacetylation of chitin (deacetylation >50%) by means of alkaline treatment [44]. Chitosan is a cationic polysaccharide with linear structure integrated by two monomers, D-glucosamine (2-amino-2-deoxy-β-glucopyranose) and *N*-acetyl-D-glucosamine (2-acetamido-2-deoxy-β-D-glucopyranose), linked by means of 1,4-glycosidic bonds [54]. Chitosan is a semi-crystalline biopolymer, nontoxic, biocompatible, biodegradable, and with antifungal, antimicrobial, and antioxidants properties [55, 56]. Due to these characteristics, chitosan is a biopolymer widely used in food packaging applications [55, 56].

Films based on chitosan have good permeability to gases (CO_2 and O_2) and acceptable mechanical properties; however, these materials are not good barrier against water vapor. The properties of chitosan films are impacted by the chitosan morphology and molecular weight, as well as by the *N*-acetylation degree, solvent type used to manufacture films, among others [54, 57, 58]. Similarly, edible coatings based on chitosan have been applied on fresh fruit and vegetables, as well as on minimally processed, decreasing the respiration rate and delaying the senescence of these products. In addition, coatings based on chitosan can be used to reduce the moisture loss and maintain the overall quality of food products [59]. Therefore, several researchers have used this macromolecule to improve the physical and mechanical properties of films and coatings, including blends with another polymers and nanoparticles. In Table 2.3, recent studies about chitosan films and coating applied in food products are listed.

2.2.4 Collagen and Gelatin

Gelatin and its precursor collagen are biopolymers obtained from protein animal sources. Collagen is considered the most abundant protein of animal kingdom,

Table 2.2 Films and coatings based on cellulose and derivatives for food packaging applications.

Components	Production approach	Main results	References
WPI[a)]/CNF[b)]	Casting	Improvement of structural properties of WPI[a)]-based films using CNF[b)] for food packaging applications	[32]
Tara gum/CNC[c)]/grape skin extract	Casting	Colorimetric pH-sensing films with positive activation test for milk spoilage	[38]
HPMC[d)]/beeswax	Dip coating	Coatings with antifungal activity against *A. alternata* on cherry tomato fruit	[39]
CMC[e)]/sodium montmorillonite/TiO_2	Casting	Biodegradable nanocomposite films with improved mechanical and light barrier properties	[40]
PLA[f)]/rosin modified CNF[b)]/chitosan	Casting	Films with antimicrobial activity against *E. coli* and *B. subtilis*	[41]
Gelatin/CNF[b)]/chitosan Starch/CNF[b)]/chitosan	Casting	Films with better structural properties and antimicrobial activity	[42]
Chitosan/gelatin/MC[g)]/tannic acid	Casting	Films with antimicrobial activity against *E. coli* and *S. aureus*	[43]
Cellulose derived from corncob	Thermo-pressing molding	Biodegradable and sustainable films applied in breads	[44]
CMC[e)]/okra mucilage/ZnO	Casting	Films with antimicrobial activity against *E. coli* and *S. aureus*	[45]
κ-Carrageenan/HPMC[d)]/*Prunus maackii* juice	Casting	Biodegradable pH-sensitive and antioxidant films for oil and lard packaging application and label for testing pork freshness	[46]

Gelatin/CMC[e)]/chitin nanofibers/*Trachyspermum ammi* (ajowan) essential oil	Casting	Films with improved mechanical and barrier properties and antimicrobial activity against *E. coli* and *S. aureus*	[47]
EC[h)]/azo indicators (methyl orange; methyl red)	Casting	Biodegradable material with pH-sensitive properties for food packaging applications	[48]
HEC[i)]/ZnO	Casting	Films with antimicrobial activity against *E. coli* and *S. aureus*	[49]
HPMC[b)]/liposome/CNF[b)]	Dip coating	Hydrophobic coatings used for hydration of fat food surfaces (slices of almonds and chocolate)	[50]

a) WPI: whey protein isolate.
b) CNF: cellulose nanofibers.
c) CNC: cellulose nanocrystals.
d) HPMC: hydroxypropyl methylcellulose.
e) CMC: carboxymethylcellulose
f) PLA: poly(lactic acid).
g) MC: methylcellulose
h) EC: ethylcellulose.
i) HEC: hydroxyethyl cellulose.

Table 2.3 Films and coatings based on chitosan for food packaging applications.

Components	Production approach	Main results	References
Chitosan/gelatin/starch/sorbitol/tween/geraniol/thymol	Casting	The coatings reduced the weight loss and delayed the physicochemical alterations of strawberries	[59]
Chitosan/black chokeberry extract/acetic acid	Casting	Colorimetric pH indicator films with high resistance to water	[60]
Chitosan/glycerol/sorbitol/acetic acid	Casting	Films with antimicrobial activity against *L. monocytogenes*	[58]
Chitosan/nisin/potassium sorbate/acetic acid	Casting	Potassium sorbate and nisin reduced the resistance and increased the flexibility and hydrophobicity of chitosan films	[61]
Chitosan/ε-polylysine (ε-PL)/TPP[a)]/acetic acid	Self-assembly	Films with low solubility in water and water vapor permeability, as well as with antimicrobial activity against *E. coli* and *S. aureus*	[62]
Chitosan/poly (acrylic acid)/sodium chloride/methanol/human plasma/fibronectin/silicone oil	Layer-by-layer	Hydrophobic films highly stable for 28 days in food simulants	[63]

Chitosan/ESs[b]/choline chloride based/malic acid/lactic acid/citric acid/glycerol	Thermo-pressing molding	Film with improved mechanical and barrier properties manufactured at industrial scale	[64]
Chitosan/montmorillonite/aromatic aldehydes/ethanol/acetic acid	Self-assembly	Hydrophobic films with good mechanical properties	[65]
Chitosan/carbon/L-(β)-lactic acid/glycerol	Radiofrequency reactive/magnetron sputtering	Films with acceptable barrier properties for food packaging applications	[66]
Chitosan/acetic acid	Casting	Films reduced the growth of mesophilic bacteria in fresh pork loins stored under vacuum, at 4 °C, for 28 days	[67]
Chitosan/sodium alginate/calcium chloride	Layer- by-layer	Chitosan coating layer-by-layer preserved the ascorbic acid content, antioxidant capacity, and firmness and avoid the fungal growth of on fruit bars during storage	[68]
Chitosan nanoparti-cles/TPP[a]/acetic acid	Ionic gelation	Coating was effective to delay the grapes ripening, reducing the weight loss and maintaining the sugar content, soluble solids, the titratable acidity, and sensory characteristics	[69]

a) TPP: tripolyphosphate.
b) ESs: eutectic solvents.

composed of around 25–35% of total proteins of the animal body and is mostly found in the tissues as bones, cartilage, ligaments, tendons, skins, blood vessels, intervertebral discs, guts, and corneas [70]. Collagen is mainly extracted from bovine, porcine, and fish [71], while gelatin is obtained after partial hydrolysis of the collagen. Gelatin is characterized by its versatility, high digestibility, and gels melting at human body temperature. Both biopolymers show interesting physicochemical and structural properties, especially for food industries [71].

Collagen is produced by connective tissue cells, and it is classified as a superelastic fibrous protein. Analyzing the deconstruction of collagen fibers (Figure 2.1a), their quaternary structure is characterized by a set of collagen fibrils composed of collagen molecules, whose protein structure is tertiary [72]. This super coiling is composed of three identical or nonidentical polypeptide chains twisted together. Each polypeptide chain constitutes the primary structure of the collagen and contains around 1000 units of amino acids, whose glycine (Gly), hydroxyproline (Hyp), and proline (Pro) are in vast majority [72]. The interactions between N—H and C=O (hydrogen bonds) from amino acids are responsible by the α-helical conformation of the collagen secondary structure [73]. On the other hand, collagen tertiary structure is stabilized by means of hydrogen bonds between C–O groups from glycine and O—H groups from hydroxyproline [73]. Finally, the collagen quaternary structure is stabilized by hydrogen bonds, intramolecular van der Waals interactions, and some covalent bonds. Each collagen molecule can have until 300 nm in length and 1.5 nm in diameter [70].

There are at least 28 types of collagen, which differ as to the arrangement of amino acids composing the primary structure. The most abundant collagens are of the types I, II, and III, which manage cell differentiation, proliferation, and migration and provide the scaffolding [70]. Because of the difficult digestion of collagen by the human body, this protein is also commercialized in its complete hydrolyzed form [74].

Gelatin is composed of collagen polypeptide fragments (Figure 2.1b), whose structure is based on α-helical conformation and its combinations (β and γ conformations) [75]. Gelatin functionality depends on raw material, which causes variations of its relative fractions of peptides and molecular mass (95–100 kDa), consequently [70]. The variation of molecular mass of gelatin peptide fractions causes changes in the gelation time (setting time), gel strength (bloom), and viscosity of the biopolymer solution [76]. Gelatin bloom depends on the number of α- and β-chains, which constitute the fractions of the largest peptides, and its viscosity depends on average molecular mass of its peptide chains [70, 77].

During the partial hydrolysis of the collagen, its cross-linking structure is preserved, but some peptide bonds between chains are broken. The cross-linking degree varies as to raw material used to the gelatin fabrication, and its pretreatment determines the type of gelatin that will be produced, type A or type B [75]. Gelatins type A and B are produced by acid and alkaline processes, and they have isoelectric points in a pH range between 6.0 and 9.0, and of around pH 5.0, respectively [77].

Beyond traditional food applications of the gelatin and collagen as emulsifiers, stabilizers, foaming and microencapsulating agents, these biopolymers are also applied as biodegradable films and coatings in order to extend the shelf life of food

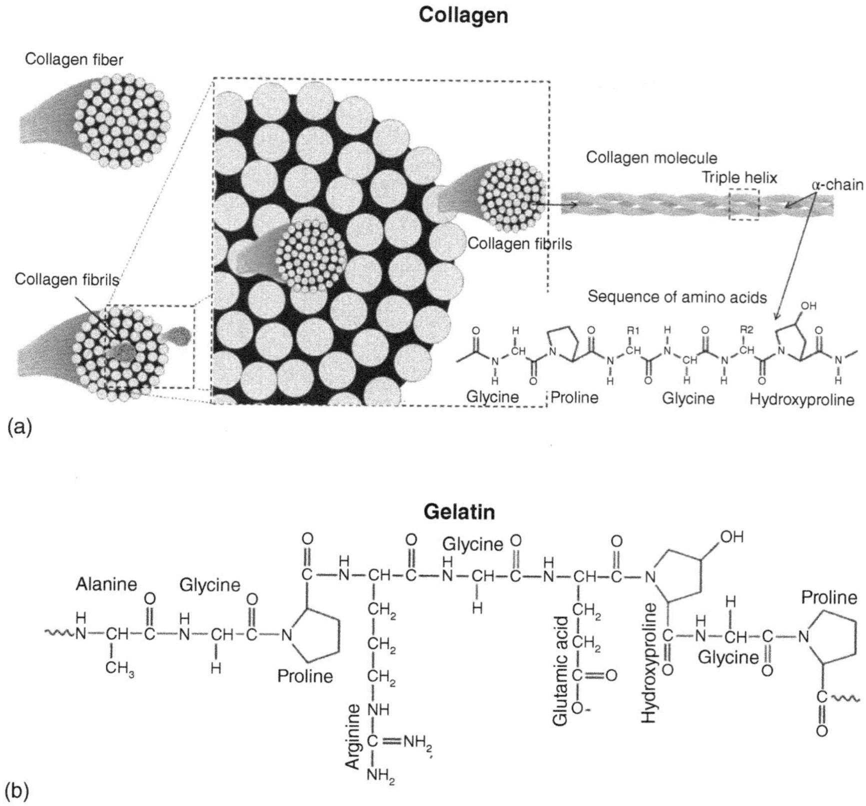

Figure 2.1 Schematic and chemical structure of collagen (a) and gelatin (b).

and as carriers of active agents [72, 78, 79]. The most important applications of films and coatings based on collagen and gelatin are presented at Table 2.4. The versatility of collagen and gelatin to form blends and composites and to carry different active ingredients has been observed, whose main characteristic is the antimicrobial activity.

2.2.5 Soybean and Derivatives

Soybean is a leguminous composed of proteins (40%), total carbohydrates (insoluble fibers and soluble saccharides) (34%), oil (21%), and minerals (ash) (4%) [93]. The proteins, water-soluble carbohydrates, and fibers are the most used soybean compounds to manufacture films and coatings, aiming food applications.

2.2.5.1 Soy Protein

The commercial soy protein is a by-product obtained from the soy oil extraction, being available as soy protein concentrate (SPC), soy protein isolate (SPI), and soy flour. SPC and SPI differ mainly by protein and carbohydrate contents. According to Koshy et al. [94], full soy flour and SPC showed carbohydrate content (34% and

Table 2.4 Films and coatings based on collagen and gelatin for food packaging applications.

Components	Production approach	Mains results	References
Gelatin/mint essential oil	Dip coating	Coatings with antimicrobial properties used to extend the shelf life of strawberries	[78]
Gelatin (PS[a])/ starch/acerola powder	Casting	Active films containing acerola compounds	[79]
Gelatin (SS[b])/TiO_2	Casting	Films with antimicrobial activity against *E. coli* and *S. aureus*	[80]
Collagen/lysozyme	Dip coating	Coatings with antimicrobial activity against psychrotrophic bacteria used to extend the shelf life of fresh salmon fillets	[81]
Collagen (FS[c])/sodium alginate	Casting	Biodegradable films with better structural organization	[82]
Gelatin (FS[c])/cellulose nanofibers/palmitic acid	Casting	Films with improved physicochemical properties	[83]
Collagen/nisin	Casting	Films with antimicrobial activity against *L. monocytogenes*, *S. aureus*, and *Lactobacillus*, used to package ready-to-eat sausage	[84]
Gelatin (PS[a])/chlorophyllin	Drying tunnel	Films with antimicrobial activity against *L. monocytogenes* and *E. coli.*, used to package bologna slices	[85]
Gelatin (FS[c])/ZnO/chitosan nanofibers	Casting	Films with antibacterial activity used to package chicken fillet and cheese	[86]

Collagen/chitosan/pomegranate peel extract	Casting	Films with antimicrobial activity against *B. saprophyticus*, *B. subtilis*, *S. typhi*, and *E. coli*	[87]
Collagen (LW[d)]) hydrolysate	Spraying nozzles	Coatings used to improve the growth of peas, soybean, broad bean, and rape seeds	[88]
Gelatin ([e)]BB and [f)]BC)/mint essential oil	Casting	Coatings with antifungal activity against *B. cinerea* and *R. stolonifera* for fruit application	[89]
Gelatin (FS [c)]) hydrolysate	Dip coating	Films were used to extend the shrimp shelf life, reducing microbial growth and lipid oxidation	[90]
Collagen from cowhide	Casting	Films with better structural organization	[91]
Gelatin (FS [c)])/squalene	Casting	Biodegradable packaging used to package fried fish crackers	[92]

a) PS: porcine skin.
b) SS: shark skin.
c) FS: fish skin.
d) LWs: leather wastes.
e) BBs: bovine bones.
f) BC: bovine cartilage.

18%, respectively) and protein content for full soy flour (56%), SPC (65%), and SPI (>90%).The full fat soy flour is produced by grinding soybeans into fine powders, while the SPC is prepared by defatting and removing of water-soluble carbohydrates from soy flour. Finally, the SPI is a highly refined soy flour and free of the most nonprotein components, fats, and carbohydrates [94].

The soy protein consists in a mixture of albumin and globulin proteins, which have an isoelectric point from pH 4.5 to 4.8 and are classified in four main categories: 2S, 7S, 11S, and 15S fractions, where S corresponds to Svedberg units [95, 96]. The 7S (β conglycinin) and 11S (glycinin) globulin proteins constitute 35% and 52% of the soy structure, respectively, while the 2S and 15S albumin proteins only correspond to 8% and 5% of the soy structure, respectively [97]. SPI has an ability to form films softer and more transparent than other soy products [96]. Its films also exhibit higher gas barrier than lipids and polysaccharides-based films [94]. However, the complexity of globular proteins decreases the flexibility of chains, which results in the formation of fragile and stiff films and coatings [97].

In order to improve the mechanical properties of soy-based biomaterials, soy proteins have been associated to other materials to form composites since their structure exhibit polar functional groups as hydroxyl, carboxyl, and amine that facilitate its chemical and physical modifications [97] (Table 2.5). These groups are distributed along 18 amino acids composing the soy protein structure including arginine, lysine, leucine, cysteine, phenylalanine, tyrosine, aspartic, and glutamic acid [94, 97].

2.2.5.2 Soybean Soluble Polysaccharide

Soybean soluble polysaccharide (SSPS) is a macromolecule extracted from cell wall material of cotyledons of soybeans [98]. Its structure is similar to pectin, being considered as an acidic carbohydrate composed mainly of rhamnogalacturonan backbone branched by β-1,4-galactan, homogalacturonan, and α-1,3- or α-1,5-arabinan chains [112]. Beyond its ability to form films with excellent mechanical properties and high thermal stability, SSPS films and coatings are extremely sensitive to water molecules (Table 2.5) [113, 114].

2.2.5.3 Soybean Fiber and Derivatives

Soybean fiber is an insoluble fraction obtained from SPC processing and recovered by centrifugation. It is composed of carbohydrates derived from soy cotyledon cell wall as pectin, cellulose, hemicellulose, and lignin [115]. Rhamnose, arabinose, galactose, and galacturonic acids form the pectin structure and are ionically linked to Ca^{2+} and Mg^{2+} ions. In the cell wall, pectin fraction is covalently linked to cellulose, phenol groups, and proteins [115, 116]. Lignin is the highest insoluble fraction of the soybean fibers due to the presence of polyphenols, which are produced by polymerization of trans-p-coumaryl, sinapyl alcohols, and coniferyl [117]. The considerable resistance of lignin is caused by the extensive cross-linking on cell wall that characterizes the high crystallinity of its chains. The significant crystallinity of soybean fibers improves oxygen and hydrocarbon barriers of soybean fibers-based materials but limits their flexibility (Table 2.5) [109].

Table 2.5 Films and coatings based on soybean and derivatives for food packaging applications.

Components	Production approach	Main results	References
SSPS[a)]/halloysite nanoclay	Casting	Biodegradable and sustainable films for food packaging applications	[98]
SSPS[a)]/ZnO	Casting	Films with antimicrobial activity against *P. expansum* and *C. albicans*	[99]
SSPS[a)]/nanorod-rich ZnO	Casting	Films with antimicrobial activity against *E. coli* and *S. aureus*	[100]
SPC[b)]	Casting/cold plasma	Films with antioxidant activity for smoked salmon packaging	[101]
SPI[c)]/soy straw	Casting	Films were used to extend the shelf life of fruits and vegetables	[102]
SSPS[a)]/nanoclay Na^+	Casting	Films with improved mechanical and barrier properties and with antimicrobial properties against *B. cereus*, *E. coli*, and *S. aureus* and *P. expansum*	[103]
SPC[b)]	Thermo-pressing molding	Films with acceptable physicochemical properties manufactured at industrial scale	[104]
SSPS[a)]/cinnamon essential oil nanoemulsion	Casting	Films with antibacterial activity against *S. pyogenes* and *S. aureus*	[105]
Soybean fibers/acid citric	Casting	Films with improved mechanical and barrier properties	[106]

Table 2.5 (Continued)

Components	Production approach	Main results	References
SPI[c]/pine (*Pinus densiflora*) bark extract	Casting	Films with antioxidant activity and improved mechanical and barrier properties	[107]
SPI[c]/chitosan	Dip coating	Coatings were used to extend the shelf life of apricot fruit (*Prunus armeniaca* L. "Katy")	[108]
Alginate/soybean fibers (cellulose)	Casting	Films with acceptable mechanical and barrier properties	[109]
SPI[c]/soybean fibers (cellulose)	Casting	Films with an improved barrier to oxygen, aromas, and lipids, as well as good mechanical properties	[110]
Cassava starch/soybean hulls and Cassava starch/microcrystalline cellulose derived from soybean	Extrusion	Films with improved mechanical and barrier properties	[111]

a) SSPS: soybean soluble polysaccharide.
b) SPC: soybean protein concentrate.
c) SPI: soybean protein isolate.

2.2.6 Starch and Derivatives

Starch is a carbohydrate, considered one of the most abundant polysaccharides in the world. This macromolecule is in the form of granules that are synthesized after the polymerization of glucose in plant [118]. Starch is composed of amylose and amylopectin [119]. Amylose is formed by glucose units linked by α-(1,4), containing sparing α-(1,6) branch points (~1%). This macromolecule has a linear structure, and it contributes mainly to the amorphous phase of the starch granule, whereas amylopectin is a branched biopolymer formed by α-(1,4) glucose, highly branched by means of α-(1,6) glucose links (~5%). This macromolecule is responsible by the crystalline structure in starch granules [120].

Starch can be isolated from different botanical sources including cereals (corn, wheat, rice, and sorghum), vegetables (pea, chickpeas, and lentils), and tubers (cassava, potato, sweet potato, and yam). Due to the overall structure of starch, botanical sources, climate conditions, geographic location for cultivation, and soil type, the ratio of amylose to amylopectin is variable [118].

Films and coatings based on starch are extensively studied due to its abundance, low cost, and excellent film-forming properties. Furthermore, starch-based films are transparent, odorless, tasteless, and good gas barrier. Hence, films and coatings based on starch have been used to extend the shelf life of fruits (Table 2.6) [120, 127]. The physical properties of starch-based materials are impacted by the starch source, manufacture time and temperature, type of plasticizer, co-biopolymers, and storage conditions [120].

The molecular weight, amylopectin/amylose ratio, and crystalline fraction (15–45%) depend on the botanical source of starch, modifying the physicochemical properties of films and coatings based on this macromolecule [128]. A higher proportion of amylopectin in the starch granules leads to manufacturing films with lower water solubility. Unfortunately, starch-based materials have poor mechanical and gas barrier properties. Aiming to improve the mechanical and barrier properties of starch-based materials, some researchers have studied the addition of different plasticizers, solvents, and macromolecules. Also, other natural compounds such as fatty acids, natural plant extracts (green tea, pomegranate), organic compounds, and essential oils have been investigated, aiming the manufacture of starch-based materials with antimicrobial activity [120, 129]. Another alternative is to blend the starch with other biopolymers such as gelatin, guar gum, and chitosan (Table 2.6) [121, 130, 131]. Thus, the interactions among biopolymer improve the mechanical and barrier properties in starch films [120]. Finally, fatty acids can be added in starch-based materials to reduce the water sensibility in these materials; however, fatty acids could induce phase separation and reduce the mechanical properties in films and coatings [132].

2.3 Main Polymers Obtained by Microbial Production

The main polymers obtained by microbial production can be classified as: polyhydroxyalkanoates (PHAs), poly(hydroxy-butyrate), and poly(hydroxy-butyrate-*co*-hydroxyvalerate) [11]. PHAs are thermoplastic polyesters composed of

Table 2.6 Films and coatings based on starch for food packaging applications.

Components	Production approach	Main results	References
Pea starch/guar gum/shellac and oleic acid	Layer-by-layer	Coatings were used to reduce the weight loss and preserve the firmness in citrus fruits during storage	[121]
Cassava starch/casein/gelatin/sorbitol	Dip coating	Coatings showed desirable optical properties and water vapor transmission rate, and these materials delayed the ripening of guava fruits in two days	[122]
Corn starch/gelatin/guabiroba pulp	Casting	Guabiroba pulp improved the mechanical properties of films	[123]
Mango kernel starch/glycerol/sorbitol	Dip coating	Coatings were used to reduce the weight loss and maintain the sensory attributes of tomatoes at 20 °C	[124]
Potato starch/glycerol/extracts: white and green tea	Casting	Films reduced the weight loss and darkening in apple slides during storage	[125]
Cassava starch/glycerol/rosemary extract	Casting	Hydrophobic films with better barrier properties against UV light with promissory applications in food packaging	[126]

hydroxyalkanoic acid as the monomeric unit. This polymer is generally classified according to the number of carbon atoms in the side chain. In this sense, side chains having between 3 and 5, 6 and 14, and more than 14 carbon atoms can be classified as short-, medium-, and high-chain-length PHAs, respectively. Depending on the R group linked to the main chain, different derivatives of PHA were found in the consulted literature [133]. In general, 150 different units of homo or combination of copolyesters of PHA can be found [134]. Beyond the PHAs, the short-chain-length PHAs such as poly(3-hydroxybutyrate) (PHB), poly(3-hydroxyvalerate) (PHV), and their copolymer poly(3-hydroxybutyrate-*co*-3-hydroxyvalerate) (PHBV) are the most popular and commercially exploited [135].

PHA can be produced by means of microbial routes using bacteria and cyanobacteria. *Bacillus*, *Cupriavidus*, *Haloferax*, *Halomonas*, and *Escherichia* have been the most used microorganism to synthetize PHA [136, 137]. Beyond the cyanobacteria, *Synechococcus*, *Spirulina*, *Nostoc*, *Phormidium*, and *Synechocystis* genus can be highlighted as producers of PHA and its derivatives [135, 138]. The microorganisms produce PHA as a carbon and energy store, and they encountered intracellular cytoplasmic inclusions when some stress culture condition is applied as nitrogen,

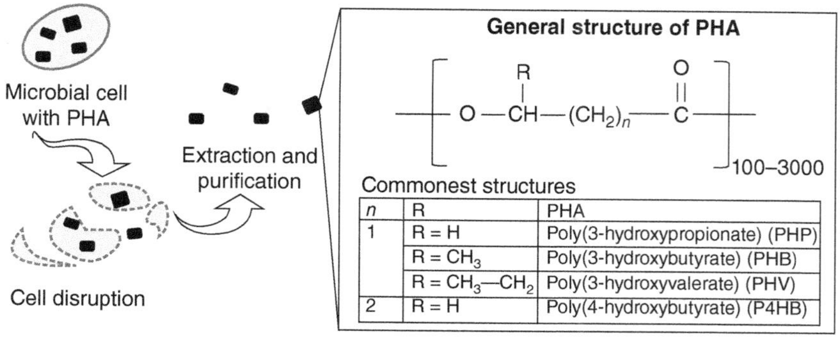

n	R	PHA
1	R = H	Poly(3-hydroxypropionate) (PHP)
	R = CH_3	Poly(3-hydroxybutyrate) (PHB)
	R = CH_3—CH_2	Poly(3-hydroxyvalerate) (PHV)
2	R = H	Poly(4-hydroxybutyrate) (P4HB)

Figure 2.2 Polyhydroxyalkanoate (PHA) extraction and its general structure, where *R* is variable, and *n* can be from 1 to 13.

phosphorus, or oxygen limitation [138, 139]. Various factors affect the PHA produced: the microorganism itself, the substrate source, the stress condition applied, the activity of the enzymes, etc. [133]. The R side chain of the PHA can vary from hydrogen to methyl tridecyl group, and this is dependent on the substrate used and the metabolism of the microorganism [138]. Another important factor to be considered is the extraction of these biopolymers, which may influence their properties and applicability [139]. Figure 2.2 presents the steps to obtain PHA, its structure, and the most common PHA derivatives. Additionally, polymerization through the opening of β-butyrolactone chains using zinc or aluminum can be performed to produce PHA derivatives [140].

Food packaging materials based on PHA are the most common application of this biopolymer. PHA has interesting physicochemical due to the monomeric composition of its copolymers. The PHA-based materials properties vary from fragile and crystalline thermoplastics (PHB) to elastomeric ones (PHV, PHBV) [141]. PHA is considered biodegradable and biocompatible [140–142]. In addition, hydrophobicity property makes PHA-based materials insoluble in water and provides good water barrier properties. PHA is a semi-crystalline material with crystallinity degree between 30% and 80%, and melting temperature oscillating between 50 and 180 °C [133, 138]. Beyond PHAs, PHB is the most studied one regarding food packaging applications. PHB has similar properties to those found in conventional plastics; hence PHB can be extruded and molded to manufacture films at industrial scale. However, since this biopolymer has high crystallinity, the films and coatings become rigid and brittle [143]; in this sense, the incorporation of 3-hydroxyvalerate, producing PHBV, or other monomers has been used to overcome its shortcoming. The blend with other biopolymers or incorporation of other types of materials to produce composites is also studied in the food packaging area [133]. Table 2.7 shows an overview of studies that applied PHA and its derivatives as food packaging materials, highlighting the main results obtained by the authors.

The PHA derivatives have been used as food packaging materials; most studies have focused on the application of this biopolymer such as film for direct or indirect food contact, as well as barrier to coat paper-based packaging (Table 2.7).

Table 2.7 Films and coatings based on polyhydroxyalkanoates for food packaging applications.

Components	Production approach	Main results	References
PHA[a]/long alkyl chain quaternary/graphene oxide Nanocomposite	Casting	Films with improved mechanical, thermal, and oxygen barrier properties. Films with antimicrobial activity against *S. aureus* and *E. coli*	[144]
PHB[b]/PEG[c]/organobentonite or organovermiculite	Casting	Increase on the processability of the PHB[b] with the improvement of the thermal resistance and crystallinity degree	[145]
PHBV[d]/PHB[b]/PLA[e]/catechin	Electrospun/thermo-pressing molding	The addition of PHB[b] increased the PLA[e]-based fibers crystallinity. The film presented antioxidant activity related to catechin release on fatty food model. The incorporation of PHB[b]/PLA[e] layer improved the mechanical properties	[146]
PHA[e]/apple extract/cellulose	Casting	The PHA[a] increased the hydrophobicity and transparency of the films whereas the tensile strength was reduced. The apple extract gives an antioxidant property to the coating	[147]
PHBV[d]/cellulose/TEC[f] or PEG[c]	Extrusion	A sufficient adhesion (cohesion break) between paper and PHBV[d] layer was obtained. The use of TEC[f] or PEG[c] produced PHBV layers with lower defects and with increase of grease barrier property	[148]

PHB[b)]/bacterial cellulose/zinc oxide nanoparticle	Thermo-pressing molding/plasma	Films with better mechanical properties and antimicrobial activity against *E. coli* and *S. aureus*	[149]
PHB[b)]/starch/montmorillonite/eugenol	Extrusion	Reinforced films with antimicrobial properties manufactured at industrial scale	[150]
PLA[e)]/PHB[b)]/cinnamaldehyde	Casting	Films with better mechanical properties and slower release of the active compound. These films were used to extend the shelf life of salmon dices	[151]
PHB[b)]/palladium nanoparticles	Electrospinning/thermo-pressing molding	Films with oxygen scavenging capability and good barrier properties	[152]
PHB[b)]/silver nanoparticles	Thermo-pressing molding	Films with antimicrobial activity against *S. enterica* and *L. monocytogenes*. In addition, nanoparticles did not influence the biodegradability of films	[153]
PHBV[d)]/PEG[c)] or PHB[b)]/zein	Electrospinning/thermos-pressing molding	Films with acceptable mechanical and thermal properties, as well as with antimicrobial activity against *L. monocytogenes*	[154]

a) PHA: polyhydroxyalkanoate.
b) PHB: poly(3-hydroxybutyrate).
c) PEG: poly(ethylene glycol).
d) PHBV: poly(3-hydroxybutyrate-*co*-3-hydroxyvalerate).
e) PLA: poly(lactic acid).
f) TEC: triethyl citrate.

2.4 Main Biodegradable Polymers Chemically Synthesized

Polylactide (PLA) also known as poly(lactic acid) are macromolecules synthetized by the polymerization of lactic acid (LA) or by the ring-opening polymerization (ROP) of lactide, referred to as polylactide [155, 156]. Both types are commonly referred as PLA; however, ROP of lactide is the most used route to obtain PLA for food packaging applications [157].

PLA is a biodegradable macromolecule, and it is considered as one of the most extensively developed commercial biomaterials. PLA is produced by lactic fermentation of polysaccharides or sugars extracted from corn, potato, sugar cane, or sugar beet. Other raw materials such as cellulosic biomass, agricultural waste, and nonfood crops (e.g. switchgrass) can be used as alternative during the fermentation [155, 156]. PLA can be processed by means of extrusion, thermoforming, injection molding, blow molding, and industrial scale [156].

Films and coating based on PLA are transparent materials with glass transition at room temperature. These materials have tensile strength of ≈60 MPa, an elongation at break of ≈5%, and a Young modulus of 3–4 GPa, and it constitutes a promising alternative to conventional plastics in food packaging. Some applications of PLA are presented in Table 2.8. However there are several factors that limit the application of PLA: low oxygen barrier (O_2) compared to poly(ethylene terephthalate) (PET) and similar to that of polystyrene (PS). The carbon dioxide permeability coefficient of PLA is lower than PS and low-density polyethylene (LDPE). The nitrogen permeability of PLA is lower than PS but higher than that informed for unplasticized poly(vinyl chloride) (PVC). PLA is a hygroscopic material and very sensitive to high relative humidity and temperature [163].

Other factors that reduce its use commercially are associated with the low thermal stability and high price [156]. Besides that, the PLA is brittle and has low toughness due to its glass transition temperature (55–60 °C) and slow crystallization rate. Thus, its use at low temperatures is limited by the low heat deflection temperature [157, 164]. To overcome the intrinsic fragility of PLA, a number of approaches have been developed such as blend modification, plasticization, copolymerization, and incorporation of natural nanofillers [165].

Some plasticizers such as polyols, citrate ester, lactic ester, and glycerol esters can be used to improve the processability of PLA and enhance its flexibility and ductility [166]. Also, in the literature, PLA has been blended with poly(ethylene glycol), poly(propylene glycol), and polymeric adipates [155, 167]. In this sense, Yang et al. [165] studied PLA films blended with ethylene-vinyl acetate-glycidyl methacrylate (NPG) and lignin microparticles to improve the toughness of PLA and performance of antioxidant for active food packing. The results showed improvement of toughness and water vapor barrier properties and demonstrated the satisfactory compatibility between lignin, PLA, and NPG.

Table 2.8 Films and coatings based on poly(lactic acid) for food packaging applications.

Components	Production approach	Main results	References
PLA[a)]/nanochitosan/*Polylophium involucratum* essential oil	Casting	Films with antibacterial and antioxidant properties were used to preserve chicken fillets at refrigeration temperature	[158]
PLA[a)]/pea starch	Casting	Films reduced the weight loss of cherry tomatoes and extended the retention of organic acids and vitamin C	[159]
PLA[a)]/GTE[b)]	Blown extrusion	Films reduced the lipid oxidation of smoked salmon during the storage (60 days)	[160]
PLA[a)]/carvacrol	Electrospinning	Films with high thermal stability and antimicrobial properties against aerobic bacteria, mold, and yeast in whole wheat bread	[161]
PLA[a)]/GO[c)]/CLO[d)]	Casting	Elastic films with antimicrobial activity against *S. aureus* and *E. coli*	[162]

a) PLA: poly(lactic acid).
b) GTE: green tea extract.
c) GO: graphene oxide.
d) CLO: clove essential oil.

2.5 Conclusions

Films and coatings based on biopolymer are probably the future materials that will be used to package food products from the economic and environmental viewpoints; however, the mechanical, barrier, and thermal properties of these materials must be improved as well as their production at industrial scale. Films and coatings with improved physicochemical properties can be obtained by blending different biopolymers, or biopolymers with nanoparticles, antioxidant, or antimicrobial compounds, among others. Most researchers have focused on developing films and coatings at laboratory scale, particularly using the casting and dip casting methodologies. It is necessary to regard other approaches such as thermo-pressing molding, extrusion, blown extrusion, reactive extrusion, and spraying, aiming films and coatings production at industrial scale. The use of alternative approaches such as electrospinning, electrospun, ionic gelation, layer-by-layer, plasma, and self-assembly could produce new biopolymer architectures with a positive impact on the physicochemical properties of films and coatings. Finally, the toxicity and migration of macromolecules and additives in films and coatings must be carefully investigated before the application of these materials by the food packaging industry.

Conflicts of Interest

The author declares no conflict of interest.

Acknowledgments

The authors gratefully acknowledge the Coordination for the Improvement of Higher Education Personnel (CAPES), National Council for Scientific and Technological Development (CNPq), and the Federal University of Santa Catarina (UFSC). G.A. Valencia gratefully acknowledges CNPq for the research grant (405432/2018-6).

References

1 Ghaani, M., Cozzolino, C.A., Castelli, G., and Farris, S. (2016). An overview of the intelligent packaging technologies in the food sector. *Trends Food Sci. Technol.* 51: 1–11. https://doi.org/10.1016/j.tifs.2016.02.008.

2 Valencia, G.A., Luciano, C.G., and Monteiro, A.R. (2019). Smart and active edible coatings based on biopolymers. In: *Polymers for Agri-Food Applications* (ed. T.J. Gutiérrez), 391–416. Gewerbestrasse: Springer.

3 Valencia, G.A. and Sobral, P.J.d.A. (2018). Recent trends on nano-biocomposite polymers for food packaging. In: *Polymers for Food Applications*, 1e (ed. T.J. Gutiérrez), 101–130. Gewerbestrasse: Springer.

4 da Silva, M.N., de Fonseca, J., M., Feldhaus, H.K. et al. (2019). Physical and morphological properties of hydroxypropyl methylcellulose films with curcumin polymorphs. *Food Hydrocolloids* 97: 105217. https://doi.org/10.1016/j.foodhyd.2019.105217.

5 Valencia, G.A., Luciano, C.G., Lourenço, R.V. et al. (2019). Morphological and physical properties of nano-biocomposite films based on collagen loaded with laponite®. *Food Packag. Shelf Life* 19: 24–30. doi: 10.1016/j.fpsl.2018.11.013.

6 Andretta, R., Luchese, C.L., Tessaro, I.C., and Spada, J.C. (2019). Development and characterization of pH-indicator films based on cassava starch and blueberry residue by thermocompression. *Food Hydrocolloids* 93: 317–324. https://doi.org/10.1016/j.foodhyd.2019.02.019.

7 Hanani, Z.A.N. (2018). Surface properties of biodegradable polymers for food packaging. In: *Polymers for Food Applications*, 1e (ed. T.J. Gutiérrez), 817. Cham: Springer.

8 Gutiérrez, T.J., Guarás, M.P., and Alvarez, V.A. (2017). Reactive extrusion for the production of starch-based biopackaging. In: *Biopackaging* (ed. M.A. Masuelli). Miami: CRC Press, Taylor & Francis Group.

9 Gruji, R., Vujadinovi, D., and Savanovi, D. (2017). Biopolymers as food packaging materials. In: *Advances in Applications of Industrial Biomaterials*, 1e (eds. E. Pellicer, D. Nikolic, J. Sort, et al.), 139–160. Springer International Publishing.

10 Shankar, S. and Rhim, J.-W. (2018). Bionanocomposite films for food packaging applications. In: *Reference Module in Food Science*, 1e, 1–10. Elsevier https://doi.org/10.1016/B978-0-08-100596-5.21875-1.

11 Vieira, M.G.A., da Silva, M.A., dos Santos, L.O., and Beppu, M.M. (2011). Natural-based plasticizers and biopolymer films: a review. *Eur. Polym. J.* 47 (3): 254–263. https://doi.org/10.1016/j.eurpolymj.2010.12.011.

12 Fox, P.F. and Brodkorb, A. (2008). The casein micelle: historical aspects, current concepts and significance. *Int. Dairy J.* 18 (7): 677–684.

13 Chevalier, E., Assezat, G., Prochazka, F., and Oulahal, N. (2018). Development and characterization of a novel edible extruded sheet based on different casein sources and influence of the glycerol concentration. *Food Hydrocolloids* 75: 182–191.

14 Ramesh, C. and Kilara, A. (2011). *Dairy Ingredients for Food Processing*. Wiley https://doi.org/10.1002/9780470959169.

15 Abd El-Salam, M.H. and El-Shibiny, S. (2019). Preparation and potential applications of casein–polysaccharide conjugates: a review. *J. Sci. Food Agric.* 100: 1852–1859.

16 Picchio, M.L., Linck, Y.G., Monti, G.A. et al. (2018). Casein films crosslinked by tannic acid for food packaging applications. *Food Hydrocolloids* 84: 424–434.

17 Ranadheera, C.S., Liyanaarachchi, W.S., Chandrapala, J. et al. (2016). Utilizing unique properties of caseins and the casein micelle for delivery of sensitive food ingredients and bioactives. *Trends Food Sci. Technol.* 57: 178–187.

18 Farhan, A. and Hani, N.M. (2017). Characterization of edible packaging films based on semi-refined kappa-carrageenan plasticized with glycerol and sorbitol. *Food Hydrocolloids* 64: 48–58.

19 Cano, A., Andres, M., Chiralt, A., and González-Martinez, C. (2020). Use of tannins to enhance the functional properties of protein based films. *Food Hydrocolloids* 100: 1–9.

20 Eghbal, N., Dumas, E., Yarmand, M.S. et al. (2019). Antimicrobial films based on pectin and sodium caseinate for the release of antifungal natamycin. *J. Food Process. Preserv.* 43 (7): 1–10.

21 Murmu, S.B. and Mishra, H.N. (2018). The effect of edible coating based on Arabic gum, sodium caseinate and essential oil of cinnamon and lemon grass on guava. *Food Chem.* 245: 820–828.

22 Andrade, M.A., Ribeiro-Santos, R., Guerra, M., and Sanches-Silva, A. (2019). Evaluation of the oxidative status of salami packaged with an active whey protein film. *Foods* 8 (9): 1–15.

23 Tavakolipour, H., Kalbasi-Ashtari, A., and Mokhtarian, M. (2020). Effects of coating pistachio kernels with mixtures of whey protein and selected herbal plant extracts on growth inhibition of *Aspergillus flavus* and prevention of aflatoxin during storage. *J. Food Saf.* 40: e12711.

24 Castro, F.V.R., Andrade, M.A., Silva, A.S. et al. (2019). The contribution of a whey protein film incorporated with green tea extract to minimize the lipid oxidation of salmon (*Salmo salar* L.). *Foods* 8 (8): 1–16.

25 Nallan Chakravartula, S.S., Cevoli, C., Balestra, F. et al. (2019). Evaluation of the effect of edible coating on mini-buns during storage by using NIR spectroscopy. *J. Food Eng.* 263 (April): 46–52.

26 Feng, Z., Wu, G., Liu, C. et al. (2018). Edible coating based on whey protein isolate nano fibrils for antioxidation and inhibition of product browning. *Food Hydrocolloids* 79: 179–188.

27 Phupoksakul, T., Leuangsukrerk, M., Somwangthanaroj, A. et al. (2017). Storage stability of packaged baby formula in poly(lactide)-whey protein isolate laminated pouch. *J. Sci. Food Agric.* 97 (10): 3365–3373.

28 Yadav, J.S.S., Yan, S., Pilli, S. et al. (2015). Cheese whey: a potential resource to transform into bioprotein, functional/nutritional proteins and bioactive peptides. *Biotechnol. Adv.*

29 de Castro, R.J.S., Domingues, M.A.F., Ohara, A. et al. (2017). Whey protein as a key component in food systems: physicochemical properties, production technologies and applications. *Food Struct.*

30 Murrieta-Martínez, C.L., Soto-Valdez, H., Pacheco-Aguilar, R. et al. (2018). Edible protein films: sources and behavior. *Packag. Technol. Sci.* 31 (3): 113–122.

31 Schmid, M. and Müller, K. (2019). Whey protein-based packaging films and coatings. In: *Whey Proteins* (eds. H.C. Deeth and N. Bansal), 407–437. Elsevier Inc.

32 Carvalho, R.A., Santos, T.A., de Azevedo, V.M. et al. (2018). Bio-nanocomposites for food packaging applications: effect of cellulose nanofibers on morphological, mechanical, optical and barrier properties. *Polym. Int.* 67 (4): 386–392.

33 Sun, B., Zhang, M., Shen, J. et al. (2019). Applications of cellulose-based materials in sustained drug delivery systems. *Curr. Med. Chem.* 26 (14): 2485–2501.

34 Kalia, S., Dufresne, A., Cherian, B.M. et al. (2011). Cellulose-based bio- and nanocomposites: a review. *Int. J. Polym. Sci.* 2011: 1–35.

35 Zaman, A., Huang, F., Jiang, M. et al. (2020). Preparation, properties, and applications of natural cellulosic aerogels: a review. *Energy Built Environ.* 1: 60–76.

36 Zhang, X., Lin, F., Yuan, Q. et al. (2019). Hydrogen-bonded thin films of cellulose ethers and poly(acrylic acid). *Carbohydr. Polym.* 215: 58–62.

37 Clasen, C. and Kulicke, W.M. (2001). Determination of viscoelastic and rheo-optical material functions of water-soluble cellulose derivatives. *Prog. Polym. Sci.* 26: 1839–1919.

38 Ma, Q. and Wang, L. (2016). Preparation of a visual pH-sensing film based on tara gum incorporating cellulose and extracts from grape skins. *Sens. Actuators, B* 235: 401–407. https://doi.org/10.1016/j.snb.2016.05.107.

39 Pérez-Gago, M.B., Fagundes, C., Monteiro, A.R., and Palou, L. (2016). Antifungal activity of food additives in vitro and as ingredients of hydroxypropyl methylcellulose (HPMC)-based edible coatings against *Alternaria alternata* on cherry tomato fruit. In: *IX Simpósio Ibérico Matur e Pós-Colheita*, Lisboa, Port 2 a 4 novembro 2016, 107–114.

40 Fathi Achachlouei, B. and Zahedi, Y. (2018). Fabrication and characterization of CMC-based nanocomposites reinforced with sodium montmorillonite and TiO_2 nanomaterials. *Carbohydr. Polym.* 199 (February): 415–425.
41 Niu, X., Liu, Y., Song, Y. et al. (2018). Rosin modified cellulose nanofiber as a reinforcing and co-antimicrobial agents in polylactic acid /chitosan composite film for food packaging. *Carbohydr. Polym.* 183 (August 2017): 102–109.
42 Noorbakhsh-Soltani, S.M., Zerafat, M.M., and Sabbaghi, S. (2018). A comparative study of gelatin and starch-based nano-composite films modified by nano-cellulose and chitosan for food packaging applications. *Carbohydr. Polym.* 189 (February): 48–55.
43 Halim, A.L.A., Kamari, A., and Phillip, E. (2018). Chitosan, gelatin and methylcellulose films incorporated with tannic acid for food packaging. *Int. J. Biol. Macromol.* 120: 1119–1126.
44 Sinaga, M.Z.E., Gea, S., Panindia, N., and Sihombing, Y.A. (2018). The preparation of cellulose nanocomposite film from isolated cellulose of corncobs as food packaging. *Orient. J. Chem.* 34 (1): 562–567.
45 Mohammadi, H., Kamkar, A., and Misaghi, A. (2018). Nanocomposite films based on CMC, okra mucilage and ZnO nanoparticles: Physico mechanical and antibacterial properties. *Carbohydr. Polym.* 181 (August 2017): 351–357.
46 Sun, G., Chi, W., Zhang, C. et al. (2019 Sep). Developing a green film with pH-sensitivity and antioxidant activity based on κ-carrageenan and hydroxypropyl methylcellulose incorporating *Prunus maackii* juice. *Food Hydrocolloids* 94: 345–353.
47 Azarifar, M., Ghanbarzadeh, B., Sowti Khiabani, M. et al. (2019). The optimization of gelatin-CMC based active films containing chitin nanofiber and *Trachyspermum ammi* essential oil by response surface methodology. *Carbohydr. Polym.* 208 (December 2018): 457–468.
48 Yadav, N. and Kaur, R. (2019 Dec). Environment friendly qualitatively responsive ethyl cellulose films as smart food packaging. *Mater. Express* 9 (7): 792–800.
49 El Fawal, G., Hong, H., Song, X. et al. (2020). Fabrication of antimicrobial films based on hydroxyethylcellulose and ZnO for food packaging application. *Food Packag. Shelf Life* 23.
50 Lopez-Polo, J., Silva-Weiss, A., Zamorano, M., and Osorio, F.A. (2020). Humectability and physical properties of hydroxypropyl methylcellulose coatings with liposome-cellulose nanofibers: Food application. *Carbohydr. Polym.* 231: 1–10.
51 Abdou, E.S., Nagy, K.S.A., and Elsabee, M.Z. (2008). Extraction and characterization of chitin and chitosan from local sources. *Bioresour. Technol.* 99: 1359–1367.
52 Marei, N.H., El-Samie, E.A., Salah, T. et al. (2016). Isolation and characterization of chitosan from different local insects in Egypt. *Int. J. Biol. Macromol.* 82: 871–877.
53 Rinaudo, M. (2006). Chitin and chitosan: properties and applications. *Prog. Polym. Sci.* 31 (7): 603–632.

54 Mujtaba, M., Morsi, R.E., Kerch, G. et al. (2019). Current advancements in chitosan-based film production for food technology; a review. *Int. J. Biol. Macromol.* 121: 889–904.

55 Yoshida, C.M.P., Borges, V., Maciel, V. et al. (2014). Chitosan biobased and intelligent films : monitoring pH variations. *LWT Food Sci. Technol.* 55 (1): 83–89.

56 Ge, J., Yue, P., Chi, J. et al. (2018). Formation and stability of anthocyanins-loaded nanocomplexes prepared with chitosan hydrochloride and carboxymethyl chitosan. *Food Hydrocolloids* 74: 23–31.

57 Elsabee, M.Z. and Abdou, E.S. (2013). Chitosan based edible films and coatings: a review. *Mater. Sci. Eng., C* 33 (4): 1819–1841.

58 Rodríguez-Núñez, J.R., Madera-Santana, T.J., Sánchez-Machado, D.I. et al. (2014). Chitosan/hydrophilic plasticizer-based films: preparation, physicochemical and antimicrobial properties. *J. Polym. Environ.* 22 (1): 41–51.

59 Badawy, M.E.I., Rabea, E.I., El-Nouby M, A.M. et al. (2017). Strawberry shelf life, composition, and enzymes activity in response to edible chitosan coatings. *Int. J. Fruit Sci.* 17 (2): 117–136.

60 Halász, K. and Csóka, L. (2018). Black chokeberry (*Aronia melanocarpa*) pomace extract immobilized in chitosan for colorimetric pH indicator film application. *Food Packag. Shelf Life* 16 (September 2017): 185–193.

61 Remedio, L.N., Silva dos Santos, J.W., Vieira Maciel, V.B. et al. (2019). Characterization of active chitosan films as a vehicle of potassium sorbate or nisin antimicrobial agents. *Food Hydrocolloids* 87 (June 2018): 830–838.

62 Wu, C., Sun, J., Lu, Y. et al. (2019). In situ self-assembly chitosan/ε-polylysine bionanocomposite film with enhanced antimicrobial properties for food packaging. *Int. J. Biol. Macromol.* 132: 385–392.

63 Gates, S.J. and Shukla, A. (2017). Layer-by-layer assembly of readily detachable chitosan and poly(acrylic acid) polyelectrolyte multilayer films. *J. Polym. Sci., Part B: Polym. Phys.* 55 (2): 127–131.

64 Galvis-Sánchez, A.C., Castro, M.C.R., Biernacki, K. et al. (2018). Natural deep eutectic solvents as green plasticizers for chitosan thermoplastic production with controlled/desired mechanical and barrier properties. *Food Hydrocolloids* 82: 478–489.

65 Chabbi, J., Jennah, O., Katir, N. et al. (2018). Aldehyde-functionalized chitosan-montmorillonite films as dynamically-assembled, switchable-chemical release bioplastics. *Carbohydr. Polym.* 183: 287–293.

66 Fernandes, C., Calderon V., S., Ballesteros, L.F. et al. (2018). Carbon-based sputtered coatings for enhanced chitosan-based films properties. *Appl. Surf. Sci.* 433: 689–695.

67 Serio, A., Chaves-López, C., Sacchetti, G. et al. (2018). Chitosan coating inhibits the growth of *Listeria monocytogenes* and extends the shelf life of vacuum-packed pork loins at 4 °C. *Foods* 7 (10): 1–10.

68 Bilbao-Sainz, C., Chiou, B.S., Punotai, K. et al. (2018). Layer-by-layer alginate and fungal chitosan based edible coatings applied to fruit bars. *J. Food Sci.* 83 (7): 1880–1887.

69 Castelo Branco Melo, N.F., de MendonçaSoares, B.L., Marques Diniz, K. et al. (2018). Effects of fungal chitosan nanoparticles as eco-friendly edible coatings on the quality of postharvest table grapes. *Postharvest Biol. Technol.* 139 (January): 56–66.

70 Gorgieva, S. and Kokol, V. (2011). Collagen- vs. gelatine-based biomaterials and their biocompatibility: review and perspectives. In: *Biomaterials Applications for Nanomedicine* (ed. R. Pignatello), 17–52. InTech.

71 Etxabide, A., Uranga, J., Guerrero, P., and de la Caba, K. (2017). Development of active gelatin films by means of valorisation of food processing waste: a review. *Food Hydrocolloids* 68: 192–198.

72 Bhagwat, P.K. and Dandge, P.B. (2018). Collagen and collagenolytic proteases: a review. *Biocatal. Agric. Biotechnol.* 15 (May): 43–55.

73 Da Silva, T.F. and Penña, A.L.B. (2012). Colágeno: Características químicas e propriedades funcionais. *Rev. Inst. Adolfo Lutz* 71 (3): 530–539.

74 Osawa, Y., Mizushige, T., Jinno, S. et al. (2018). Absorption and metabolism of orally administered collagen hydrolysates evaluated by the vascularly perfused rat intestine and liver in situ. *Biomed Res.* 39 (1): 1–11.

75 Correia, D.M., Padrão, J., Rodrigues, L.R. et al. (2013). Thermal and hydrolytic degradation of electrospun fish gelatin membranes. *Polym. Test.* 32 (5): 995–1000.

76 Lin, L., Regenstein, J.M., Lv, S. et al. (2017). An overview of gelatin derived from aquatic animals: properties and modification. *Trends Food Sci. Technol.* 68: 102–112.

77 Lv, L.C., Huang, Q.Y., Ding, W. et al. (2019). Fish gelatin: the novel potential applications. *J. Funct. Foods* 63: 1–14.

78 Aitboulahsen, M., Zantar, S., Laglaoui, A. et al. (2018). Gelatin-based edible coating combined with *Mentha pulegium* essential oil as bioactive packaging for strawberries. *J. Food Qual.* 2018.

79 dos Garcia, V.A., S., Borges, J.G., Osiro, D. et al. (2020). Orally disintegrating films based on gelatin and starch pregelatinized: new carriers of active compounds from acerola. *Food Hydrocolloids* 101: 1–12.

80 He, Q., Zhang, Y., Cai, X., and Wang, S. (2016). Fabrication of gelatin-TiO_2 nanocomposite film and its structural, antibacterial and physical properties. *Int. J. Biol. Macromol.* 84: 153–160.

81 Wang, Z., Hu, S., Gao, Y. et al. (2017). Effect of collagen-lysozyme coating on fresh-salmon fillets preservation. *LWT Food Sci. Technol.* 75: 59–64.

82 Wang, Z., Hu, S., and Wang, H. (2017). Scale-up preparation and characterization of collagen/sodium alginate blend films. *J. Food Qual.* 2017: 1–11.

83 Wang, W., Liu, Y., Jia, H. et al. (2017). Effects of cellulose nanofibers filling and palmitic acid emulsions coating on the physical properties of fish gelatin films. *Food Biophys.* 12 (1): 23–32.

84 Batpho, K., Boonsupthip, W., and Rachtanapun, C. (2017). Antimicrobial activity of collagen casing impregnated with nisin against foodborne microorganisms associated with ready-to-eat sausage. *Food Control* 73: 1342–1352.

85 López-Carballo, G., Hernández-Muñoz, P., and Gavara, R. (2018). Photoactivated self-sanitizing chlorophyllin-containing coatings to prevent microbial contamination in packaged food. *Coatings* 8 (9): 1–14.

86 Amjadi, S., Emaminia, S., Nazari, M. et al. (2019). Application of reinforced ZnO nanoparticle-incorporated gelatin bionanocomposite film with chitosan nanofiber for packaging of chicken fillet and cheese as Food models. *Food Bioprocess Technol.* 12 (7): 1205–1219.

87 Bhuimbar, M.V., Bhagwat, P.K., and Dandge, P.B. (2019). Extraction and characterization of acid soluble collagen from fish waste: development of collagen-chitosan blend as food packaging film. *J. Environ. Chem. Eng.* 7 (2): 1–7.

88 Ławińska, K., Lasoń-Rydel, M., Gendaszewska, D. et al. (2019). Coating of seeds with collagen hydrolysates from leather waste. *Fibres Text. East. Eur* 27 (4): 59–64.

89 Scartazzini, L., Tosati, J.V., Cortez, D.H.C. et al. (2019). Gelatin edible coatings with mint essential oil (*Mentha arvensis*): film characterization and antifungal properties. *J. Food Sci. Technol.* 56 (9): 4045–4056.

90 Kouhdasht, A.M. and Nasab, M.M. (2020). Shelf-life extension of whole shrimp using an active coating containing fish skin gelatin hydrolysates produced by a natural protease. *Food Sci. Nutr.* 8: 214–223.

91 Ma, Y., Teng, A., Zhao, K. et al. (2020). A top-down approach to improve collagen film's performance: the comparisons of macro, micro and nano sized fibers. *Food Chem.* 309.

92 Moula Ali, A.M., de la Caba, K., Prodpran, T., and Benjakul, S. (2020). Quality characteristics of fried fish crackers packaged in gelatin bags: effect of squalene and storage time. *Food Hydrocolloids* 99.

93 Medic, J., Atkinson, C., and Hurburgh, C.R. (2014). Current knowledge in soybean composition. *J. Am. Oil Chem. Soc.* 91: 363–384.

94 Koshy, R.R., Mary, S.K., Thomas, S., and Pothan, L.A. (2015). Environment friendly green composites based on soy protein isolate - a review. *Food Hydrocolloids* 50: 174–192.

95 Nishinari, K., Fang, Y., Guo, S., and Phillips, G.O. (2014). Soy proteins: a review on composition, aggregation and emulsification. *Food Hydrocolloids* 39: 301–318.

96 Abaee, A., Mohammadian, M., and Jafari, S.M. (2017). Whey and soy protein-based hydrogels and nano-hydrogels as bioactive delivery systems. *Trends Food Sci. Technol.* 70 (October): 69–81.

97 Tian, H., Guo, G., Fu, X. et al. (2018). Fabrication, properties and applications of soy-protein-based materials: a review. *Int. J. Biol. Macromol.* 120: 475–490.

98 Alipoormazandarani, N., Ghazihoseini, S., and Mohammadi, N.A. (2015). Preparation and characterization of novel bionanocomposite based on soluble soybean polysaccharide and halloysite nanoclay. *Carbohydr. Polym.* 134: 745–751.

99 Salarbashi, D., Mortazavi, S.A., Noghabi, M.S. et al. (2016). Development of new active packaging film made from a soluble soybean polysaccharide incorporating ZnO nanoparticles. *Carbohydr. Polym.* 140: 220–227.

100 Akbariazam, M., Ahmadi, M., Javadian, N., and Mohammadi, N.A. (2016). Fabrication and characterization of soluble soybean polysaccharide and nanorod-rich ZnO bionanocomposite. *Int. J. Biol. Macromol.* 89: 369–375.

101 Oh, Y.A., Roh, S.H., and Min, S.C. (2016). Cold plasma treatments for improvement of the applicability of defatted soybean meal-based edible film in food packaging. *Food Hydrocolloids* 58: 150–159.

102 Martelli-Tosi, M., Assis, O.B.G., Silva, N.C. et al. (2017). Chemical treatment and characterization of soybean straw and soybean protein isolate/straw composite films. *Carbohydr. Polym.* 157: 512–520.

103 Salarbashi, D., Noghabi, M.S., Bazzaz, B.S.F. et al. (2017). Eco-friendly soluble soybean polysaccharide/nanoclay Na+bionanocomposite: properties and characterization. *Carbohydr. Polym.* 169: 524–532.

104 Ciannamea, E.M., Espinosa, J.P., Stefani, P.M., and Ruseckaite, R.A. (2018). Long-term stability of compression-molded soybean protein concentrate films stored under specific conditions. *Food Chem.* 243 (August 2017): 448–452.

105 Ghani, S., Barzegar, H., Noshad, M., and Hojjati, M. (2018). The preparation, characterization and in vitro application evaluation of soluble soybean polysaccharide films incorporated with cinnamon essential oil nanoemulsions. *Int. J. Biol. Macromol.* 112: 197–202.

106 Ma, W., Rokayya, S., Xu, L. et al. (2018). Physical-chemical properties of edible film made from soybean residue and citric acid. *J. Chem.* 2018: 1–8.

107 Han, Y., Yu, M., and Wang, L. (2018). Bio-based films prepared with soybean by-products and pine (*Pinus densiflora*) bark extract. *J. Cleaner Prod.* 187: 1–8.

108 Zhang, L., Chen, F., Lai, S. et al. (2018). Impact of soybean protein isolate-chitosan edible coating on the softening of apricot fruit during storage. *LWT* 96: 604–611.

109 Barbut, S. and Harper, B.A. (2019). Dried Ca-alginate films: effects of glycerol, relative humidity, soy fibers, and carrageenan. *LWT* 103: 260–265.

110 González, A., Gastelú, G., Barrera, G.N. et al. (2019). Preparation and characterization of soy protein films reinforced with cellulose nanofibers obtained from soybean by-products. *Food Hydrocolloids* 89: 758–764.

111 Merci, A., Marim, R.G., Urbano, A., and Mali, S. (2019). Films based on cassava starch reinforced with soybean hulls or microcrystalline cellulose from soybean hulls. *Food Packag. Shelf Life* 20.

112 Liu, Q.R., Qi, J.R., Yin, S.W. et al. (2017). Preparation and stabilizing behavior of octenyl succinic esters of soybean soluble polysaccharide in acidified milk beverages. *Food Hydrocolloids* 63: 421–428.

113 Chivero, P., Gohtani, S., Ikeda, S., and Nakamura, A. (2014). The structure of soy soluble polysaccharide in aqueous solution. *Food Hydrocolloids* 35: 279–286.

114 Yuan, C., Chen, M., Luo, J. et al. (2017). A novel water-based process produces eco-friendly bio-adhesive made from green cross-linked soybean soluble polysaccharide and soy protein. *Carbohydr. Polym.* 169: 417–425.

115 Fátima Seibel, N. and Beléia, A.D.P. (2008). Carboidratos das fibras de cotilédones e proteínas de produtos derivados de soja (*Glycine max* (L.) Merril). *Cienc. Tecnol. Aliment.* 28 (3): 607–613.

116 Willats, W.G.T., Mccartney, L., Mackie, W., and Knox, J.P. (2001). Pectin: cell biology and prospects for functional analysis. *Plant Mol. Biol.* 47: 9–27.

117 Beckham, G.T., Johnson, C.W., Karp, E.M. et al. (2016). Opportunities and challenges in biological lignin valorization. *Curr. Opin. Biotechnol.* 42: 40–53.

118 Ogunsona, E., Ojogbo, E., and Mekonnen, T. (2018). Advanced material applications of starch and its derivatives. *Eur. Polym. J.* 108: 570–581.

119 Copeland, L., Blazek, J., Salman, H., and Tang, M.C. (2009). Form and functionality of starch. *Food Hydrocolloids* 23: 1527–1534.

120 Thakur, R., Pristijono, P., Scarlett, C.J. et al. (2019). Starch-based films: major factors affecting their properties. *Int. J. Biol. Macromol.* 132: 1079–1089.

121 Saberi, B., Golding, J.B., Marques, J.R. et al. (2018). Application of biocomposite edible coatings based on pea starch and guar gum on quality, storability and shelf life of 'Valencia' oranges. *Postharvest Biol. Technol.* 137: 9–20.

122 Pellá, M.C.G., Silva, O.A., Pellá, M.G. et al. (2020). Effect of gelatin and casein additions on starch edible biodegradable films for fruit surface coating. *Food Chem.* 309.

123 Malherbi, N.M., Schmitz, A.C., Grando, R.C. et al. (2019). Corn starch and gelatin-based films added with guabiroba pulp for application in food packaging. *Food Packag. Shelf Life* 19: 140–146.

124 Nawab, A., Alam, F., and Hasnain, A. (2017). Mango kernel starch as a novel edible coating for enhancing shelf- life of tomato (*Solanum lycopersicum*) fruit. *Int. J. Biol. Macromol.* 103: 581–586. https://doi.org/10.1016/j.ijbiomac.2017.05.057.

125 Pająk, P., Socha, R., Łakoma, P., and Fortuna, T. (2017). Antioxidant properties of apple slices stored in starch-based films. *Int. J. Food Prop.* 20 (5): 1117–1128.

126 Piñeros-Hernandez, D., Medina-Jaramillo, C., López-Córdoba, A., and Goyanes, S. (2017). Edible cassava starch films carrying rosemary antioxidant extracts for potential use as active food packaging. *Food Hydrocolloids* 63 (February): 488–495.

127 Hassan, B., Chatha, S.A.S., Hussain, A.I. et al. (2018). Recent advances on polysaccharides, lipids and protein based edible films and coatings: a review. *Int. J. Biol. Macromol.* 109: 1095–1107.

128 Cazón, P., Velazquez, G., Ramírez, J.A., and Vázquez, M. (2017). Polysaccharide-based films and coatings for food packaging: a review. *Food Hydrocolloids* 68: 136–148.

129 Sothornvit, R. and Krochta, J.M. (2005). Plasticizers in edible films and coatings. *Innov. Food Packag.*: 403–433.

130 Santacruz, S., Rivadeneira, C., and Castro, M. (2015). Edible films based on starch and chitosan. Effect of starch source and concentration, plasticizer, surfactant's hydrophobic tail and mechanical treatment. *Food Hydrocolloids* 49: 89–94.

131 Podshivalov, A., Zakharova, M., Glazacheva, E., and Uspenskaya, M. (2017). Gelatin/potato starch edible biocomposite films: correlation between morphology and physical properties. *Carbohydr. Polym.* 157: 1162–1172.

132 Sánchez-Ortega, I., García-Almendárez, B.E., Santos-López, E.M. et al. (2016). Characterization and antimicrobial effect of starch-based edible coating suspensions. *Food Hydrocolloids* 52: 906–913.

133 Masood, F. (2017). Polyhydroxyalkanoates in the food packaging industry. In: *Nanotechnology Applications in Food* (eds. A.E. Oprea and A.M. Grumezescu), 153–177. Elsevier Inc. https://doi.org/10.1016/B978-0-12-811942-6.00008-X.

134 Steinbüchel, A. and Lütke-eversloh, T. (2003). Metabolic engineering and pathway construction for biotechnological production of relevant polyhydroxyalkanoates in microorganisms. *Biochem. Eng. J.* 16: 81–96.

135 Costa, J.A.V., Moreira, J.B., Lucas, B.F. et al. (2018). Recent advances and future perspectives of PHB production by cyanobacteria. *Ind. Biotechnol.* 14 (5): 249–256.

136 Favaro, L., Basaglia, M., Casella, S., and Food, A. (2018). Improving polyhydroxyalkanoate production from inexpensive carbon sources by genetic approaches: areview. *Biofuels, Bioprod. Biorefin.*: 1–20.

137 Rodriguez-Perez, S., Serrano, A., Pantión, A.A., and Alonso-Fariñas, B. (2018). Challenges of scaling-up PHA production from waste streams. A review. *J. Environ. Manage.* 205: 215–230.

138 Costa, S.S., Miranda, A.L., de Morais, M.G. et al. (2019). Microalgae as source of polyhydroxyalkanoates (PHAs) — a review. *Int. J. Biol. Macromol.* 131: 536–547. https://doi.org/10.1016/j.ijbiomac.2019.03.099.

139 Pavan, F.A., Junqueira, T.L., Watanabe, M.D.B. et al. (2019). Economic analysis of polyhydroxybutyrate production by *Cupriavidus necator* using different routes for product recovery. *Biochem. Eng. J.* 146: 97–104. https://doi.org/10.1016/j.bej.2019.03.009.

140 Rosa, D.S., Lotto, N.T., Lopes, D.R., and Guedes, C.G.F. (2004). The use of roughness for evaluating the biodegradation of poly-β-(hydroxybutyrate) and poly-β-(hydroxybutyrate-*co*-β-valerate). *Polym. Test.* 23: 3–8.

141 Koller, M. (2014). Poly(hydroxyalkanoates) for food packaging: application and attempts towards implementation. *Appl. Food Biotechnol.* 1 (1): 3–15.

142 Fukada, E. and Ando, Y. (1986). Piezoelectric properties of poly-β-hydroxybutyrate and copolymers of β-hydroxybutyrate and β-hydroxyvalerate. *Int. J. Biol. Macromol.* 8: 361–366.

143 Khosravi-darani, K. and Bucci, D.Z. (2015). Application of poly (hydroxyalkanoate) in Food packaging: improvements by nanotechnology. *Chem. Biochem. Eng.* 29 (2): 275–285.

144 Xu, P., Yang, W., Niu, D. et al. (2020). Multifunctional and robust polyhydroxyalkanoate nanocomposites with superior gas barrier, heat resistant and inherent antibacterial performances. *Chem. Eng. J.* 382: 122864. https://doi.org/10.1016/j.cej.2019.122864.

145 Miguel, R., Júnior, S., Araújo, T. et al. (2019). Thermal behavior of biodegradable bionanocomposites: influence of bentonite and vermiculite clays. *J. Mater. Res. Technol.* 8 (3): 3234–3243. https://doi.org/10.1016/j.jmrt.2019.05.011.

146 Arrieta, M.P., Alberto, D., Daniel, L. et al. (2019). Antioxidant bilayers based on PHBV and plasticized electrospun PLA-PHB fibers encapsulating catechin. *Nanomaterials* 9 (346): 1–14.

147 Urbina, L., Eceiza, A., Gabilondo, N. et al. (2019). Valorization of apple waste for active packaging: multicomponent polyhydroxyalkanoate coated nanopapers with improved hydrophobicity and antioxidant capacity. *Food Packag. Shelf Life* 21: 100356. https://doi.org/10.1016/j.fpsl.2019.100356.

148 Sängerlaub, S., Brüggemann, M., Rodler, N. et al. (2019). Extrusion coating of paper with poly(3-hydroxybutyrate-*co*-3-hydroxyvalerate) (PHBV)—packaging related functional properties. *Coatings* 457: 1–28.

149 Panaitescu, D.M., Ionita, E.R., Nicolae, C. et al. (2018). Poly(3-hydroxybutyrate) modified by nanocellulose and plasma treatment for packaging applications. *Polymers (Basel)* 10: 1–24.

150 Garrido-Miranda, K.A., Rivas, B.L., Pérez, M.A. et al. (2018). Antioxidant and antifungal effects of eugenol incorporated in bionanocomposites of poly (3-hydroxybutyrate)-thermoplastic starch. *LWT Food Sci. Technol.* 98 (January): 260–267. https://doi.org/10.1016/j.lwt.2018.08.046.

151 Ma, Y. and Wang, Y. (2018). Development of PLA - PHB - based biodegradable active packaging and its application to salmon. *Packag. Technol. Sci.* 31: 739–746.

152 Cherpinski, A., Gozutok, M., Sasmazel, H.T., and Lagaron, J.M. (2018). Electrospun oxygen scavenging films of poly (3-hydroxybutyrate) containing palladium nanoparticles for active packaging applications. *Nanomaterials* 8: 1–19.

153 Castro-Mayorga, J.L., Freitas, F., Reis, M. et al. (2018). Biosynthesis of silver nanoparticles and polyhydroxybutyrate nanocomposites of interest in antimicrobial applications. *Int. J. Biol. Macromol.* 108: 426–435. https://doi.org/10.1016/j.ijbiomac.2017.12.007.

154 Fabra, M.J., Sánchez, G., López-Rubio, A., and Lagaron, J.M. (2014). Microbiological and ageing performance of polyhydroxyalkanoate-based multilayer structures of interest in food packaging. *LWT Food Sci. Technol.* 59: 760–767.

155 Murariu, M. and Dubois, P. (2016). PLA composites: from production to properties. *Adv. Drug Delivery Rev.* 107: 17–46.

156 Gerometta, M., Rocca-Smith, J.R., Domenek, S., and Karbowiak, T. (2019). Physical and chemical stability of PLA in food packaging. Reference module in food science. *Elsevier* 1: 1–11.

157 Nofar, M., Sacligil, D., Carreau, P.J. et al. (2019). Poly(lactic acid) blends: processing, properties and applications. *Int. J. Biol. Macromol.* 125: 307–360.

158 Javaherzadeh, R., Tabatabaee Bafroee, A.S., and Kanjari, A. (2020). Preservation effect of Polylophium involucratum essential oil incorporated poly lactic acid/nanochitosan composite film on shelf life and sensory properties of chicken fillets at refrigeration temperature. *LWT Food Sci. Technol.* 118 (October 2019): 108783.

159 Zhou, X., Yang, R., Wang, B., and Chen, K. (2019). Development and characterization of bilayer films based on pea starch/polylactic acid and use in the cherry tomatoes packaging. *Carbohydr. Polym.* 222 (January): 1–7.

160 Martins, C., Vilarinho, F., Sanches Silva, A. et al. (2018). Active polylactic acid film incorporated with green tea extract: development, characterization and effectiveness. *Ind. Crops Prod.* 123: 100–110.

161 Altan, A., Aytac, Z., and Uyar, T. (2018). Carvacrol loaded electrospun fibrous films from zein and poly(lactic acid) for active food packaging. *Food Hydrocolloids* 81: 48–59.

162 Arfat, Y.A., Ahmed, J., Ejaz, M., and Mullah, M. (2018). Polylactide/graphene oxide nanosheets/clove essential oil composite films for potential food packaging applications. *Int. J. Biol. Macromol.* 107: 194–203.

163 Domenek, S., Nassar-Fernandes, S., and Ducruet, V. (2017). Rheology, mechanical properties, and barrier properties of poly(lactic acid). *Adv. Polym. Sci.* 279: 303–341.

164 Rasal, R.M., Janorkar, A.V., and Hirt, D.E. (2010). Poly(lactic acid) modifications. *Prog. Polym. Sci.* 35: 338–356.

165 Yang, W., Weng, Y., Puglia, D. et al. (2020). Poly(lactic acid)/lignin films with enhanced toughness and anti-oxidation performance for active food packaging. *Int. J. Biol. Macromol.*

166 Jin, F.L., Hu, R.R., and Park, S.J. (2019). Improvement of thermal behaviors of biodegradable poly(lactic acid) polymer: a review. *Compos. Part B Eng.* 164: 287–296.

167 Liu, H. and Zhang, J. (2011). Research progress in toughening modification of poly(lactic acid). *J. Polym. Sci., Part B: Polym. Phys.* 49 (15): 1051–1083.

3

Nanostructures Based on Starch, Their Preparation, Processing, and Application in Packaging

Cristian C. Villa

Universidad del Quindio, Programa de Quimica, Facultad de Ciencias Básicas y Tecnologias, Cra 15 Calle 12 630004 Armenia, Colombia

3.1 Introduction

Starch is a polysaccharide used by plants as an energy reserve that is often considered the second most common biomass on Earth [1–4]. It is found as granules of different morphologies (depending on the botanical source) in plant tissues, mainly seed, roots, tubers, leaves, and fruits [5]. Starch structure is composed by α-D-glucopyranosyl units that can be linked in either α-D-(1–4) and/or α-D-(1–6) linkages. Likewise, these linkages give rise to two types of molecules: the linear amylose formed by approximately 1000 glucose units linked in α-D-(1–4) manner and the branched amylopectin, formed by approximately 4000 glucose units, branched through α-D-(1–6) linkages [2, 3, 6]. The union of both amylose and amylopectin forms a semi-crystalline structure arranged as small granules with diameters between 1 and 100 μm. Most unmodified starches have an amylose content around 20–30%, with amylopectin ranging from 70% to 80%. Some starches can have very low amylose contents, below 1%, such as waxy corn starches. The size and morphology of starch granules are dependent on their botanical source, as their shape can vary from spherical, oval, polygonal, lenticular, and kidney shapes and their size can range from <1 to 100 μm [7].

3.1.1 Starch Nanoparticles and Nanocrystals

Nanomaterials based on starch can be classified according to their nature in two main groups: starch nanoparticles (SNPs) and starch nanocrystals (SNCs). SNPs are almost completely amorphous particles that are commonly obtained by the controlled nanoprecipitation of gelatinized starch, while SNC synthetized through the hydrolysis of the amorphous phase of the starch granule, removing mostly amylose until nanosized particles are achieved [8, 9].

As mentioned before the synthesis of SNP materials is achieved mostly through nanoprecipitation of gelatinized starch, and most methods involve either hot

Nanotechnology-Enhanced Food Packaging, First Edition. Edited by Jyotishkumar Parameswaranpillai, Radhakrishnan Edayileveettil Krishnankutty, Aswathy Jayakumar, Sanjay Mavinkere Rangappa, and Suchart Siengchin.

or cold (addition of NaOH) gelatinization, which is followed by the addition of an anti-solvent, mostly ethanol, methanol, or acetone [10–13]. The anti-solvent methods involve the nucleation of the amylose molecules that precipitate once the anti-solvent is slowly added; this causes supersaturation in the solution and particle growth [14]. After the critical overlapping concentration is reached, nucleation starts and particles are formed. Furthermore, as nuclei are formed and growth supersaturation decreases, quickly a controlled process of growing nuclei is allowed [15]. In general, the anti-solvent method is considered simple, and the physicochemical properties of the SNP can be controlled by small changes in the process, such as the addition of surfactants [16] and changing the anti-solvent or the amylopectin/amylose ratios of the starch [16, 17]. Furthermore, aspects like the concentration of starch and process temperature are the important parameters affecting the yield of SNP [15, 18].

Extrusion has also been used as methods to synthetize SNP. In this method starch granules with limited amounts of water are subjected to high temperatures, pressures, and mechanical forces, thus undergoing changes such as melting, fragmentation, and incomplete gelatinization [15]. This process allows the disruption of the molecular bonds in the starch granules, forming smaller particles with less crystallinity [19]. Likewise, in reactive extrusion native starches are mixed with different reactants like plasticizers (glycerol, sorbitol, etc.) and cross-linkers and subjected to the extrusion process [20–22]. Starch fragments formed due to the extreme conditions are then cross-linked forming nanoparticles of different sizes. Furthermore, SNPs size can be reduced after addition of cross-linkers, as they increase the shear forces and torque, which facilitate size reduction [15].

Another mechanical technique used in SNPs formation is high-pressure homogenization [20]. This process, usually used in emulsion formation and microorganism inactivation, uses a continuous flow of liquid, which is subjected to external pressure sources and passed through microfabricated channels [23]. The mechanical forces reduce the size of particles breaking molecular bonds, obtaining very narrow size distributions [24]. Finally, gamma irradiation has also been used as a technique to reduce the starch granule size. In this technique a homogeneous starch paste is irradiated with gamma rays inducing breaking in the bonds of the amorphous regions, thus reducing particle size while maintaining crystallinity [25, 26]. Other methods include successive ultrasound treatments [27] and ball milling [28].

The methods for SNCs synthesis involve the hydrolysis of the amorphous, reducing particle size while increasing crystallinity. Most methods use acid hydrolysis, either with HCl or H_2SO_4; the later allows faster hydrolysis rates and higher yields, while HCl hydrolysis is more time consuming [29]. Reports have shown that SNCs can be obtained after 5–7 days using H_2SO_4, while using HCl this can take up to 15 days [11, 15, 30]. However, one of the main drawbacks of H_2SO_4 hydrolysis is that sulfate groups are incorporated into the SNCs surface; this can lead to increase stability in aqueous solution but can limit their applications in several fields [31].

In recent years, there has been an increasing need for green methods in SNCs synthesis that reduce the high amounts of hydrolyzing agent [14, 32]. Among this methods, enzymatic hydrolysis followed by acid hydrolysis has become one of the most important alternatives. Most examples use both α and β amylases as the hydrolyzing agent; the first one is a type of endo-amylase that hydrolyses internal α-1,4 bonds, while β-amylase is an exo-amylase that causes cleavage of alternated α-1,4 bonds from nonreducing ends [14]. Although amylases are an important part in starch digestion, it seems that using them by themselves is not enough for obtaining nano-sized particles. Studies have shown that using only α-amylose as the hydrolyzing agent for waxy maize and rice starch results in starch particles of 11 and 3.6 μm, respectively [33, 34]. On the other hand, it has been shown that combined treatments of both enzymatic and acid hydrolysis can reduce synthesis time from seven to three days [6, 35]. Finally some physical methods have been used in combination with acid hydrolysis, such as ball milling and ultrasonication [36, 37].

3.1.2 Starch Nanomaterials in Food Packaging

Over the last decades, starch-based nanomaterials have been proposed as fillers in composite polymeric films, as they have shown the capability to improve mechanical, barrier, and electrical properties of the films [6]. They have been used in order to improve properties of biodegradables films made from biodegradables polymers, such as starch, and other carbohydrate polymers, proteins, and lipids; furthermore, some examples can be found of their use in nonbiodegradables composite polymers [38]. Both SNCs and SNPs have been used in order to reinforce mechanical and barrier properties of polymeric films.

According to Le Corre and Angellier-Coussy [6], two types of nanocomposite formation can be distinguished. In the first case aqueous systems and hydrosoluble and hydro-dispersable polymers are grouped. The second group is formed by nonaqueous systems that use organic solvents.

Examples in the first group include the inclusion of SNCs in films formed by natural rubber and latex [39–41]. Likewise, one of the main applications of starch nanomaterials as fillers in polymeric packaging has been in the development of starch-based films. Biodegradable (sometimes edible) films made from starch have been of great interest as they are odorless, tasteless, colorless, nontoxic, and semipermeable to moisture, gases (carbon dioxide and oxygen), and flavor components [42]. However, they have shown high water solubility and poor water vapor barrier due to their hydrophilicity; furthermore, their mechanical properties are poor, with low tensile strength and elongation values [43–45]. Studies have shown that due to their small size, SNCs can interact with the polymeric matrix by forming strong hydrogen bonds. This interaction allows the transfer of stress from the matrix to the nanoparticles that carries the load and enhances the film's strength. Furthermore, water vapor barrier values decreased due to the tortuous path created by the SNCs on the polymeric film, stopping water vapor transmission though the film [26, 31, 46–48]. These behaviors have also been observed in protein films reinforced with starch-based nanomaterials. Soy protein films reinforced with citric acid modified SNPs showed an increase on their tensile strength and water resistance, as nanoparticles created a hydrophobic surface [49, 50]. This was also observed for amaranth protein films [51–53].

3.1.3 Starch Nanomaterials as Carriers of Bioactive Molecules

One of the most interesting applications of starch-based nanomaterials is in nanocarriers of bioactive molecules, as they are biodegradable, nontoxic, and biocompatible [54]. Most studies use SNPs as carriers, as they have shown an increase in both solubility and bioavailability of the bioactive molecules [54].

A study by Farrag et al. [55] presented results of the use of SNPs made from potato, pea, and corn starch to encapsulate quercetin. This molecule is a polyphenol found in many leaves, fruits, and vegetables, which is well-known for its antioxidant and anticancer activities, as well as for its low water solubility [56, 57]. It was reported that starch's amylopectin content has a significant effect on the molecule encapsulation. SNPs obtained from potato starch (highest amylopectin content) encapsulated higher quercetin molecules due to their higher amylopectin than SNPs obtained from corn starch; this as the branched regions produced by the amylopectin molecules allowed a better "accommodation" of the quercetin molecules [55]. Furthermore, they reported that quercetin antioxidant activity is preserved by encapsulation in the SNPs and is related to the SNPs loading capacity; thus, the higher loading percentage of quercetin leads to higher radical scavenging activity [55].

Another polyphenol that has nanoencapsulated in starch-based nanomaterials is curcumin, a molecule present in the rhizomes of turmeric, and it is well-known for its anticancer, antioxidant, anti-inflammatory, antimicrobial, and antiviral activity. Even curcumin has shown great potential in several biological applications, its use has been limited by its low water solubility and low bioavailability [58–66]. Chin et al. [67] used SNPs made from sago as carriers of curcumin, reporting particle sizes around 50–80 nm and loading capacity reaching around 78% [16].

In general, encapsulation of the bioactive molecules in SNP is mostly done using the anti-solvent synthesis method. In this method, an organic solvent containing the bioactive molecules is slowly added to gelatinized starch at room temperature. The molecules are then encapsulated in interior of the SNPs, due to its hydrophobic nature. In this method, it is possible to increase molecule encapsulation by reducing starch polarity through chemical modifications like acetylation or cross-linking with other molecules [10, 68]. This was considered by Pang et al. [69] in order to increase loading efficiency of curcumin into SNPs made from sago starch. They modified the SNPs by using maleate ester modified sago starch, reaching a loading capacity of 85%. Increasing loading capacity by chemical modification of the SNPs was also observed by Acevedo-Guevara et al. [10], using native and acetylated banana starch SNPs with higher encapsulation capacity and particle size in the modified SNPs than in their native counterpart. In both cases, it was explained that modifications such as acetylation and esterification lead to higher hydrogen bond interactions between the bioactive molecule and ester groups incorporated into the SNPs structure, leading to higher incorporation into the nanoparticle. Finally, Sadeghi et al. [17] studied the effect of the amylose/amylopectin ratios in different corn starch SNPs in curcumin encapsulation. It was reported that high amylose content leads to increasing SNPs sizes; furthermore, they observed that nanoencapsulation can be used to protect curcumin from photodegradation, as more than 83% of the encapsulated polyphenol remained after 10 days of storage.

Other molecules with high antioxidant activity has been encapsulated in SNPs. Report by de Oliveira et al. [70] showed that chemical such as acetylation increases loading capacities of both molecules. This was explained as the formation of specific interactions (hydrogen bonds) between the SNPs and the antioxidant molecules drives more molecules into the starch-based nanomaterial. Ahmad et al. [71] used SNPs made from horse chestnut, water chestnut, and lotus stem starch as carriers of catechin, while Shabana et al. [72] used SNCs obtained from potato starch to encapsulated antioxidant molecules (ascorbic and oxalic acid). Results showed that mechanical methods, such as ultrasound, lead to smaller particle sizes while increasing antioxidant loading capacity

Starch-based nanomaterials have also been used in order to protect vitamins, like vitamin A and vitamin D, due to their high lipophilic nature and easy degradation by light and oxygen [73–75]. Firstly, Hasanvand et al. [76] reported that SNPs made from high amylose maize can be used as carriers of vitamin D. They reported encapsulation values up to 78% and through the use of several techniques (DSC, FTIR, and XRD) observed the formation of hydrogen bonds between vitamin D molecules and the starch-based nanovehicle as the encapsulation main drive. More recently, Santoyo-Aleman et al. [68] used citric acid modified SNPs made from banana starch to encapsulate vitamin A. They reported that hydrogen bond formation between the citric acid molecules in the modified vehicle and the vitamin allowed higher encapsulation values.

Due to the high biocompatibility of starch and their edible nature, SNPs can be used as vehicles for controlled release of molecules in the gastrointestinal track. SNPs can be used to shield bioactive molecules from the harsh environment of the stomach (low pH and enzymatic attacks) while increasing release in the intestines [14, 32, 54]. Acevedo-Guevara et al. [10] showed that curcumin release from native and acetylated banana SNPs in simulated gastric and intestinal conditions occurred mostly in intestinal conditions, while the molecule was protected in the stomach simulation. Santoyo-Aleman et al. [68] used a native citric acid modified banana starch SNPs for β-carotene release and encapsulation, reporting that the resistant nature of banana starch to acid hydrolysis led to an increasing stability of the carotene molecule in the gastric conditions. Finally, Ahmad et al. [71] reported that SNPs extracted from horse chestnut, water chestnut, and lotus can be used for the controlled release of catechin in gastric conditions. As explained starch-based nanoencapsulation is an efficient tool in the preservation of catechin bioactive properties in the harsh conditions of the stomach.

3.1.4 Perspectives and Outlook

Starch nanomaterials have generated great interest due to their relatively easy synthesis and the fact that they can be easily modified, through chemical, physical, and enzymatic methods to tailor their properties as needed. Although there have been several studies about the application of starch nanomaterials in food packaging and as nanovehicles for bioactive molecules, there is still plenty of room for research, as starch from every botanical source has different properties that can be used in order to obtain different types of nanovehicles. Furthermore, starch nanomaterials can be modified to improve mechanical, barrier, and optical properties of packaging materials, while other modifications can be used to improve encapsulation of bioactive molecules in either active food packaging or nanomedicine.

References

1 Ai, Y. and Jane, J.-l. (2015). Gelatinization and rheological properties of starch. *Starch - Stärke* 67 (3–4): 213–224. https://doi.org/10.1002/star.201400201.

2 Magallanes-Cruz, P.A., Flores-Silva, P.C., and Bello-Perez, L.A. (2017). Starch structure influences its digestibility: a review. *J. Food Sci.* 82 (9): 2016–2023. https://doi.org/10.1111/1750-3841.13809.

3 Pérez, S. and Bertoft, E. (2010). The molecular structures of starch components and their contribution to the architecture of starch granules: a comprehensive review. *Starch - Stärke* 62 (8): 389–420. https://doi.org/10.1002/star.201000013.

4 Wang, S., Li, C., Copeland, L. et al. (2015). Starch retrogradation: a comprehensive review. *Compr. Rev. Food Sci. Food Saf.* 14 (5): 568–585. https://doi.org/10.1111/1541-4337.12143.

5 Odeku, O.A. (2013). Potentials of tropical starches as pharmaceutical excipients: a review. *Starch - Stärke* 65 (1–2): 89–106. https://doi.org/10.1002/star.201200076.

6 Le Corre, D. and Angellier-Coussy, H. (2014). Preparation and application of starch nanoparticles for nanocomposites: a review. *React. Funct. Polym.* 85: 97–120. https://doi.org/10.1016/j.reactfunctpolym.2014.09.020.

7 Jane, J.-l. (2009). Structural features of starch granules II, Chapter 6. In: *Starch*, 3e (eds. J. BeMiller and R. Whistler), 193–236. San Diego: Academic Press.

8 LeCorre, D., Bras, J., and Dufresne, A. (2011). Influence of botanic origin and amylose content on the morphology of starch nanocrystals. *J. Nanopart. Res.* 13 (12): 7193–7208. https://doi.org/10.1007/s11051-011-0634-2.

9 LeCorre, D.S., Bras, J., and Dufresne, A. (2012). Influence of the botanic origin of starch nanocrystals on the morphological and mechanical properties of natural rubber nanocomposites. *Macromol. Mater. Eng.* 297 (10): 969–978. https://doi.org/10.1002/mame.201100317.

10 Acevedo-Guevara, L., Nieto-Suaza, L., Sanchez, L.T. et al. (2018). Development of native and modified banana starch nanoparticles as vehicles for curcumin. *Int. J. Biol. Macromol.* 111: 498–504. https://doi.org/10.1016/j.ijbiomac.2018.01.063.

11 Kim, H.-Y., Park, S.S., and Lim, S.-T. (2015). Preparation, characterization and utilization of starch nanoparticles. *Colloids Surf., B* 126: 607–620. https://doi.org/10.1016/j.colsurfb.2014.11.011.

12 Ma, X., Jian, R., Chang, P.R., and Yu, J. (2008). Fabrication and characterization of citric acid-modified starch nanoparticles/plasticized-starch composites. *Biomacromolecules* 9 (11): 3314–3320. https://doi.org/10.1021/bm800987c.

13 Qin, Y., Liu, C., Jiang, S. et al. (2016). Characterization of starch nanoparticles prepared by nanoprecipitation: influence of amylose content and starch type. *Ind. Crops Prod.* 87 (Supplement C): 182–190. https://doi.org/10.1016/j.indcrop.2016.04.038.

14 Qiu, C., Hu, Y., Jin, Z. et al. (2019). A review of green techniques for the synthesis of size-controlled starch-based nanoparticles and their applications as nanodelivery systems. *Trends Food Sci. Technol.* 92: 138–151. https://doi.org/10.1016/j.tifs.2019.08.007.

15 Kumari, S., Yadav, B.S., and Yadav, R.B. (2020). Synthesis and modification approaches for starch nanoparticles for their emerging food industrial applications: a review. *Food Res. Int.* 128: 108765. https://doi.org/10.1016/j.foodres.2019.108765.

16 Chin, S.F., Pang, S.C., and Tay, S.H. (2011). Size controlled synthesis of starch nanoparticles by a simple nanoprecipitation method. *Carbohydr. Polym.* 86 (4): 1817–1819. https://doi.org/10.1016/j.carbpol.2011.07.012.

17 Sadeghi, R., Daniella, Z., Uzun, S., and Kokini, J. (2017). Effects of starch composition and type of non-solvent on the formation of starch nanoparticles and improvement of curcumin stability in aqueous media. *J. Cereal Sci.* 76: 122–130. https://doi.org/10.1016/j.jcs.2017.05.020.

18 Hebeish, A., El-Rafie, M.H., El-Sheikh, M.A., and El-Naggar, M.E. (2014). Ultra-fine characteristics of starch nanoparticles prepared using native starch with and without surfactant. *J. Inorg. Organomet. Polym. Mater.* 24 (3): 515–524. https://doi.org/10.1007/s10904-013-0004-x.

19 Liu, D., Wu, Q., Chen, H., and Chang, P.R. (2009). Transitional properties of starch colloid with particle size reduction from micro- to nanometer. *J. Colloid Interface Sci.* 339 (1): 117–124. https://doi.org/10.1016/j.jcis.2009.07.035.

20 Escobar-Puentes, A.A., Rincón, S., García-Gurrola, A. et al. (2019). Preparation and characterization of succinylated nanoparticles from high-amylose starch via the extrusion process followed by ultrasonic energy. *Food Bioprocess Technol.* 12 (10): 1672–1682. https://doi.org/10.1007/s11947-019-02328-5.

21 Kaur, J., Kaur, G., Sharma, S., and Jeet, K. (2018). Cereal starch nanoparticles—a prospective food additive: a review. *Crit. Rev. Food Sci. Nutr.* 58 (7): 1097–1107. https://doi.org/10.1080/10408398.2016.1238339.

22 Song, D., Thio, Y.S., and Deng, Y. (2011). Starch nanoparticle formation via reactive extrusion and related mechanism study. *Carbohydr. Polym.* 85 (1): 208–214. https://doi.org/10.1016/j.carbpol.2011.02.016.

23 Homayouni, A., Sohrabi, M., Amini, M. et al. (2019). Effect of high pressure homogenization on physicochemical properties of curcumin nanoparticles prepared by antisolvent crystallization using HPMC or PVP. *Mater. Sci. Eng., C* 98: 185–196. https://doi.org/10.1016/j.msec.2018.12.128.

24 Tester, R.F., Karkalas, J., and Qi, X. (2004). Starch—composition, fine structure and architecture. *J. Cereal Sci.* 39 (2): 151–165. https://doi.org/10.1016/j.jcs.2003.12.001.

25 Akhavan, A. and Ataeevarjovi, E. (2012). The effect of gamma irradiation and surfactants on the size distribution of nanoparticles based on soluble starch.

Radiat. Phys. Chem. 81 (7): 913–914. https://doi.org/10.1016/j.radphyschem.2012.03.004.

26 Lamanna, M., Morales, N.J., García, N.L., and Goyanes, S. (2013). Development and characterization of starch nanoparticles by gamma radiation: potential application as starch matrix filler. *Carbohydr. Polym.* 97 (1): 90–97. https://doi.org/10.1016/j.carbpol.2013.04.081.

27 Minakawa, A.F.K., Faria-Tischer, P.C.S., and Mali, S. (2019). Simple ultrasound method to obtain starch micro- and nanoparticles from cassava, corn and yam starches. *Food Chem.* 283: 11–18. https://doi.org/10.1016/j.foodchem.2019.01.015.

28 Ahmad, M., Gani, A., Masoodi, F.A., and Rizvi, S.H. (2020). Influence of ball milling on the production of starch nanoparticles and its effect on structural, thermal and functional properties. *Int. J. Biol. Macromol.* 151: 85–91. https://doi.org/10.1016/j.ijbiomac.2020.02.139.

29 Le Corre, D., Bras, J., and Dufresne, A. (2010). Starch nanoparticles: a review. *Biomacromolecules* 11 (5): 1139–1153. https://doi.org/10.1021/bm901428y.

30 Putaux, J.-L., Molina-Boisseau, S., Momaur, T., and Dufresne, A. (2003). Platelet nanocrystals resulting from the disruption of waxy maize starch granules by acid hydrolysis. *Biomacromolecules* 4 (5): 1198–1202. https://doi.org/10.1021/bm0340422.

31 Angellier, H., Molina-Boisseau, S., Dole, P., and Dufresne, A. (2006). Thermoplastic starch–waxy maize starch nanocrystals nanocomposites. *Biomacromolecules* 7 (2): 531–539. https://doi.org/10.1021/bm050797s.

32 Villa, C.C., Sanchez, L.T., and Rodriguez-Marin, N.D. (2019). Starch nanoparticles and Nanocrystals as bioactive molecule carriers. In: *Polymers for Agri-food Applications* (ed. T.J. Gutierrez), 91–98. Springer Nature https://doi.org/10.1007/978-3-030-19416-1_6.

33 Foresti, M.L., Williams, M.d.P., Martínez-García, R., and Vázquez, A. (2014). Analysis of a preferential action of α-amylase from *B. licheniformis* towards amorphous regions of waxy maize starch. *Carbohydr. Polym.* 102: 80–87. https://doi.org/10.1016/j.carbpol.2013.11.013.

34 Kim, J.-Y., Park, D.-J., and Lim, S.-T. (2008). Fragmentation of waxy rice starch granules by enzymatic hydrolysis. *Cereal Chem.* 85 (2): 182–187. https://doi.org/10.1094/CCHEM-85-2-0182.

35 Hao, Y., Chen, Y., Li, Q., and Gao, Q. (2018). Preparation of starch nanocrystals through enzymatic pretreatment from waxy potato starch. *Carbohydr. Polym.* 184: 171–177. https://doi.org/10.1016/j.carbpol.2017.12.042.

36 Boufi, S., Bel Haaj, S., Magnin, A. et al. (2018). Ultrasonic assisted production of starch nanoparticles: structural characterization and mechanism of disintegration. *Ultrason. Sonochem.* 41: 327–336. https://doi.org/10.1016/j.ultsonch.2017.09.033.

37 Dai, L., Li, C., Zhang, J., and Cheng, F. (2018). Preparation and characterization of starch nanocrystals combining ball milling with acid hydrolysis. *Carbohydr. Polym.* 180: 122–127. https://doi.org/10.1016/j.carbpol.2017.10.015.

38 Chen, G., Wei, M., Chen, J. et al. (2008). Simultaneous reinforcing and toughening: new nanocomposites of waterborne polyurethane filled with low loading

level of starch nanocrystals. *Polymer* 49 (7): 1860–1870. https://doi.org/10.1016/j.polymer.2008.02.020.

39 Angellier, H., Molina-Boisseau, S., and Dufresne, A. (2005). Mechanical properties of waxy maize starch nanocrystal reinforced natural rubber. *Macromolecules* 38 (22): 9161–9170. https://doi.org/10.1021/ma0512399.

40 Angellier, H., Molina-Boisseau, S., Lebrun, L., and Dufresne, A. (2005). Processing and structural properties of waxy maize starch nanocrystals reinforced natural rubber. *Macromolecules* 38 (9): 3783–3792. https://doi.org/10.1021/ma050054z.

41 Angellier, H., Molina-Boisseau, S., and Dufresne, A. (2006). Waxy maize starch nanocrystals as filler in natural rubber. *Macromol. Symp.* 233 (1): 132–136. https://doi.org/10.1002/masy.200690009.

42 Alberto Jiménez, María José Fabra, Pau Talens, Amparo Chiralt, Effect of re-crystallization on tensile, optical and water vapour barrier properties of corn starch films containing fatty acids, *Food Hydrocolloids*, Volume 26, Issue 1, 2012, Pages 302-310.

43 Campos, C.A., Gerschenson, L.N., and Flores, S.K. (2011). Development of edible films and coatings with antimicrobial activity. *Food Bioprocess Technol.* 4: 849–875. https://doi.org/10.1007/s11947-010-0434-1.

44 Pinzon, M.I., Garcia, O.R., and Villa, C.C. (2018). The influence of *Aloe vera* gel incorporation on the physicochemical and mechanical properties of banana starch-chitosan edible films. *J. Sci. Food Agric.* 98: 4042–4049. https://doi.org/10.1002/jsfa.8915.

45 Sánchez-Ortega, I., García-Almendárez, B.E., Santos-López, E.M. et al. (2016). Characterization and antimicrobial effect of starch-based edible coating suspensions. *Food Hydrocolloids* 52: 906–913.

46 Li, X., Qiu, C., Ji, N. et al. (2015). Mechanical, barrier and morphological properties of starch nanocrystals-reinforced pea starch films. *Carbohydr. Polym.* 121: 155–162. https://doi.org/10.1016/j.carbpol.2014.12.040.

47 Nieto-Suaza, L., Acevedo-Guevara, L., Sánchez, L.T. et al. (2019). Characterization of aloe vera-banana starch composite films reinforced with curcumin-loaded starch nanoparticles. *Food Struct.*: 100131. https://doi.org/10.1016/j.foostr.2019.100131.

48 Silva, A.P.M., Oliveira, A.V., Pontes, S.M.A. et al. (2019). Mango kernel starch films as affected by starch nanocrystals and cellulose nanocrystals. *Carbohydr. Polym.* 211: 209–216. https://doi.org/10.1016/j.carbpol.2019.02.013.

49 Tian, H. and Xu, G. (2011). Processing and characterization of glycerol-plasticized soy protein plastics reinforced with citric acid-modified starch nanoparticles. *J. Polym. Environ.* 19 (3): 582–588. https://doi.org/10.1007/s10924-011-0304-6.

50 Zheng, H., Ai, F., Chang, P.R. et al. (2009). Structure and properties of starch nanocrystal-reinforced soy protein plastics. *Polym. Compos.* 30 (4): 474–480. https://doi.org/10.1002/pc.20612.

51 Condés, M.C., Añón, M.C., Mauri, A.N., and Dufresne, A. (2015). Amaranth protein films reinforced with maize starch nanocrystals. *Food Hydrocolloids* 47: 146–157. https://doi.org/10.1016/j.foodhyd.2015.01.026.

52 Dai, L., Zhang, J., and Cheng, F. (2019). Effects of starches from different botanical sources and modification methods on physicochemical properties of starch-based edible films. *Int. J. Biol. Macromol.* 132: 897–905. https://doi.org/10.1016/j.ijbiomac.2019.03.197.

53 Mukurumbira, A.R., Mellem, J.J., and Amonsou, E.O. (2017). Effects of amadumbe starch nanocrystals on the physicochemical properties of starch biocomposite films. *Carbohydr. Polym.* 165: 142–148. https://doi.org/10.1016/j.carbpol.2017.02.041.

54 Rostamabadi, H., Falsafi, S.R., and Jafari, S.M. (2019). Starch-based nanocarriers as cutting-edge natural cargos for nutraceutical delivery. *Trends Food Sci. Technol.* 88: 397–415. https://doi.org/10.1016/j.tifs.2019.04.004.

55 Farrag, Y., Ide, W., Montero, B. et al. (2018). Preparation of starch nanoparticles loaded with quercetin using nanoprecipitation technique. *Int. J. Biol. Macromol.* 114: 426–433. https://doi.org/10.1016/j.ijbiomac.2018.03.134.

56 Bose, S., Du, Y., Takhistov, P., and Michniak-Kohn, B. (2013). Formulation optimization and topical delivery of quercetin from solid lipid based nanosystems. *Int. J. Pharm.* 441 (1): 56–66. https://doi.org/10.1016/j.ijpharm.2012.12.013.

57 Jeszka-Skowron, M., Krawczyk, M., and Zgoła-Grześkowiak, A. (2015). Determination of antioxidant activity, rutin, quercetin, phenolic acids and trace elements in tea infusions: influence of citric acid addition on extraction of metals. *J. Food Compos. Anal.* 40: 70–77. https://doi.org/10.1016/j.jfca.2014.12.015.

58 Anand, P., Kunnumakkara, A.B., Newman, R.A., and Aggarwal, B.B. (2007). Bioavailability of curcumin: problems and promises. *Mol. Pharmaceutics* 4 (6): 807–818. https://doi.org/10.1021/mp700113r.

59 Maghsoudi, A., Yazdian, F., Shahmoradi, S. et al. (2017). Curcumin-loaded polysaccharide nanoparticles: optimization and anticariogenic activity against *Streptococcus* mutans. *Mater. Sci. Eng., C* 75: 1259–1267. https://doi.org/10.1016/j.msec.2017.03.032.

60 Mai, Z., Chen, J., He, T. et al. (2017). Electrospray biodegradable microcapsules loaded with curcumin for drug delivery systems with high bioactivity. *RSC Adv.* 7 (3): 1724–1734. https://doi.org/10.1039/C6RA25314H.

61 Menon, V.P. and Sudheer, A.R. (2007). Antioxidant and anti-inflammatory properties of curcumin. In: *The Molecular Targets and Therapeutic Uses of Curcumin in Health and Disease* (eds. B.B. Aggarwal, Y.-J. Surh and S. Shishodia), 105–125. Boston, MA: Springer US.

62 Mirzaei, H., Shakeri, A., Rashidi, B. et al. (2017). Phytosomal curcumin: a review of pharmacokinetic, experimental and clinical studies. *Biomed. Pharmacother.* 85: 102–112. https://doi.org/10.1016/j.biopha.2016.11.098.

63 Nelson, K.M., Dahlin, J.L., Bisson, J. et al. (2017). The essential medicinal chemistry of curcumin. *J. Med. Chem.* 60 (5): 1620–1637. https://doi.org/10.1021/acs.jmedchem.6b00975.

64 Oliveira, A.S., Sousa, E., Vasconcelos, M.H., and Pinto, M. (2015). Curcumin: a natural lead for potential new drug candidates. *Curr. Med. Chem.* 22: 4196.

65 Sahu, A.K., Mishra, J., and Mishra, A.K. (2020). Introducing tween-curcumin niosomes: preparation, characterization and microenvironment study. *Soft Matter* 16 (7): 1779–1791. https://doi.org/10.1039/C9SM02416F.

66 Stanić, Z. (2017). Curcumin, a compound from natural sources, a true scientific challenge – a review. *Plant Foods Hum. Nutr.* 72 (1): 1–12. https://doi.org/10.1007/s11130-016-0590-1.

67 Chin, S.F., Mohd Yazid, S.N.A., and Pang, S.C. (2014). Preparation and characterization of starch nanoparticles for controlled release of curcumin. *Int. J. Polym. Sci.* 2014: 8. https://doi.org/10.1155/2014/340121.

68 Santoyo-Aleman, D., Sanchez, L.T., and Villa, C.C. (2019). Citric-acid modified banana starch nanoparticles as a novel vehicle for β-carotene delivery. *J. Sci. Food Agric.* https://doi.org/10.1002/jsfa.9918.

69 Pang, S.C., Tay, S.H., and Chin, S.F. (2014). Facile synthesis of curcumin-loaded starch-maleate nanoparticles. *J. Nanomater.* 2014: 7. https://doi.org/10.1155/2014/824025.

70 de Oliveira, N.R., Fornaciari, B., Mali, S., and Carvalho, G.M. (2017). Acetylated starch-based nanoparticles: synthesis, characterization, and studies of interaction with antioxidants. *Starch - Stärke* 70 (3–4): 1700170. https://doi.org/10.1002/star.201700170.

71 Ahmad, M., Mudgil, P., Gani, A. et al. (2019). Nano-encapsulation of catechin in starch nanoparticles: characterization, release behavior and bioactivity retention during simulated in-vitro digestion. *Food Chem.* 270: 95–104. https://doi.org/10.1016/j.foodchem.2018.07.024.

72 Shabana, S., Prasansha, R., Kalinina, I. et al. (2018). Ultrasound assisted acid hydrolyzed structure modification and loading of antioxidants on potato starch nanoparticles. *Ultrason. Sonochem.* https://doi.org/10.1016/j.ultsonch.2018.07.023.

73 Ballard, J.M., Zhu, L., Nelson, E.D., and Seburg, R.A. (2007). Degradation of vitamin D3 in a stressed formulation: the identification of esters of vitamin D3 formed by a transesterification with triglycerides. *J. Pharm. Biomed. Anal.* 43 (1): 142–150. https://doi.org/10.1016/j.jpba.2006.06.036.

74 Mahmoodani, F., Perera, C.O., Fedrizzi, B. et al. (2017). Degradation studies of cholecalciferol (vitamin D3) using HPLC-DAD, UHPLC-MS/MS and chemical derivatization. *Food Chem.* 219: 373–381. https://doi.org/10.1016/j.foodchem.2016.09.146.

75 Walia, N., Dasgupta, N., Ranjan, S. et al. (2017). Fish oil based vitamin D nanoencapsulation by ultrasonication and bioaccessibility analysis in simulated gastro-intestinal tract. *Ultrason. Sonochem.* 39: 623–635. https://doi.org/10.1016/j.ultsonch.2017.05.021.

76 Hasanvand, E., Fathi, M., Bassiri, A. et al. (2015). Novel starch based nanocarrier for vitamin D fortification of milk: production and characterization. *Food Bioprod. Process.* 96: 264–277. https://doi.org/10.1016/j.fbp.2015.09.007.

4

Cellulose Nanostructures and Its Application as Effective Food Packaging Systems

Girilal M and Jyothi K. Nair

Saintgits College of Engineering (Autonomous), Department of Food Technology, Kottayam - 686532, Kerala, India

4.1 Introduction

Cellulose $(C_6H_{10}O_5)_n$ is an abundant natural polymer present in all plants, algae, certain marine organisms, and in some microorganisms. It is capable of overcoming challenges related to renewability, material biodegradability, energy, and cost. Cellulose is both a renewable and a multifunctional raw material. It is having potential to replace several nonrenewable materials. Structurally, cellulose consists of thousands of monosaccharide units of glucose. Majority of cellulose is found inside plants that amount to almost 50% of the dry mass of wood, and it is the basic component of the plant cell wall. Its capacity for tension resistance is competent to that of steel. Even if humans cannot digest cellulose, the unique and distinct properties of the biopolymer are useful for many applications like making paper, linen, cotton, paint, rayon, explosives (nitrocellulose), biofilms (cellulose acetate), biofuel, etc. and in variety of applications in food industry, pharmaceutical industries, textile industries, chemical industries, construction industry, etc. [1, 2].

In short, cellulose polysaccharide is nontoxic, is biodegradable, has ductile and durable networks, has relatively high cellulose purity and good physical properties, and has high tensile and compressive strength, which make it an asset to a variety of fields like food packaging industry, textile and paper industry, pharmaceutical industry, medicine, etc. Cellulose is a hydrocolloid that can be treated from agricultural waste raw materials. Solid or liquid waste may be used. Certain agricultural by-products, which are the source of cellulose, include bagasse sugarcane, corn husk, paper, wood, and other agri-waste materials. Especially in a developing as well as agriculture-based country like India, almost half of the wastes generated are rich in cellulose; this can be effectively modified and used for production of cellulose-based materials with better biocompatibility, stability, flexibility, and at the same time cost-effective [1].

Although cellulose cannot be digested by humans and have zero calorific value, it can be used as better food additive to serve purposes including inducing a sensation

Nanotechnology-Enhanced Food Packaging, First Edition. Edited by Jyotishkumar Parameswaranpillai, Radhakrishnan Edayileveettil Krishnankutty, Aswathy Jayakumar, Sanjay Mavinkere Rangappa, and Suchart Siengchin.

of fullness. Other than the application in food industry, it also can be extensively used as a construction material for the production of cellulose-based threads, biofilms, and other derivatives. It is also popular for its strong cohesion between chains and high tenacity. Another new area where cellulose could be exploited is in sustainable fuel production. Cellulose can be exploited for production of biofuel, which is environment friendly with significant reduction of carbon emissions. The production cost also can be optimized on further extensive research. The availability, sustainability, and simple conversion of cellulose into cellulose nanoparticle has helped it gain interest of the food and packaging industry as well as other industries.

Chemical modification is performed to cellulose material by a variety of chemical reactions to produce various cellulose derivatives. These chemical modifications make the cellulose a better candidate by the enhancement in strength, biocompatible, biodegradable, eco-friendly, and at the same time cost-effective. Cellulose and its derivatives, like methyl cellulose (MC), hydroxyl propyl cellulose (HPC), hydroxyl propyl methyl cellulose (HPMC), ethyl cellulose (EC), oxy cellulose (OC), microcrystalline cellulose (MCC), hydroxyethyl cellulose (HEC), silicified microcrystalline cellulose (SMCC), and sodium carboxymethyl cellulose (SCMC), are widely used in different industries mentioned earlier [3, 4]. Another area of prime importance is nanocellulose, nanostructure of cellulose, which will be discussed in detail in this chapter.

4.2 Source of Cellulose

Cellulose can be made either naturally or synthetically based on the application for which it is used for. Wood or generally plants are the common and abundant source of cellulose. A biological source of pure cellulose is cotton fiber. Some other sources of cellulose are corn, jute, flasks, hemp, rice, sisal, and wheat straw. Another source of cellulose is microorganisms and majorly from bacteria. Through specific chemical reactions on the cellulose material, various cellulose derivatives are made having different application in food, pharmaceutical, cosmetics, and paper, textile, and agriculture sectors. Detailed sources of cellulose, its derivatives, and its applications are consolidated in Table 4.1 [3, 5].

4.3 Cellulose Structure

Cellulose is a linear homopolymer made up of β(1-4)-linked glucose residues [1, 3]. Cellobiose, which is the two-glucose residue, is the structural unit of cellulose. Hundreds and thousands of these repetitive cellobiose units make the complex cellulose and its derivative structure [1]. The basic structure of cellulose is demonstrated in Figure 4.1.

Cellulose polysaccharide has ductile and durable networks, has relatively high cellulose purity, has good physical properties, and has high tensile and compressive strength, which make it an asset to various fields such as paper, food and textile industries, pharmaceutical industry, medicine, etc.

Table 4.1 Classification, source, and application of cellulose and its derivatives.

	Category	Source/type	Applications
Natural fiber	Vegetable fiber	Seed fibers, e.g. cotton fiber	Manufacture of paper and cloth, food industry, and pharmaceutical industry
		Fruit fibers, e.g. coconut	
		Leaf fiber, e.g. sisal and agave	
		Stalk fiber, e.g. bamboo and grass	
		Skin/bast fibers, e.g. flax, hemp, and jute	
Synthetic fiber	Microbial	*Acetobacter xylinum, Gluconacetobacter, Cladophora, Sarcina, Achromobacter, Agrobacterium, Aerobacter, Azotobacter, Alcaligenes, Rhizobium, Pseudomonas, Rhodobacter*, and *Salmonella*	Clothing, food industry, and pharmaceutical industry
	Chemical reactions	Oxycellulose from cotton	Pharmaceutical, cosmetics, and agricultural industry
		Microcrystalline cellulose (MCC)	Food industry
		Silicified MCC	Food and pharmaceutical industry
		Ethyl cellulose	Pharmaceutical industry and drug delivery
		Methyl cellulose	Pharmaceutical and chemical industry
		Hydroxyethyl cellulose	Pharmaceutical, textile and paper industry, cosmetics, and agriculture
		Hydroxy propyl cellulose	Food and pharmaceutical industry
		Carboxy methyl cellulose	Food and textile industry
		Sodium carboxy methyl cellulose	Pharmaceutical, paper, and agriculture industry
	Natural	Modal, bamboo, sea weed, sugarcane, and lyocell	Food and pharmaceutical industry

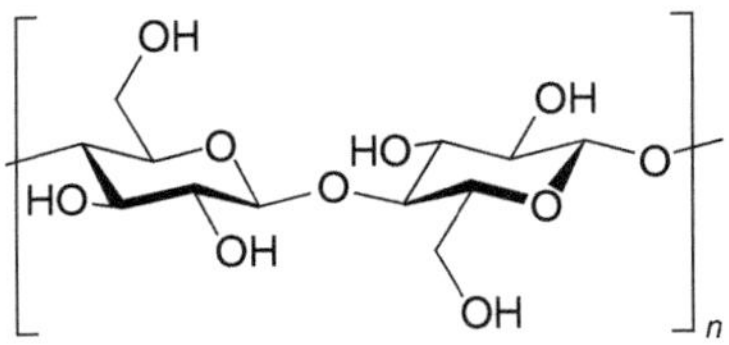

Figure 4.1 Basic cellulose structure.

4.4 Properties of Cellulose

Cellulose is a biopolymer that is not soluble in water or any other common solvents, which make them a suitable food additive. There exists a strong hydrogen bonding between the cellulose chain, which makes it insoluble in water or similar conventional solvents. Various chemical modifications are made for the production of cellulose derivatives, which make them better candidates in various industries. When the chemical changes are introduced, the durability and strength get enhanced, and at the same time it becomes more environment friendly and cost-effective. The purity of cellulose in each of the derivatives makes them more durable and efficient in having a unique property [6, 7].

Cellulose is one of the hydrocolloids that are made as a result of processing from agricultural waste raw materials. Some potential waste that may be used is solid or liquid waste. The agricultural by-products from which cellulose can be extracted includes bagasse sugarcane, corn husks, wood, paper, and other waste materials. In many developing countries, almost half of the waste generated have 50% of cellulose, which can be effectively modified and used for producing structures with better flexibility, biocompatibility, stability, film-forming ability, and emulsifying ability. As cellulose is indigestible by humans, its calorific value is zero and can be used as food additive to serve purposes including inducing a sensation of fullness [8, 9].

The stiffness of crystalline cellulose is about 140–220 GPa, which is comparable and better than that of glass fiber. Films made of nanocellulose possess high strength and high stiffness. The strength-to-weight ratio of nanocellulose is almost eight times than that of stainless steel. These mechanical properties make it a suitable raw material for making packaging materials. When considering the comparative strength, strong cohesion between chains, high tenacity, and flexibility, cellulose and its derivatives have potential application in construction industry as a multipurpose starting material for production of cellulose-based threads and other derivatives. They are also being widely used for production of biofuel, which is cost-effective as well as useful in significant reduction of carbon emissions [10].

4.5 Nanocellulose

As the size of a particle reduces, the physical, chemical, and biological property of a material when compared with its bulk material changes substantially. Nanosize is the size in which the particle is less than 100 nm in at least one dimension and is the smallest size possible in which the particle is having at least some similar properties in common with that of its bulk material. The unique properties of nanoparticles are mainly due to increased surface-area-to-volume ratio where the specificity and accuracy of the reaction will be increased [11].

The invention of cellulose nanocellulose is a big milestone in the field of material science and engineering, which attributes to the estimated high potential of nanocellulose. Cellulose materials can be self-assembled at multiple scales. The cellulose required for conversion to nanostructures can be acquired from a wide range of sources such as bacteria, algae, tunicates, and plants. Bacterial cellulose, or microbial cellulose, can be made extracellularly by gram-negative bacteria of different genera, e.g. *Achromobacter*, *Acetobacter*, *Aerobacter*, *Alcaligenes*, *Azotobacter*, *Agrobacterium*, *Rhizobium*, *Pseudomonas*, *Rhodobacter*, *Sarcina*, *Salmonella*, and especially *Gluconacetobacter*, which is the most efficient producer. The properties, size, and energy consumption of the extraction process for producing nanocellulose depend on the source used for the extraction of cellulose [3, 5].

4.5.1 Types of Nanocellulose

Based on the morphology nanostructures are categorized into nanofibrils, nanofibers, nanowhiskers, nanocrystals, nanorods, and nanoballs as shown in Figure 4.2 [12, 13].

4.5.1.1 Cellulose Nanofibrils

Nanofibrils are generally produced either from bacterial cellulose or from plants. It is available in the form of hydrogel in bacterial cellulose, which is a microbial polysaccharide. Due to the high purity, *in situ* change of flavor and color, and its potential to form into various textures and shapes, it can be used as a food ingredient. This can be used for gelling and thickening, water-binding, stabilizing, and as a packing material. Nanofibrils, which are produced from bacterial cellulose, will have a diameter from 70 to 140 nm, whereas those produced from wood-derived cellulose will

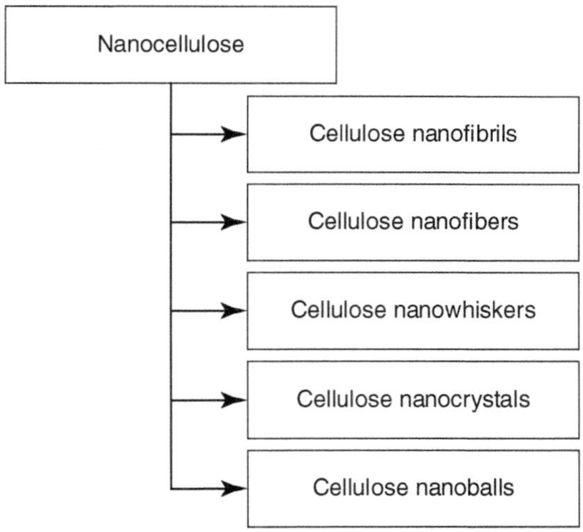

Figure 4.2 Types of nanocellulose.

have a diameter of 3–5 nm. From plants, nanofibrils can be obtained by oxidation, hydrolysis, and mechanical disintegration from wood [14].

Nanofibrils are further classified into two categories. They are two-dimensional and three-dimensional micro- and macrostructures.

4.5.1.2 Cellulose Nanofibers

They are thicker and longer in structure than nanofibrils. They are produced by electrospinning of cellulose, which can be done in combination with natural/artificial polymers or without the addition of additives. They have a diameter in the range of 10–100 nm and is having a very complex structure, which consist of elementary nanofibrils having alternating amorphous and crystalline domains. For converting fibers into CNF, plant cells have to be processed through mechanical disintegration. Among the variety of CNF, bacterial cellulose is having the highest purity in CNF form.

4.5.1.3 Cellulose Nanowhiskers

Cellulose nanowhiskers, nanorods, and nanocrystals are normally short fiber arrangements when compared with nanofibrils. Both nanocrystals and nanorods are rodlike structures. Sometimes nanocrystals take the shape of needle structure also. Cellulose nanofibers have lower crystallinity index than the cellulose nanocrystals because during the preparation of nanostructures, acid hydrolysis of cellulose will destroy the amorphous region, and the crystalline segment of the cellulose will be intact in nature [12]. Cellulose nanowhiskers are having needlelike structure. The diameter of this crystal is 4–25 nm and length is 100–1000 nm. It is produced by the process in which bleaching and acid hydrolysis take place.

4.5.1.4 Cellulose Nanoballs

Cellulose nanomaterials, which have round-shaped structure, are called nanocellulose balls for cellulose nanoballs. Nanoballs derived from wood have a diameter ranging from 80 to 85 nm [15].

4.5.1.5 Cellulose Nanocrystals

In case of cellulose nanocrystals, crystalline regions are isolated from CNF, and it is combined together to form CNC. Due to this reason, CNC is having greater crystallinity than CNF. It is normally produced by chemical treatment methods. The size of nanocellulose rages from 3 to 30 nm in thickness and a few hundred nm in length. This varies depending upon the type of plant sources. It is produced from algae, tunicates, and bacteria [16].

4.5.2 Properties of Nanocellulose

Due to their renewability, abundance, eco-friendliness, low weight, high strength, and stiffness, cellulose nanomaterials isolated from various cellulosic sources are finding applications in food packaging recently. While comparing with conventional cellulose, the nanocellulose has many advantages such as better surface area,

Young's modulus, and higher aspect ratio, and in its pure form it is harmless to people and is biocompatible. Nanocellulose has better advantages for its application when compared with that of cellulose fibers [11].

Recent studies throwing light on cellulose nanostructures (CNS) have attracted industries for its nontoxicity, biodegradability, low density, and exceptional mechanical and chemical properties. The tests conducted on CNS display its possibilities to be used in advanced manufacturing, energy efficiency, rural economic development, and accelerated technology transfer. It can also be applied for commercial purposes such as components for automobiles, applications in aerospace, electronics, cosmetic applications, food packaging materials, special papers, polymer reinforcement, etc. They have also gained fame for their recyclability, biodegradability, and renewability. It is also strong, stiff, and light in weight. Nanocellulose has a broad range of productive chemical, physical, and biological properties. Its increased surface area allows the adsorption of different molecules and microbial cells and porous nanocellulose materials, which can isolate different molecules.

4.5.2.1 Mechanical Properties

Nanocellulose incorporated thin films have tensile strength similar to that of metals and polymer materials. It is due to the substantial amount of hydrogen bonding and high density of the nanocellulose. It also increases the fiber-to-fiber bond strength. The crystals of nanocellulose show elastic modulus and breaking strength characteristics for common materials.

4.5.2.2 Barrier Properties

Barrier properties include oxygen/oil/grease barrier properties and are essential in increasing the shelf life of a food product. To increase the shelf life of the food product, packaging material should have a good barrier against oxygen and other gases and volatile components. Nanocellulose incorporated films have better efficiency in preventing the entry of oxygen through it. Hydrogen bonds in the cellulose-based films will give greater barriers toward the oxygen, oil, and grease as they cannot interact tightly with the bonds. This property can be enabled by making the nanocellulose-based films into oleophobic. Nanocellulose also finds application as additive to increase retention and dry and wet strength in paper and paper board products.[16].

4.5.2.3 Water Vapor Properties

Nanocellulose-based films do not have much resistance toward water vapor. When the film is subjected to moist environment, it becomes wet and loses its structure. The moisture sensitivity can be improved by adding cellulose reinforcing materials at suitable levels or by adding cellulose nanocrystals into rubber matrix or in poly acetic acid or by adding nanocellulose in a sodium caseinate-type protein film [16].

4.5.2.4 Other Properties

Nanocellulose incorporated packaging materials have another application in controlled release of drugs. In this type of packaging materials, the nanocellulose will be a barrier against the particular drug component. This type of packaging materials can even be utilized for acting against microbial agents. In order to achieve this, the packaging material can be treated along with food grade components and bio-based cationic polymers. This will escalate the shelf life of the food product by acting against microbes. Biodegradable packaging materials have greater importance nowadays. Nanocellulose incorporated packaging materials are biodegradable in nature. As it is produced from nature, these types of packaging materials are assumed to be edible in nature [16].

4.5.3 Synthesis of Nanocellulose

Cellulose nanomaterials are categorized into nano-object cellulose and nanostructures according to the size and structure. Nano-object cellulose has 10–100 μm size, whereas and nanostructured cellulose has size of 1–50 nm. Nano-object cellulose was further classified into cellulose micro crystal (CMC) and cellulose microfiber, and nanostructure are classified into cellulose nanocrystal (CNC) and cellulose nanofiber. Classification is represented in Figure 4.3.

The different methods for synthesis of nanocellulose are as on Figure 4.4.

4.5.3.1 Electrospinning

The technique is popularly used for manufacturing nanofibers from synthetic and natural polymers. Electrospinning is a technique that is used for spinning natural or synthetic fibers into nanofibers. Electrospinning technology is based on electrostatic forces. It is a flexible, simple, and an affordable technology. This method is suitable for producing nanofibers with an increased surface-to-volume ratio. For the maintenance of the bioactivity of biological components delivered by nanofibers, coaxial electrospinning is used. It maintains the function of the delivery factor of components such as proteins, drugs, bacteria, viruses, enzymes, and various other biomolecules. Food additives like antioxidants, volatile, or unstable flavors are generally susceptible to oxygen, heat, or light. This is where nanofibers can come into play as they usually display high stability against increased temperature and can be used for the stabilization in the functional additives. Applications of CNS can be developed in nutraceutical delivery in food systems, embedding of food compounds in nanofibers through coaxial electrospinning, etc. CNS produced using electrospinning offers high encapsulation efficiency, tunable release of the encapsulated component, high thermal, storage and light stability, and gives bioactive compounds an increased protection. These nanofibers can be exploited for food applications, applications in tissue repair, tissue engineering and wound healing, bioactive compound delivery, can be used in biosensors, and can be used as membranes in filtration for food and beverage processing [17].

There are three major components. They are pump, high-voltage power supply, and a drum. The voltage can vary from 1 to 30 kV. The solution is pumped from a needle and reaches its end and comes out in the shape of Taylor cone when the voltage reaches to the required level. At this voltage level, the electrostatic force overcomes the surface tension, and the solution pumps out in the form of nanofiber. This will

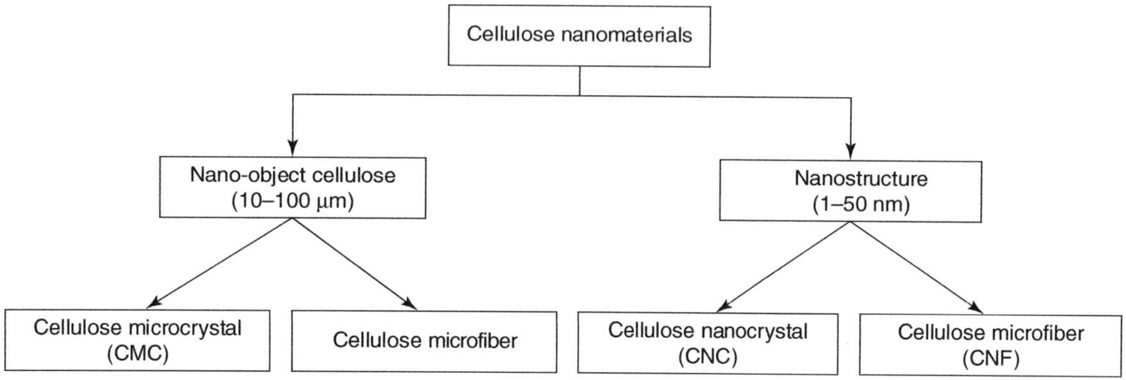

Figure 4.3 Classification of cellulose nanomaterials.

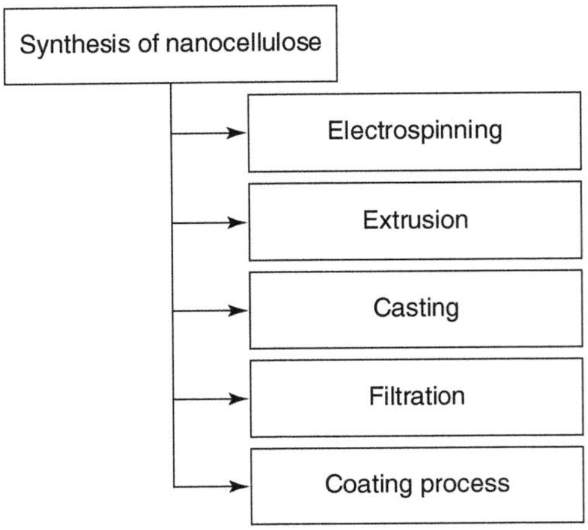

Figure 4.4 Different methods for synthesis of nanocellulose.

be deposited in a collector. The diameter of the nanofiber produced by this method has range between 40 and 2000 nm [18].

4.5.3.2 Extrusion

In extrusion, the product is passed out through a die at high temperature and pressure. When the samples that contains nanocellulose is passed though the extruder, the parameters like breakage, thermal degradation, alignment of nanocellulose, and high viscosity have to be noticed for the smooth flow of the sample and for the formation of the final packaging material. Nanocellulose is used in very small quantities in thermoplastic resin. The addition of nanocellulose will improve the modulus of elasticity in such materials. After extrusion, the product is then cooled below the melting point of the polymer. This will prevent thermal damage to the cellulosic compound. This method is mainly to include nanocellulose in different polymeric matrices and thereby improving various extra properties favorable for packaging material [16, 19].

4.5.3.3 Casting

The extrusion is not possible if the packaging material contains high levels of nanocellulose. Casting can be done in two methods. The first method is casting from aqueous solution or pure water. The hydrogen bond in the water will bond together with cellulose nanoparticles during drying. The second method is casting from a nonaqueous solution. In this method, it will allow dissolution of water insoluble matrix material. This will affect the properties of the film produced [16].

4.5.3.4 Paper Making Process Using Filtration

Two pretreatments are done prior to isolation of CNS in the paper industry, i.e. mercerization followed by bleaching. CNS is produced by two methods – chemical and mechanical process. Acid hydrolysis is used under chemical method. In mechanical method, fibrillation process is done in a grinder using shear force, and individualization of pulp to nanoscale is done. For the manufacturing of paper, a suspension of cellulosic fibrous material is kept on a screen and it is dried. This will produce inter fiber hydrogen bonding. The suspension can be prepared from mixing nanocellulose and zeolites. This film is having a higher capacity to absorb odors. To improve the strength of the paper, nanofibrillated cellulose can be added to the pulp. As surface area is more in nanocellulose, the dewatering process is very slow, and it will have more capacity to hold the water than normal cellulose. To increase the dewatering process, residual lignin and added cationic polymer can be included in the pulp [20].

4.5.3.5 Coating Process

Coating process is done by applying slurry of coating agent onto a porous surface. The solid content in the slurry will be adhered into the film and converted as a film. Most of the content will be absorbed into the material, and some portion will be evaporated. The nanocellulose is having higher dispersibility in water. This property makes this an excellent aqueous coating. Nowadays the thickness of oil-based plastic films is reduced by using high performing bio-based and thin layer of functional material. In such cases, nanocellulose acts an outstanding alternative owing to its high crystallinity, among other properties [16, 19].

4.6 Nanocellulose as Packaging Material

Food packaging is mainly intended to keep the packaged product fresh, safe, and stable, and to avoid harm to product during transport and storage. Packaging can also increase food products' shelf life by regulating the movement of heat, chemicals, tastes, and taints, thereby playing a major role in eliminating food waste worldwide. A vital application of nanoparticles and nanofibers is in the encapsulation of bioactive components, in particular hydrophobic components. Material used for packaging should not in any way harm the consumer, bring about composition change in the food, or bring any change to the organoleptic properties of the food. Cellulose nanomaterials isolated from different cellulosic sources are finding applications in food packaging due to their unique properties [17].

When the properties of packaging materials are concerned, barrier properties are important especially that of the food packaging material, which are crucial in reducing gas and water vapor exchanges between the food and the surroundings. This in turn reduces the rate of chemical, physical, and microbiological changes. Cellulose nanocrystals, bacterial cellulose, and cellulose nanofibrils are the major types of cellulose used in food packaging materials. Cellulose nanocrystals with a crystallinity of 85% are commonly used in the packaging materials. When native cellulose has undergone hydrolysis, the resultant cellulose nanocrystals formed are rigid and straight in structure. This will provide a possible thickness of thin film structures, which contains cellulose nanocrystals. Cellulose nanofibrils with diameter of 20–30 nm and crystallinity of 60–70% are used in food packaging material. Highly fibrillated cellulose is more flexible in wet state because it contains elevated levels of noncrystalline cellulose matter. Highly fibrillated cellulose with a large number of fibrils attached together is called nanocellulose aggregate. Bacterial cellulose is produced from bacteria through polymerization and crystallization. It is free from lignin, pectin, and hemicellulose. Certain purified bacterial celluloses are graded as food grade and is used directly into fruits, vegetables, meat, etc. [16, 21].

Cellulosic nanofibers have the advantage of favorable mechanical properties, which enable its applications in manufacturing of packaging materials with barrier coatings in active as well as intelligent food packaging products, manufacturing of strong polymer nanocomposites with lightweight, and manufacturing reinforcing agents used in foam and food packaging. The crystalline regions of nanocellulose are regarded as gas impermeable. Nanocellulose can be used as a barrier material due to its high crystallinity nature, along with its capacity to form a solid network bond due to the strong interfibrillar bonds. Nanocellulose is quite minor in dimension but hold greater surface area with increased porosity. This is responsible for limiting internal diffusion and quantum size effects. Nanocellulose may also provide barrier property against grease and oil. Cellulosic nanofibers have good potential to be used in food and beverage production plants in separation technology [21, 22].

Cellulose nanofibers could be considered as a potential food packaging material because of the following:

- They provide good vapor barrier.
- They improve freshness and reduce odor.
- They can be used in food items such as beverages, sauces, dairy, and meat products as thickeners as well as texturizing agents.
- They can be also used in foods to mask undesirable flavors.

CNS has great prospects to be profitable for its countless properties and applications. The residues from paper industries contain high cellulose, which makes it useful as a resource for commercial production of CNS. The properties of nanocellulose vary depending on their origin, type, processing, pretreatments, and functionalization. The strong thermal tolerance of cellulosic nanofibers substantiates its use in foods exposed to high temperatures, and it is also capable of protecting sensitive bioactive compounds. Their surface area is so high that they can detect microorganisms and adulteration in food and identify food components. Fiber-based and tamperproof packaging is also enabled by the use of cellulosic nanofibers [23].

Nanostructure composites are formed when other biological components or derivatives are incorporated into the conventional CNS, which helps incredibly to enhance the existing properties of the same. Plant-derived nanocellulose can improve their antibacterial properties by various chemical modifications or by addition of ions, other nanoparticles, and synthetic or other nature-derived molecules. Cellulosic nanofibers are integrated into the films to form a composite film with enhanced barrier properties. In plant proteins, when cellulosic nanofibers are added, it was noted that their mechanical strength as well as Young's modulus were increased. The fusion of nanocellulose with other magnetic nanomaterials has proved to be a composite adsorbent with magnetic properties. For instance, a polymeric ionic liquid magnetic biosorbent (Fe_3O_4) core–shell cellulose magnetite was developed and used for Congo red adsorption. In other words, the practice of merging two or more natural materials to form new biomaterials or by functionalizing nanocellulose is being used as an interesting technique for building materials with combined properties that the individual material alone could not achieve [24].

Nanocellulose can be utilized in the preparation of flexible, optically transparent paper, which is usually a desirable property of packaging materials. It acts as a carrier of certain active substances such as antioxidants and antimicrobials, which improve the food quality as well as prevent spoilage of food products. Cellulose nanofibers functionalized with active groups could be used for active food packaging applications since it exhibits antimicrobial properties against different bacterial groups [25, 26].

Aerogels/foams can be made from nanocellulose either homogeneously or in composite formulations, which helps in replacing the use of polystyrene-based foams. The poor physical properties of bio-based materials are responsible for their limited use in food packaging. To overcome this problem nanocellulose can be incorporated as a reinforcing structure into those materials, which may improve their properties. In food packaging applications, nanocellulose can be used as matrix for a several materials including films [27].

One of the purest forms of cellulose is bacterial nanocellulose, which is produced by non-photosynthetic organisms by the help of enzymatic polymerization of organic substrates like glycerol and sugar. It has a high water holding capacity and is also free from hemicellulose and lignin, which helps decrease its purifying cost and prevents need of harmful chemicals. Bacterial nanocellulose, when compared with plant-derived cellulose nanofibers, does not contain certain functional groups (except alcohol) and polymers like pectin, hemicellulose, and lignin. It is, therefore, recognized for its high purity in CNF form as well as high capacity to retain water and crystallinity nature, which can be exploited in food packaging. CNS can be impregnated in dispersions containing other polymers, and these can be used for manufacture of packaging materials with the ultimate function of increasing stability of the food product and maintaining its quality during shelf life. The various aspects that need to be considered while developing a packaging material are reduction of gas and water vapor exchange between food items; decreasing the rates of physical, chemical, and microbial reactions; water and oxygen permeability; decreased water solubility; transparency; etc. [25, 28].

Recent studies suggest that the sonication of gelatin – CNC – glycerol films forms a dispersion, which significantly enhances elongation, modulus, water vapor barrier, and transparent nature of the films, which shows the compatibility of CNC dispersions to ultrasonic processing. It displays the capability of CNS to be used in food packaging systems. CNCs have broad chemical modifying capacity, high mechanical resistance, and low cytotoxicity. CNCs are generally hydrophilic because of their –OH group, but it is found to be effective in resisting water in hydrophilic matrices like alginate, chitosan, soy protein isolate, gelatin, and PHB. CNCs are also known to have improved thermal stability of biopolymers like PLA and alginate. Emulsion bionanocomposites produced from chitosan, glycerol, and CNCs were found to amplify the water resistance and water vapor barrier of films. Limonene is used for its citrus flavor contribution, and this can be exploited for application in packaging of citrus fruits [23].

Coatings of nanocellulose can be applied over conventional polymers; this will allow a reduction in the size of plastic films, which in turn will help in reducing the impact that these packaging materials may have on the environment. Coatings help prevent several bacteria and other microorganisms and also extend shelf life and heat stability. The presence of silver nanoparticles reduced coalescence of CNFs due to the limited solvation and hydrogen bond between nearby CNFs, because of the carboxylate–silver binding. This functionalization of CNFs resulted in exceptional antibacterial properties of different bacterial groups, making them a potential candidate for food packaging systems [25].

There are certain beneficial properties of nanocellulose such as high water-absorbing ability and adequate mechanical strength, which allows moisture to stay within the food item, making it suitable for food packaging applications. Various properties of nanocellulose, like wettability and electrical charge, which are significant for tissue designing and other medical utilizations of this material, can be enhanced by manipulating nanocellulose with different substance gatherings and by readiness of nanocellulose from different sources. Reinforcement of materials like cellulose acetate butyrate, acrylic latex films, and poly-L-lactic acid can be achieved using nanocellulose. Detailed study on biodegradable films from nanocellulose reinforced methylcellulose, and chitosan composite films reinforced by nanocellulose, as edible films or coatings were studied, and they proved to improve the shelf life of foods. Incorporating CNCs into chitosan also increased antibacterial properties of the membranes. This property of the membrane was further increased by a hydrophobic stearic acid coating, which hampers adhesion of bacterial contaminants. Synthetic and nature-derived molecules often used or incorporated into CNS with hopes to increase their antibacterial property are antiseptics, antibiotics, antimicrobial peptides, alkannin, coumarin, shikonin, isoliquiritigenin, and curcumin [29].

Nanocarbon materials when combined with nanocellulose deliver nanocarbon functionality using an eco-friendly, strong, low-cost, non-melting, dimension-stable, nontoxic, and nonmetal matrix or carrier with flexible applications in manufacturing, biotechnology, and biomedicine alone. Addition of carbon nanoparticles into cellulose nanostructures can also significantly increase their mechanical strength. Other properties of this combination that make it favorable for food packaging and other related applications include the thermal stability, adjustable optical transparency, thermal conductivity, photothermal activity, intrinsic fluorescence and luminescence, nanoporous character, hydrolytic stability, and increased adsorption capacity. Cellulose–graphene composites displayed that fire retardant, super insulating, mechanically strong anisotropic foams could be created by the technology of freeze casting suspensions of cellulose nanofibers, and these were proven to have better performance than conventional polymer-based insulating material. Graphene oxide–nanocellulose composites also display potential of food packaging with their shape memory-based thin films and membranes [23].

The incorporation of nanocellulose in food packaging is mainly due to the need of replacement of petroleum-based materials with biodegradable products and due to the specific properties of nanocellulose as packaging materials. It has good barrier properties, improves physical properties and appearance, improves freshness, and reduces odor. The nanocellulose can act as piezoelectric material, and hence it can act as a sensor to indicate the condition of the packaged contents [16].

4.7 Comparison of Nanocellulose and Cellulose

Cellulose as we know is the most abundant, sustainable, renewable, and biodegradable polymer on this planet and is extensively used for production of paper, textiles, and pharmaceuticals, for its low-cost, inertness, lightweight, flexibility, etc. The properties of cellulose include chirality, hydrophilicity, and wide chemical variability. Cellulose has remarkable properties, which make it an environmental friendly and biocompatible material. In spite of these, cellulose could not meet the needs of the next generation as the natural form of cellulose cannot be used in food delivery systems due to its low solubility in water and large dimensions. The combination of mechanical and chemical treatment along with pretreatment can be used for the modification of the fibers, which are initially about 20–50 μm and consisting of thousands of microfibrils of diameter of a few or tens of nanometers, into nanofibrils or nanocrystals of cellulose, which helps remove all the existing defects [24].

It has been found out that nanocellulose is advantageous over conventional cellulose due to its higher aspect ratio, surface area, and Young's Modulus. One of the other advantages of using nanocellulose is that, in its pure form it is harmless to people and is biocompatible. This nanocellulose exhibits unique mechanical and optical properties, which are suitable for a broader range of applications. Films made using cellulose, films for food packaging, have barriers for commercialization such as mechanical characteristics, poor resistance to moisture, oxygen, and carbon dioxide moisture, etc. These limitations can be circumvented using nanotechnology, and this improves tensile strength and reduces permeability to water vapor and elasticity. Several studies on the implementation of nanofiber cellulose (CNF), cellulose nanocrystal (CNC), and bacterial nanocellulose (BNC) are used as alternative for fat, stabilizing agent, thickener, water binding properties, etc. [24, 30].

A transition occurs from micro- to nanoscale when the mechanical disintegration of fibers takes place, which leads to a significant reduction in thermal stability, viscosity, and crystallinity. Aqueous counter collision, blending, ball milling, electrospinning, cryo-crushing, grinding, extrusion, refining, homogenization, ultrasonication, steam explosion, or a combination of any of them is used as the principal mechanical treatments. A comparison of the nanofibers created through two different ways, i.e. via enzymatic treatment using endoglucanase and those produced from acid hydrolysis, reveals that the former has better structure with regard to average molar mass and larger aspect ratio [8, 13].

CNS offer higher aspect ratio, surface area, and Young's modulus than conventional cellulose fibers, which in turn makes them more advantageous than the latter. Cellulose needs to be explicitly modified to offer selective and enhanced properties for getting the product of our wish, since untailored cellulose normally does not show good adsorption performance. The basic properties of cellulose comprise its ability for functionalization, hydrophilicity, and ease of manipulation of various properties including aspect ratio, surface area, chemical accessibility, quantum effects, etc. [11].

Nanocellulose plays an important role as a better natural biomaterial used for wastewater treatment because of its capability of adsorption resistance to organic contaminants. Although the nanometer-sized cellulose is significantly smaller in dimensional size unlike micrometer-sized counterparts, they have large surface area with enhanced porosity that restricts internal diffusion and effects of quantum size. The enhancement in the nanocellulose adsorption removal efficiency could be due to the improved crystalline nature, surface area, and quantity of functional moieties available. On conversion of cellulose to nanocellulose, there occurs an exponential increase in the surface area, alongside its crystalline nature and the quantity of surface hydroxy groups (–OH). The liberal outside hydroxy (–OH) groups go about as hydrophilic locales to speed up nucleation and development of inorganic materials, which results in determining the particle size, morphology, and crystallinity. Such promptly accessible surface –OH groups offer a perfect base for the formation of hybrid nanocomposites. In addition, a number of chemical modifications can shape these –OH groups to selective and enhanced interactions. The hydrophilic nature of the nanocellulose can be strengthened by introducing hydrophilic groups into its polymer chains, such as alcohols, carboxylic acids, and amides. Reactive –OH groups have been shown to improve the adsorption behavior of revised nanocellulose, in particular to organic contaminants.

4.8 Disadvantages of Using Nanocellulose in Food Packaging

Although nanocellulose can be used as a promising packing material, there exist several challenges and shortcomings that affect the application. A major disadvantage is that the cellulose nanocrystals from different sources have different properties. This drawback presents a need to do a wide research on each nanocellulose obtained from different sources. Also, to ensure easy commercialization and marketability of nanocellulose, it is important that the application needs to be easily accessible, have low-cost, and produce valuable end products. But the studies on production of nanocellulose are still carried only on a small scale. To produce nanocellulose that is both environmentally and economically feasible, continuous examination of production strategies is needed. Although cellulose-based materials are considered nontoxic, nanocellulose materials have to be examined for any potential toxic effects since they are minor in dimension, which may make it possible for them to penetrate into cells and persist in the system, resulting in undesirable effects on living cell sustainability and propagation especially at higher concentrations [16].

Another disadvantage is that cellulose nanomaterial is prone to degradation by photocatalysis as it is sensitive to UV irradiation. So, it is often necessary to coat nanocellulose materials with materials absorbing UV radiation in order to shield the cellulosic fibers from UV degradation. Also, cellulosic fibers having various defects or dislocation pores, cracks, compression failures, and other sites of damages are prone to chemical attack and mechanical forces.

Though the study and understanding of nanotechnology has developed greatly, still safety, toxicological, and eco-toxicological questions exist and need to be addressed, and these challenges act as barriers to the commercialization of CNS in food packaging applications, as well as in other fields. The lack of awareness of this low-cost alternative and its many advantageous properties constitutes another challenge. In order to be successful in the market, there needs to be a better understanding about the product. Potential investors of the product will also require relevant information about the availability, technology required, and cost of production of the material. For the commercialization of the material in large volumes, producers require simultaneous increase in production to make sure the adequate supply and minimal cost. This demands technology for large scale production, for which budget friendly process equipment was not available until recently; hence awareness of the production and processing methods and protocol to ensure a uniform, reliable, and cost-effective production of cellulose nanomaterials, particularly in huge volumes, which inhibits marketing, needs to be popularized [11].

Technology transfer from the research labs to industries for commercialization needs to be strengthened. There is also a need for enhanced process control technology to ensure efficiency, including capabilities for measuring the nanoscale and manipulation. There was only very limited information about the environment, health, and safety impact that nanostructures used for food, medical, or pharmaceutical industries, which had influenced the same. The safety concerns regarding the use of the material are important in all industries to gain trust of the consumers as well. Hence, data about the health, wellness, and nutritional aspects need to reach the potential producers and consumers. The link between researchers, producers, and potential consumers also needs to be built up well to understand the setbacks, what needs to be introduced or improved, the cost/benefit ratio, etc. One of the technical challenges associated with CNS is lack of technology for the removal of water from cellulose and converting into a usable form for incorporating them into another material.

Applications particularly nonaqueous ones require a cost-effective, energy efficient drying process of CNS. Cellulose nanomaterials also raise resource and transportation costs when suspended in low concentrations in aqueous mixtures. They also show an undesirable irreversible bonding between fibers upon drying. There are also other concerns regarding how these structures may impact the flavor of the food items that are packed using them. Other demerits include the change in behavior of CNC depending upon the source from which they are extracted, for example, coconut husk when used as the source, the films produced displayed inferior properties of tensile strength due to hydrophobicity of lignin, and low compatibility within the polar matrix. CNF tangles and percolates, which results in quite high fiber agglomeration that impaired elongation and also affected transparency.

The price and availability of nanocellulose or other bio-based materials, their manufacturing process, and packaging performance will determine the incorporation of such materials into a viable food packaging system. Engineering properties of such nanocellulose incorporate packaging materials like durability properties, recyclability, and sustainability have to be considered for the manufacture of such packaging materials. Life cycle assessment of packaging material needs to be studied thoroughly. Such studies are conducted for existing products, but it is greatly missing in case of new products [27].

4.9 Conclusions

The major benefits of using nanocellulose for the design of ideal food packaging systems are its abilities to prevent food spoilage, prevent the entry of oxygen, potential to replace the use of nonbiodegradable materials like polystyrene-based foams and plastics, their ability to serve as carriers of active substances like antimicrobials and antioxidants that can improve the quality of food, and finally their ability to extend food shelf life. The unique properties of CNS such as biodegradability, renewability, sustainability, and presence in abundance make the exploitation more economical as well as reliable in many industries. In the case of food packaging systems, the biocompatibility, mechanical property, superior barrier as well as antimicrobial property, flexibility, and transparency make it a potential alternative for many of the packaging materials currently using now. The amount of preservatives and cost for extra processing techniques can be minimized, and as a result more organoleptic properties can be retained in the final product.

References

1 Reginaldo, A.F., Wagner, C.O., and Chandrashekhar, P.J. (2007). Structure, organization, and functions of cellulose synthase complexes in higher plants. *Braz. J. Plant Physiol.* 19 (1): 1–13.

2 Ahmed, M.A.H., Saim, A., and Ekrem, D. (2016). Evaluation of the possibilities for cellulose derivatives in food products. *J. Forest. Faculty* 16 (2): 383–400.

3 Devabaktuni, L., Kulkarni, P.K., Mudit, D. et al. (2011). Sources of cellulose and their applications – A review. *Int. J. Drug Formul. Res.* 2: 19–38.

4 Paunonen, S. (2013). Strength and barrier enhancements of cellophane and cellulose derivatives films: a review. *Cellulose Derivatives Rev. Bio Resour.* 8 (2): 3098–3121.

5 Sherif, M.K. (2014). Bacterial cellulose production and its industrial applications. *J. Bioprocess. Biotech.* 4 (2): 1–10.

6 Klemm, D., Brigitte, H., Hans-Peter, F., and Andreas, B. (2005). Cellulose: Fascinating biopolymer and sustainable raw material. *Angew. Chem. Int. Ed.* 44: 3358–3393.

7 Saxena, I.M. and Bown, R.M. (2005). Cellulose biosynthesis: current views and evolving concepts. *Ann. Bot.* 96: 9–21.

8 Sonam, J., Premjeet, S.S., Reetesh, M., and Babita, G. (2013). Cellulose derivatives as thermoresponsive polymer: an overview. *J. Appl. Pharmaceut. Sci.* 3 (12): 139–144.

9 Murray, J.C.F. (2000). Cellulosics. In: *Handbook of Hydrocolloids* (eds. G.O. Phillips and P.A. Williams), 219–230. Cambridge, U.K.: Wood head Publishers.

10 Xiong, R., Zhang, X., Tian, D. et al. (2012). Comparing microcrystalline with spherical nanocrystalline cellulose from waste cotton fabrics. *Cellulose* 19 (4): 1189–1198.

11 Katrina, P.Y.S., Yean, L.P., and Shee, K.M. (2018). Nanocellulose: Recent advances and its prospects in environmental remediation. *Beilstein J. Nanotechnol.* 9: 2479–2498.

12 Lucie, B., Julia, P., Maria, T. et al. (2020). Applications of nanocellulose/nanocarbon composites: focus on biotechnology and medicine. *Nanomaterials* 10: 196–228.

13 Hou-Yong, Y., Heng, Z., Mei-Li, S. et al. (2017). From cellulose nanospheres, nanorods to nanofibers: various aspect ratio induced nucleation/reinforcing effects on polylactic acid for robust-barrier food packaging. *ACS Appl. Mater. Interfaces* 9 (50): 43920–43938.

14 Zhijun, S., Yue, Z., Glyn, O.P., and Guang, Y. (2014). Utilization of bacterial cellulose in food. *Food Hydrocolloids* 35: 539–545.

15 Irina, K., Herve, B., Patricia, B. et al. (2013). Cellulosic nanorods of various aspect ratios for oil in water Pickering emulsions. *Soft Matter.* 9: 952–959.

16 Martin, A.H., Ana, F., Preeti, T. et al. (2017). Nanocellulose in thin film, coatings, and piles for packaging applications: a review. *Bioresources* 12 (1): 2143–2233.

17 Talita, S.F. and Graciela, I.B. (2018). Nanocellulose in food science and technology – potential, advantages and gaps of research. *Novel Tech. Nutr. Food Sci.* 1: 73–74.

18 Atefe, R., Ali, N., and Milad, F. (2015). Application of cellulosic nanofibers in food science using electrospinning and its potential risk. *Compr. Rev. Food Sci. Food Saf.* 14: 269–284.

19 Ana, F., Lokendra, P., and Martin, H. (2016). Nanocellulose in packaging: advances in barrier layer technologies. *Ind. Crops Prod.* 95: 574–582.

20 Alana, G., Fabiany, S.K., Jean, J.B., and Derval dosSantos, R. (2017). Cellulose nanostructures obtained from waste paper industry: a comparison of acid and mechanical isolation methods. *Mater. Res.* 20: 209–214.

21 Lucie, B., Julia, P., Marketa, B. et al. (2019). Versatile application of nanocellulose: from industry to skin tissue engineering and wound healing. *Nanomaterials* 9: 164–203.

22 Henriette, M.C.A., Morsyleide, F.R., and Luiz, H.C.M. (2017). Nanocellulose in bio-based food packaging applications. *Ind. Crops Prod.* 97: 664–671.

23 Kontturi, E., Laaksonen, P., Linder, M.B. et al. (2018). Advanced materials through assembly of nanocelluloses. *Adv. Mater.* 30: 1703779.

24 Klemm, D., Kramer, F., Moritz, S. et al. (2011). Nanocelluloses: a new family of nature-based materials. *Angew. Chem. Int. Ed.* 50: 5438–5466.

25 Xiao, Y., Rong, L., Wang, B. et al. (2018). A light-weight and high-efficacy antibacterial nanocellulose-based sponge via covalent immobilization of gentamicin. *Carbohydr. Polym.* 200: 595–601.

26 Xin, S., Li, X., Ma, Z. et al. (2013). Cytotoxicity and antibacterial ability of scaffolds immobilized by polysaccharide/layered silicate composites. *Carbohydr. Polym.* 92: 1880–1886.

27 Abdul Khalil, H.P.S., Davoudpour, Y., Chaturbhuj, K.S. et al. (2016). A review on nanocellulosic fibres as new material for sustainable packaging: process and applications. *Renewable Sustainable Energy Rev.* 64: 823–836.

28 Fu, L., Zhang, J., and Yang, G. (2013). Present status and applications of bacterial cellulose-based materials for skin tissue repair. *Carbohydr. Polym.* 92: 1432–1442.

29 Picheth, G.F., Pirich, C.L., Sierakowski, M.R. et al. (2017). Bacterial cellulose in biomedical applications: a review. *Int. J. Biol. Macromol.* 104: 97–106.

30 Dos Reis, E.M., Berti, F.V., Colla, G., and Porto, L.M. (2018). Bacterial nanocellulose-IKVAV hydrogel matrix modulates melanoma tumor cell adhesion and proliferation and induces vasculogenic mimicry in vitrol. *J. Biomed. Mater. Res. Part B App. Biomater.* 106: 2741–2749.

5

Chitosan-Based Nanoparticles and Their Applications in Food Industry

Basant E. Elsaied and Ahmed A. Tayel

Kafrelsheikh University, Faculty of Aquatic and Fisheries Sciences, El-Geish St. 33516, Kafr El Sheikh, Egypt

5.1 Introduction

The concept of nanotechnology is "understanding and controlling material at dimensions of approximately 1–100 nm where unique phenomena allow new applications." It has now got more coverage in multiple applications such as agriculture, pharmaceuticals, medicine, cosmetics, water treatment, and food industry. Food industry is a practice of manufacturing and preserving food to reach a final consumable state. Further, the role of nanotechnologies in enhancing foods can be determined in terms of improving nutritional and organoleptic values and extending shelf life of foods [1]. Commercially, in the food industry there are several nano-based items such as 24Hr Microactive® CoQ10, Aquanova NovaSol®, Asia Food beverage®, and Curcosome® [2].

Ranked second after cellulose, chitin is the natural polymer that most abounds. Henri Braconnot noticed chitin in mushroom for the first time in 1811. In 1823 Odier then discovered it in the cuticles of the insects and named it "chitin" from the Greek word "chiton" meaning a cover or a packet. Gradually, in 1859, Rouget coined an "altered chitin" could be prepared by a thermochemical treatment of chitin, and it was identified as "chitosan." Thereafter, in the 1930s and 1940s, both Rouget and Ledderhose granted about 50 patents about chitosan. Till the 1950s, advance diffraction techniques, including X-ray diffraction (XRD), were developed, and hence, it has been shown that the cell walls, mycelia, and septa of the fungi consisted of chitin and chitosan [3].

In the 1960s, chitosan has been investigated in wound healing, water treatment, and detoxification applications. It has lately been named a "fat magnet" since it binds fats and hence beneficial in weight management regimens [4]. The global market size of chitosan was estimated at USD 6.8 billion in 2019 and is projected to rise at a 24.7% compound annual growth rate (CAGR) based on sales between 2020 and 2027. Generally, chitosan nanoforms are some dispersions of particulates or solid materials usually between 1 and 100 nm in diameter [5].

Nanotechnology-Enhanced Food Packaging, First Edition. Edited by Jyotishkumar Parameswaranpillai, Radhakrishnan Edayileveettil Krishnankutty, Aswathy Jayakumar, Sanjay Mavinkere Rangappa, and Suchart Siengchin.

These chitosan-based nanoforms are usually incorporated directly to food and/or as packages. Mainly, chitosan-based nanoparticles can be used in a form of nanocapsules and/or nanoemulsions to enhance flavors, aromas, and color of food products [6, 7]. It also eventually incorporated in food packages in a form of nanofilms, nanocoatings, and nanosachets [8]. However, these chitosan-based nanoforms address many challenges regarding their applicability, scale-up productivity, and marketability [9].

Although much of research and development (R&D) has been done to improve these materials, an obvious global regulation is needed to achieve their actual potential in food industry. In this regard, this chapter provides a new investigation about the development and practical applications of nanoparticles based on chitosan in the food production industry.

5.2 Chitosan

5.2.1 Chitosan Precursor: Chitin Origins

Chitin is the only known precursor for chitosan. It is typically generated from natural origins, and it can be subdivided into three parts: (i) terrestrial sources, (ii) aquatic sources, and (iii) microbiological sources.

5.2.1.1 Terrestrial Sources

From a commercial perspective, these sources are mainly exploited for chitin extraction, due to easy accessibility and efficient processing. Terrestrial species usually comprises arthropods, nematodes, insects, silkworms, mosquitoes, honeybees, and others (Figure 5.1). The content of chitin is mainly dependent on the structure of these life forms, which is pretty diverse; thus, a remarkable difference in chitin content is observed. For example, cockroaches contain a chitin content of 30%–37%

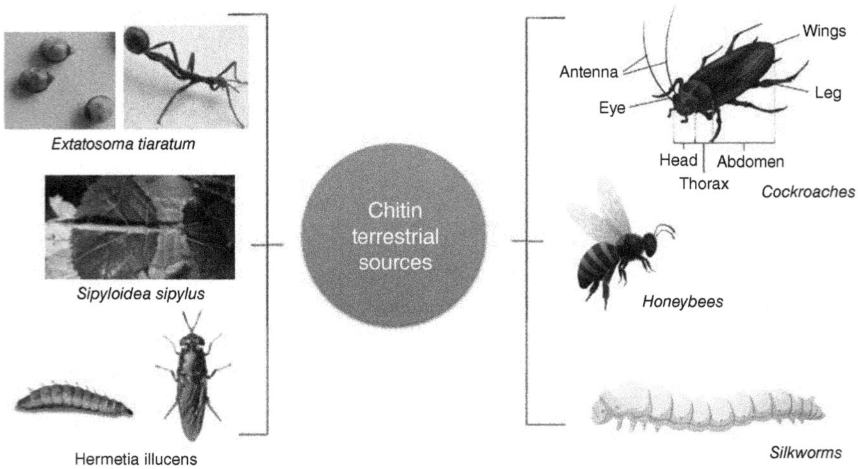

Figure 5.1 Possible terrestrial sources of chitin.

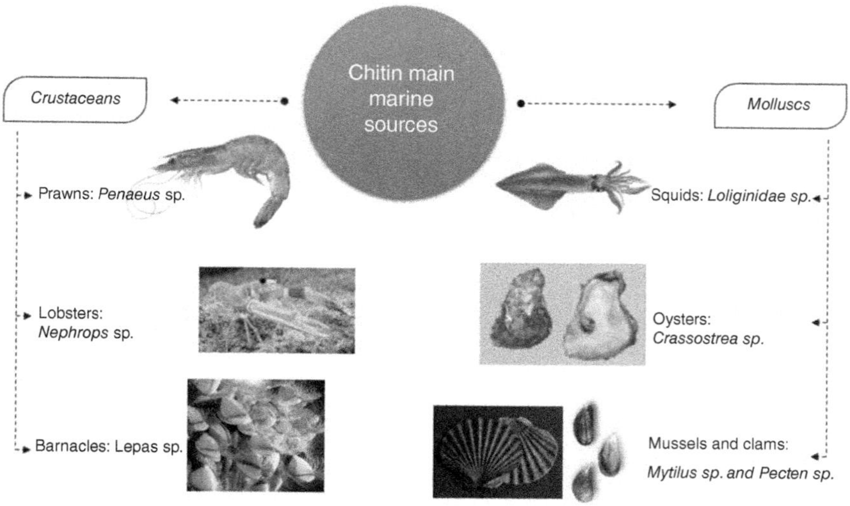

Figure 5.2 Possible aquatic sources of chitin.

(abdomen, head, and legs), 25%–29% (genitalia, cerci, thorax, and antennae), and 19% (fore and back wings). On the other hand, *Extatosoma tiaratum* has 24%, 3.4%, and 3.8% for exuviae, eggs, and eggshells, respectively. Other models are *Sipyloidea sipylus*, *Hermetia illucens*, honeybees, and silkworms, which possess a chitin content of 14%, 10–15%, 23–32%, and 20%, respectively [10].

5.2.1.2 Aquatic Sources

Chitins/chitosans are extracted from aquatic organisms such as diatoms, sponges, lobsters, barnacles, prawns, crabs, shrimps, krill, and squids. Conventionally, food processing plants release enormous waste materials consisting of heads, shells, and claws of shellfishes, which are exploited by chitin/chitosan producers as raw materials. Chitin proportions from these life forms may depend on species and season, but generally, exoskeletons produce 15–40% chitin, 20–40% protein, and 20–50% calcium carbonate ($CaCO_3$), which form complexes with pigments, fats, and other salts of metals. The main marine sources of chitin are Mollusca and crustaceans (Figure 5.2). Based on seasonal and unpredictable availability of raw materials, large-scale manufacture of chitosan from marine biowaste has faced limitations, in addition to challenges in the extraction process conditions [11].

5.2.1.3 Microbiological Sources

The microbial environment primarily contains molds, fungi, and yeasts. Mainly, Ascomycota, Basidiomycota, Deuteromycota, and Zygomycota comprise 2–65% of chitins. Particularly, Mucorales such as *Syncephalastrum racemosum* cell wall showed maximum yield of chitins/chitosans of 7% per dry weight of mycelia under optimal conditions [12]. From a market perspective, a number of companies around the world not only produce fungal chitosan on industrial scale but also develop bioproducts based on microbial sources of chitosan such as Kitozyme and

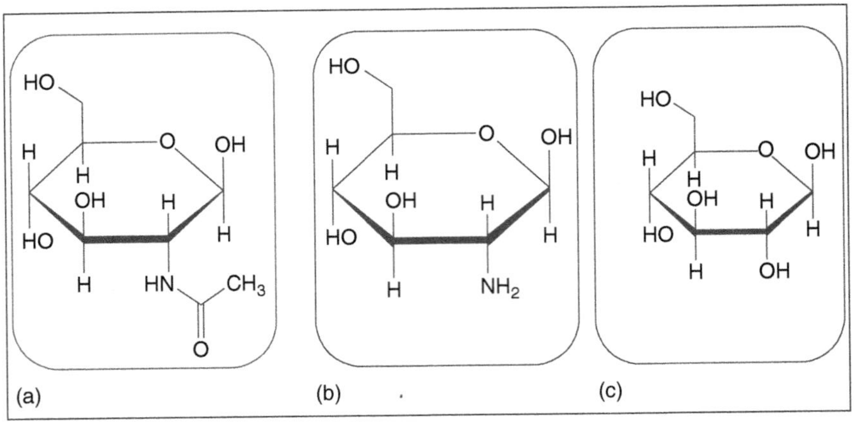

Figure 5.3 Structure of (a) *N*-acetyl-glucosamine, monomer of chitin; (b) glucosamine, monomer of chitosan; and (c) glucose, monomer of cellulose.

Chibiotech, which intend to satisfy the needs of vegan, halal, kosher, or gluten-free consumers.

5.2.2 Chemical Composition and Properties

Structurally, chitin and cellulose are known to be polymers consisting of β-(1,4)-2-acetamido-2-deoxy-β-D-glucopyranose and β-(1,4)-2-deoxy-β-D-glucopyranose units, respectively (Figure 5.3). Chitin is commonly described as a linear long-chain homopolymer composed of *N*-acetyl glucosamine units, [poly(*N*-acetyl-β-D-glucosamine)]. Based on its origin, chitin is available in three separate polymorphic forms, α-chitin, β-chitin, and γ-chitin, which can be distinguished by XRD and nuclear magnetic resonance (NMR). The isomorph of α-chitin is definitely the most abundant. These isomorphs vary in their macromolecular chain configuration [13].

Undeniably, partially deacetylated chitins results are commercial chitosans, which are polysaccharides composed of *N*-acetyl glucosamine and D-glucosamine units with variable lengths of chains. Owing to the risks of side reactions and chain depolymerization, completely deacetylated chitosans are rarely acquired. Actually, raw chitosans usually correspond to polymers with an acetyl content of less than 40%. Thus, the average percent of deacetylation degree (DDC) can characterize each commercial chitosan (Figure 5.4) [14].

5.2.2.1 Structural Properties

Chemically, due to the presence of primary and secondary hydroxyl (OH) groups on each repeated unit and the amine (NH_2) groups on each deacetylated unit, chitosan is more active than chitin. To modify the mechanical and physical properties of chitosan, the NH_2 and OH groups are readily active in chemical modifications. Compared with chitin, chitosan has an affinity for additional solvents. Extensive intra-and intermolecular reactions, such as hydrogen bonding, contribute to a rigid chitin structure that is mirrored in its insolubility in water and even in virtually all

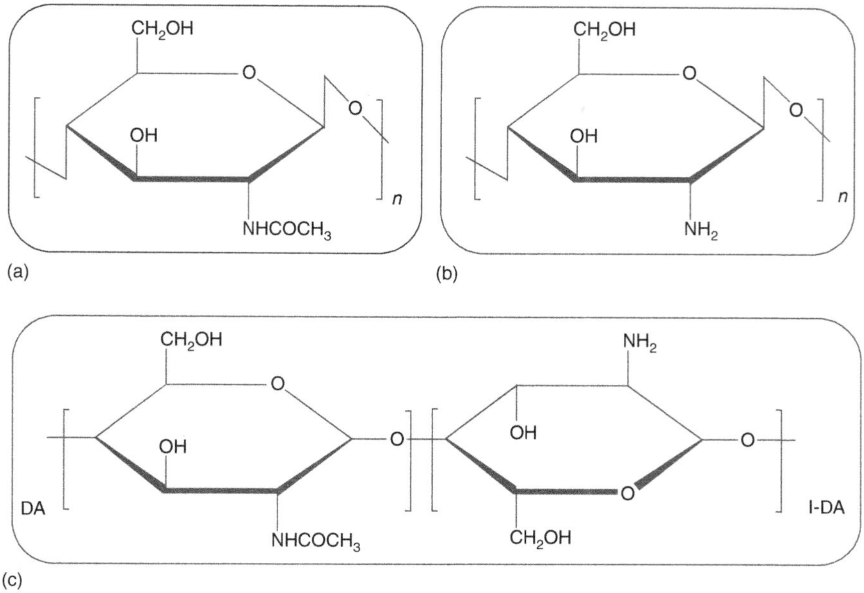

Figure 5.4 Schematic illustration of fully acetylated chitin [poly(*N*-acetyl-β-D-glucosamine] (a), fully deacetylated chitosan [poly(D-glucosamine] (b), and commercial chitosan (c).

organic solvents, wherein chitosan is readily soluble in diluted acids such as acetic acid. Positively charged groups in chitosan give it exciting chelating properties, owing to the ability of such groups to bind and adsorb various ions, which enabled chitosan to be used as coagulant and/or flocculant in wastewater treatment [15].

5.2.2.2 Physicochemical Properties

Degree of Deacetylation (DDC) DDCs are an extremely vital properties of chitosan as they influence nearly all the other properties to a significant degree. Essentially, the DDC is the ratio between the units of glucosamine and *N*-acetylated glucosamine, and the higher the DDC, the higher the purity. It has been commonly known that the alkaline solution concentration, temperature, and reaction time can adjust the characteristic patterns of chitosan deacetylation that results in modifications in its biological as well as physicochemical behaviors, since increasing those parameters will increase the DDC [16].

Molecular Weight (MW) Chitosan Molecular weight (MW) highly controls its properties and thus defines its probable applications. Considering MW, chitosan is distinguished to three main types: (i) low MW, (ii) high MW, and (iii) oligos. Based on the literature review, chitosan's antimicrobial activity depends on its MW. Chitosans of high MW cannot penetrate inside the cells. Instead, they have a tendency to accumulate on bacterial surfaces, and, consequently, they limit essential nutrients transportation. In contrast, low MW chitosans (<5000 kDa) are able to penetrate inside the cells, reach nucleus, and disturb the synthesis of mRNA. Such reactivity opened a lot of opportunities in miscellaneous applications, e.g. food preservation [17] and wound dressings [18].

Viscosity Generally, viscosity of chitosan solutions can be affected by several factors such as temperature, pH, concentration, MW, DDC, and extraction method. As already known, there is an inverse relationship between chitosan solution temperature and viscosity. Conversely, a proportional relationship is obvious between chitosan solution concentration and viscosity; the same occurs with MW and viscosity. In acidic PH, chitosan is an extraordinary enhancer of viscosity, which makes the choice of solvent type an important criterion. Thus, chitosan viscosity impacts biomedical applications such as wound dressing and also biodegradation-based technologies like bioplastics [19].

5.2.2.3 Diverse Properties

Besides the previously mentioned physicochemical properties, chitosan has countless biological properties such as being nontoxic, gelable, biocompatible, and biodegradable, in addition to being a chelating, antimicrobial, hemostatic, anticancer, immunoadjuvant, and drug carrying agent [20].

5.2.3 Preparation Methods and Manufacturing

A thousand research papers have discussed the methods of chitin and chitosan preparation [21]. The chemical and biological manufacturing processes of chitosan and derivatives thereof have been uncountably patented (EP2274337B1; US9846163B2). Three main well-known steps are required to prepare chitin, known as deproteinization, demineralization, and decolorization. The chitin/chitosan production is performed through a conventional protocol (Figure 5.5). The protocol parameters such as concentration of acid, reaction time, and temperature must be cautiously measured so as to lessen the hydrolytic depolymerization and deacetylation of chitin and to avoid thermal degradation [22].

A typical process applies a deproteinization by strong alkaline solution (e.g. sodium hydroxide (NaOH)) 40%–60% v/v at 50–60 °C, a demineralization by strong acid solution (e.g. HCl) 1 M at room temperature, a decolorization by oxidants (e.g. sodium hypochlorite (NaOCl)) to get crude colorless chitin product, and finally a deacetylation by strong alkaline solution 40%–60% v/v at 80–120 °C to get crude chitosan. The side products of carotenoids, inorganic salts, and proteins can be used as fish food additives, rubber processing, and feedstuff, respectively. It is worth mentioning that the deacetylation final solution, which contains sodium acetate, NaOH, and some impurities, can be retreated to regenerate acetate-free NaOH that can be reused in the process. This will eventually decrease the manufacturing costs of chitosan (WO1999038893A1).

However, this chemical method has a lot of drawbacks. To overcome the defects, biological treatments, such as microbial production and enzymatic hydrolysis, are alternatives to the chemical production of chitin and chitosan [23]. A comparison between these two methods is shown in Table 5.1.

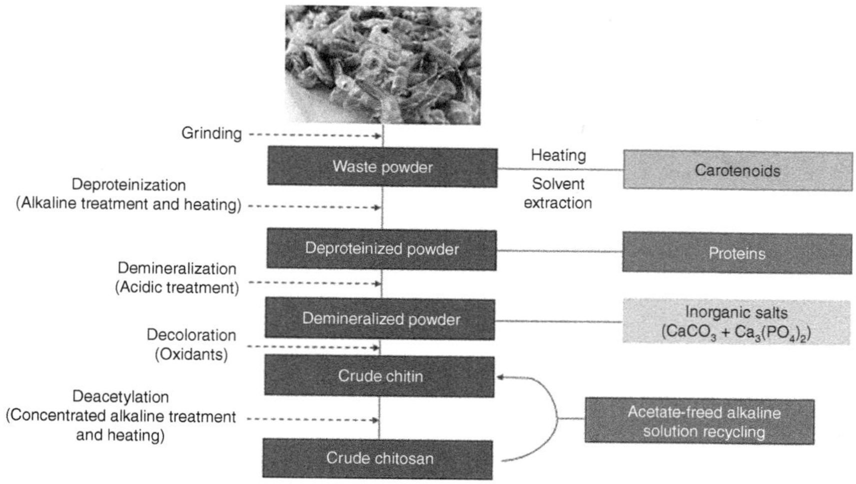

Figure 5.5 Conventional protocol of chitin and chitosan production.

Table 5.1 Advantages and disadvantages of chemical and biological treatments of chitin/chitosan.

	Chemical process	**Biological process**
Cost	Low cost	High cost
Scalability	Scalable	Scalable
Water and energy consumption	High	Low
Reagents toxicity	Toxic	Nontoxic
Time	Long	Short
Environment conservation	Not ecological	Ecological
Product quality	Excellent	Fair
Product yield	High	Low

5.2.4 Chitosan Modifications

Many studies have considered the modifications of chitosan structure to overcome its disadvantaged solubility in organic solvents and water. Mainly, chitosan possesses three reactive sites: the OH groups on C3 and C6 in each unit and the amino group on C2 in each deacetylated unit (Figure 5.6). Such groups permit substitution reactions, resulting in new derivatives. The key modifications are quaternization, which is beneficial to improve chitosan solubility in water, acetylation, acylation, reductive amination, Schiff's bases, phosphorylation, and cross-linking (Figure 5.7). This will bring pendant groups into chitosan, destroying its crystal structure and, therefore, improving its solubility. An updated review has been compiled by Negm et al., which conferred the newest advances in strategies of chitosan modification [24].

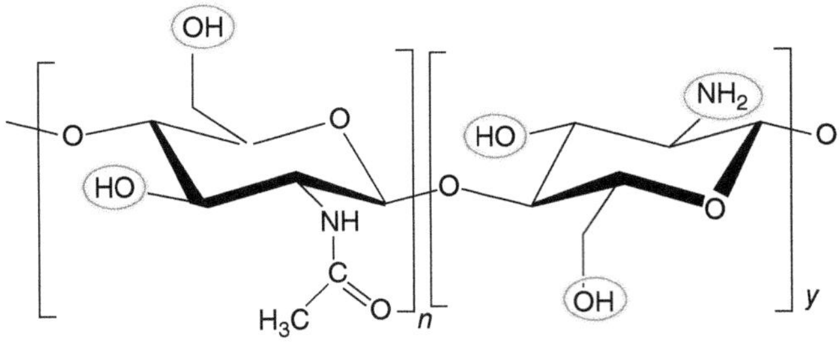

Figure 5.6 Main reactive sites of chitosan.

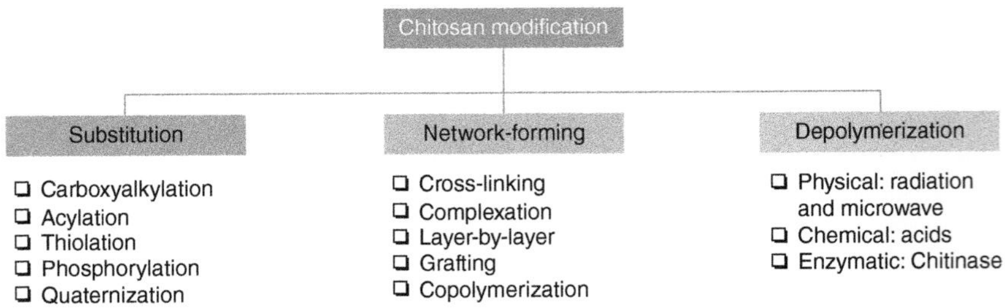

Figure 5.7 Key chemical modifications of chitosan.

Chitosan derivatives gained a lot of focus in food industry application such as being involved in food additives, e.g. stabilizing agents; nutritional quality enhancement, e.g. dietary supplement; and food packaging, e.g. antibacterial agents [24–26] (Table 5.2).

5.2.5 Overview of Chitosan Applications

5.2.5.1 Food and Beverage Industry

In the global markets of Asia, the food and beverage industry is one of the largest chitosan field, e.g. the North America, Europe, Japan, Korea, and China. Chitosan products also can be used as animal feed additive, and this is likewise a growing market. In wine production, chitosan can be made for clarification, stabilization, removal of ochratoxin A, enzymes, and other unwanted substances such as metals and pesticides [38]. As an eco-friendly flocculant, chitosan is also used for beer clarity [39]. Recently, Rocha et al. published a summary of the chitosan-based items used for different forms of beverages, such as alcoholic and nonalcoholic drinks, for clarity, protection, encapsulation, and productive and intelligent packaging [33].

5.2.5.2 Aquaculture

Alishahi and Aïder [40] described the use of chitosan and its derivatives in aquaculture. It may be used as functional food, dietary supplements, bioactive compound carriers, drug release, and water/wastewater pollutant treatment. In aquaculture, there is also an ongoing need to create successful vaccines to prevent infectious diseases [41].

Table 5.2 Derivatives of chitosan and their possible applications.

	Chitosan derivatives	Properties	Application	References
Quaternized derivatives	• *N*-substituted quaternary ammonium • *N*,*N*,*N*-trimethyl chitosan chloride • *N*-(2-hydroxyl) propyl-3 trimethyl ammonium chitosan chloride	– Cationic derivatives – Soluble in a wide scope of pH – Interactions with negative species – Mucoadhesion – Improved moisture holding and absorption	Water treatment; paper industry; dental care; drug delivery; gene delivery	[27–31]
Carboxymethyl chitosan	• *N*-carboxymethyl chitosan • *N*,*N*-carboxymethyl chitosan • *n*-Lauryl carboxymethyl chitosan • *O*,*N*-carboxymethyl chitosan • *O*-carboxymethyl chitosan	– The most studied chitosan derivatives – Amphoteric – Soluble in a wide scope of pH – Gelable – Clarifying agent	Food preservation; plant protection; excipients; drug delivery systems; dental; cosmetics	[31–34]
Chitosan 6-*O*-sulfate	• *O*-sulfated chitosan • *N*-sulfated chitosan • 2-*N*,6-*O*-sulfated chitosan • *N*-butyl-*O*-sulfate chitosan • *N*-octyl-*O*-sulfate chitosan • *N*-palmitoyl-*O*-sulfate chitosan	– Anticoagulant activity – Antimicrobial agent – Hemostatic agent – Antiviral activity	Drug delivery; hemocompatibility; and superplasticizer	[33, 35]

(Continued)

Table 5.2 (Continued)

	Chitosan derivatives	Properties	Application	References
Alkylated derivatives	• *N*-acyl derivatives • *N*-carboxyacyl chitosan • *N*-lauryl derivatives • Hydroxyethyl chitosan • Hydroxypropyl chitosan • Alkyl-glycosides branched chitosan • Hydroxybutyl chitosan	– Super substituted derivatives – Modifications need numerous protection – Capability to produce physical gels, sponges – Rapid gelation kinetics – Micelle formation – High rate of moisture absorption – Well-suited with neutral and cationic surfactants	Surfactant agents; papermaking; hemostatic agents; gene transfection; and oil recovery	[33, 35, 36]
N-methylene phosphonic chitosans	• Bifunctional derivatives • Grafted with alkyl chains	– Anionic and amphoteric character – Complexing properties – Film forming	Cosmetology; anticorrosion protection; chelating agents	[35–37]

5.2.5.3 Pharmacy and Cosmetics

Due to managed drug delivery, mucoadhesion, *in situ* gelling, and permeation enhancing properties, chitosan is predominantly used in the pharmaceutical industry. Like other polysaccharides, chitosan also exhibits efflux pump inhibitory properties. In various formulations, such as solutions, gels, tablets, capsules, fibers, films, and sponges, chitosan and its derivatives can be used [40]. A variety of cosmeceutical products based on chitosan are commercially available, such as Curasan™, Chitoseen™-K, Zenvivo™, Ritachitosan®, Chitosan MM222, etc. It is possible to apply these items as creams, lotions, and ointments [42].

5.2.5.4 Dentistry

Due to its bioactivity, anti-inflammatory, wound-healing practices, and bone repair, chitosan is predominantly used in dental treatment. In multiple formulations, such as solutions, microspheres, hydrogels, and toothpastes along with additives such as calcium phosphate, amelogenin, and quinic acid, chitosan is used [43].

5.3 Nanoforms of Chitosan

Generally, chitosan nanoforms are typically particulate dispersions or stable materials varying in diameter from 1 to 100 nm. Ionic gelation, emulsification, reverse micelles, self-assembly, and coacervation are the preparation methods widely known for chitosan nanoforms synthesis [5, 44].

5.3.1 Chitosan Nanocomposites

Fundamentally, scattered nanofillers such as nanoparticles, nanoplates, and nanofibers are made of nanocomposites, usually with a mean size of <100 nm. The strong mechanical property and thermal stability of chitosan nanocomposites makes them a viable substitute for the pharma and food processing and packaging industries [40].

5.3.2 Chitosan Nanocarriers

Nanocarriers identified as a nanomaterial used as module for transporting of another substance, such as a drug, range from 1 to 1000 nm diameter sizes. Much of the research on nanocarriers, especially in chemotherapy, have focused on their possible use in drug delivery. Since nanocarriers can be used directly to target tiny pores, lower pHs, and higher tumor temperatures, they can reduce the toxicity of many chemotherapy drugs [45].

Chitosan nanocarriers can be subdivided into (i) chitosan nanocapsules and (ii) chitosan nanoemulsions. Chitosan nanocapsules are defined as nanoparticles that consist of shell and core, where chitosan can be located or are polymeric capsules surrounded by a water or oil core prepared via interfacial polymer deposition followed by an interfacial reaction with chitosan [46]. Chitosan nanoemulsions are classified as nano-sized emulsions developed to facilitate the delivery of the active ingredients. Nanoemulsions are also used to distribute various anticancer medications, photosensitizers, medicinal agents, and food processing enhancers for selective drug distribution [47].

5.3.3 Preparation Methods

5.3.3.1 Ionic Gelation Method

This technique is very clear, quick, and is being applied by the majority of scientists in this field. By the interaction of oppositely charged molecules, chitosan nanoparticles can be synthesized. Since tripolyphosphate (TPP) is nontoxic, multivalent, and gelable, TPP has also been used to synthesize chitosan nanoparticles. The protocol can be regulated by TPP and chitosan charge density, which is dependent on the solution's pH [48].

5.3.3.2 Reverse Micellar Method

With this technique, ultrafine polymeric nanoparticles with a small size range can be prepared rather than emulsion-based methods, which produce large nanoparticles (> 200 nm). A microemulsion is formulated using a lipophilic surfactant that is dissolved in an effective organic solvent, such as n-hexane. A reverse micellar process to prepare chitosan nanoparticles was used by Kafshgari et al. [49]. Nevertheless, problems with the separation of nanoparticles and greater levels of solvent usage have been reported [50].

5.3.3.3 Emulsion-Based Solvent Evaporation Method

This approach is a minor variation of the reverse micellar method but prevents high shearing forces. In this process, a polymer solution is emulsified in aqueous phase, accompanied by evaporation of a solvent that induces the precipitation of polymers as nanospheres. The nanospheres produced were filtered, washed using suitable solvents such as petroleum ether, and gradually dried. Metformin-loaded chitosan nanospheres were effectively synthesized using this process [51, 52].

5.3.3.4 Coprecipitation Method

Highly monodispersed chitosan nanoparticles can be prepared by applying an acidic chitosan solution to an alkaline solution, such as ammonium hydroxide, in a coprecipitation process. The number of various particle sizes in this approach may be a weak point. A coprecipitation method was used to prepare lactic acid-grafted chitosan nanoparticles, which were spherical in shape and uniformly distributed [53] (Figure 5.8).

5.3.4 Characterization Techniques

5.3.4.1 UV–Visible Spectroscopy (Spectroscopic Analysis)

UV/visible spectroscopy is a method used to measure light absorption and dispersion in a sample. Only between the light source and the photodetector, a sample is placed, and the amplitude of the UV beam is measured before and after the sample passes through. At each wavelength, these measurements are compared to measure the

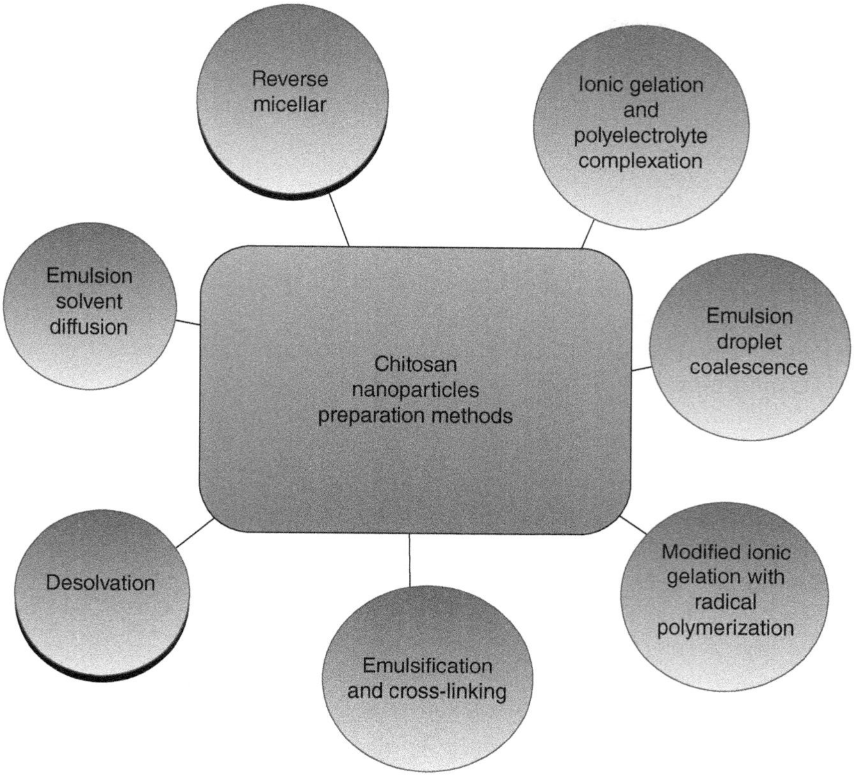

Figure 5.8 Main methods for preparation of chitosan nanoparticles.

sample wavelength of the sample. Usually, the results are plotted as a function of wavelength. Each spectrum is corrected by a buffer blank to make sure the spectral features in the sample extinction spectrum are not included [54].

5.3.4.2 Electron Microscopy (EM)

There are two major electron microscopy (EM) techniques for characterizing the morphology and homogeneity of nanoparticles, namely, transmission electron microscopy (TEM) and scanning electron microscopy (SEM). TEM provides information on the precise structure, size, and consistency of the nanoparticles, while SEM provides information on the surface topography and composition of the sample [55].

5.3.4.3 Dynamic Light Scattering (DLS)

Dynamic light scattering (DLS) records Brownian motion of nanoparticles, which in turn gives data about size distribution of the nanostructures through Stokes–Einstein equation [56].

5.3.4.4 Zeta Potential (ZP)

Zeta potential (ZP) reflects the degree of the repulsion between particles, which is the physical property in suspension of any particle. Briefly, the ZPs of 0 to ±30 mV suggest the instability of nanoparticles, while ZPs higher than ±30 mV suggest the stability of nanoparticles [56].

5.3.5 Overview of Applications

Chitosan nanoparticles are organic materials of outstanding physicochemical, antimicrobial, and biological properties, rendering them a special bioactive eco-friendly substance [57]. Hence, some of chitosan nanoparticles applications are illustrated in the following section.

5.3.5.1 Tissue Engineering

The goal of tissue engineering is to restore, substitute, preserve, or improve the function of a specific tissue or organ [58, 59]. Chitosan–gelatin/nanophase hydroxyapatite composite scaffolding showed better cell binding, proliferation, and spread of bone-like MG-63 cells relative to plain chitosan [60].

5.3.5.2 Water Treatment

Serious concerns have been raised about water pollution because of the inefficiency of traditional means of water treatment. The chitosan functional reactive groups make it an outstanding flocculant and/or coagulant for purification of wastewater for pesticide and metal contaminants elimination [61]. Chitosan nanoforms like nanobeads were tested effectively for adsorbing of heavy metals like Pb(II), Cr(VI), Cd(II), etc. [62].

5.3.5.3 Agriculture

Agrinanotechnology has attained a great attention recently due to its capability to offer simple and fast plant disease identification and effective fertilizer and pesticide distribution systems. However, nanoparticles should be carefully chosen for agricultural purposes. Chitosan has a significant influence on the development of many plants such as corn, coffee, wheat, and strawberry [63]. The biophysical features of coffee seedlings have been affected by chitosan nanoparticles through increased absorption of nutrients, photosynthesis levels, the quality of pigments, etc. [64].

5.3.5.4 Drug Delivery

The development of several colloidal drug delivery vehicles was made based on the use of chitosan nanocarriers. Chitosan nanoparticles can cross biological obstacles and supply drugs or macromolecules to a targeted site through controlled release. It is also effective in the interaction with the cell membrane since the limited volume of chitosan nanoparticles is picked up by cells by endocytosis [65]. The capabilities of chitosan nanoparticles to boost bioavailability of drugs, alter their pharmacokinetics, and stabilize encapsulated drugs have been described in several studies (Figure 5.9) [66].

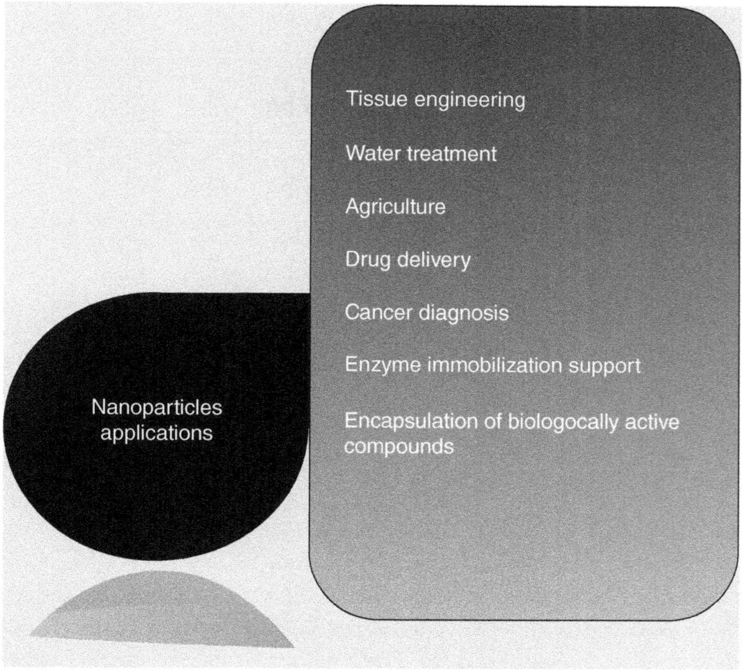

Figure 5.9 Schematic illustration of various applications of nanoparticles.

5.4 Chitosan-Based Nanoforms Applications in Food Industry

5.4.1 Opportunities in Food Processing

Food processing is a method of food preservation by turning it into a consumable form. Processing techniques are intended to keep the taste and consistency of the food intact but also to protect it from the invasion of microorganisms that contributes to spoilage of food. Three of the conventional food processing techniques are irradiation, heating, and high pressure [67].

Fresh food is not the only target of the food manufacturing industry, but the production of nutritious foods is also part of the issue. Food processing approaches that involve the nanomaterials include integration of gelling and viscosifying agents, nutrients delivery, e.g. minerals and vitamins, and nanoencapsulation of colors and flavors [68].

In addition, the role of nanotechnology in food enhancement can be measured on the basis of food texture, quality, taste, nutritional content, and shelf life. It is remarkable that nanotechnology not just addressed all of the areas mentioned earlier but also introduced dramatic improvements in foods that gave them new qualities [1].

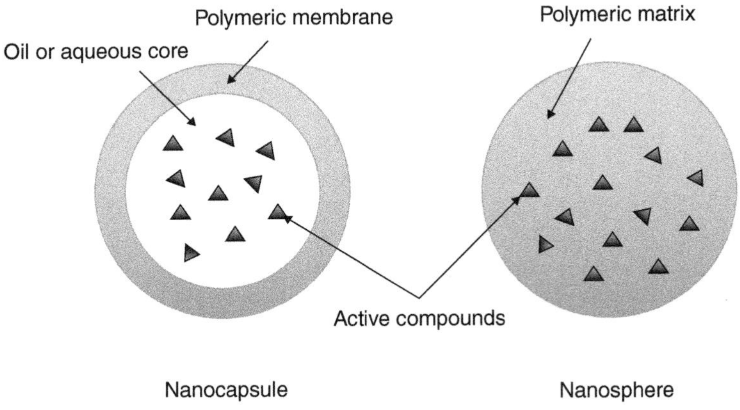

Figure 5.10 Main types of polymeric nanoparticles.

5.4.1.1 Chitosan-Based Nanoparticles: Enhancing Food Taste and Appearance

With the aid of nanocapsules, nanoencapsulation is done. They have a range of advantages, such as increased bioavailability, handling and durability, oxidation safety, preservation and distribution of volatile ingredients, taste concealing, and managed release caused by moisture and pH. These nanovesicular structures can be described as having a standard core–shell configuration in which the active ingredient is confined to a reservoir or to a cavity enclosed by a polymeric membrane or coating. (Figure 5.10). The active ingredient in the liquid, solid, or dispersion phase may be stored in the cavity. Nanocapsules transport dietary supplements through the gastrointestinal tract (GIT) of the biological system and penetrate deeply into tissues, which leads to increased delivery efficiency and bioavailability [69].

Coloring Agents TPP-low MW chitosan and poly(ethylene oxide)-4-methoxycinnamoylphthaloyl chitosan (PCPLC), respectively, have been used to encapsulate carotenoids like astaxanthin and lutein. The method of ionotropic gelation was employed to produce lutein-charged chitosan nanoparticles and assayed lutein bioavailability *in vivo* using mice and *in vitro* using simulated gastrointestinal digestion. Chitosan nanoencapsulation greatly enhanced lutein bioavailability by ~50% [6]. Moreover, PCPLC-chitosan nanospheres encapsulated astaxanthin with substantial encapsulation efficiency (EE) of 98% and protected astaxanthin from thermal pressures [70].

In 2015, Ge et al. reported β-carotene encapsulation with self-assembled nanomicelles based on chitosan, which improved its water solubility and stability [71]. Moreover, the stability of crocin, a saffron carotenoid, was improved when encapsulated into chitosan–sodium alginate nanoparticles where EE and LC were 38.16% and 30.96%, respectively [72]. Dhiman et al. concluded that chitosan nanoparticles showed a lycopene entrapment efficiency of 89.4%. Chitosan nanoparticles have enhanced the antioxidant function and color stability of anthocyanin by nanoencapsulation at elevated temperature [73, 74].

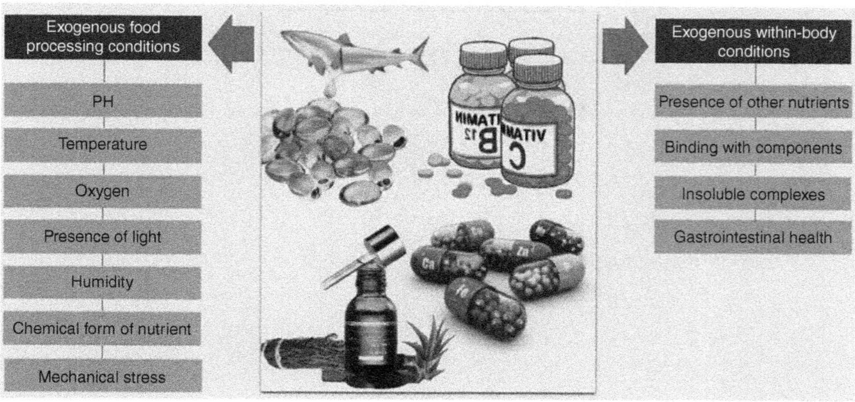

Figure 5.11 Factors affecting bioavailability and stability of nutrients.

Flavoring Agents Flavors are essential ingredients in foods that make food attractive and can satisfy consumers. Since the most flavors are unstable, nanoencapsulation seems a rational approach to store them. A double-shelled chitosan-coated silica nanocapsules possessed EE of 95.5% and sustained release behaviors for vanillin flavor [75].

In hot chili peppers, the pungent vanilloid compound, capsaicin, was successfully encapsulated in chitosan nanocapsules, suggesting the employment of chitosan nanocapsules in flavor delivery in food systems [7].

Eugenol-loaded spherically formed chitosan nanoparticles that have an LC and EE of 12% and 20%, respectively, were developed by an emulsion-ionic gelation cross-linking method. Compared with free eugenol, the extrusion of eugenol-charged chitosan nanoparticles guaranteed the increased thermal stability of encapsulated eugenol up to 155 °C [76].

5.4.1.2 Chitosan-Based Nanoparticles: Maintaining Nutritional Value

Many active components, such as lipids, proteins, carbohydrates, and vitamins, are extremely susceptible to the acidic and enzymatic action of the stomach and duodenum condition. Nanoencapsulation of these substances helps them to withstand certain brutal environments and also enables them to integrate readily in food products, which are difficult to do in nonencapsulated types due to the poor solubility in water. Figure 5.11 summarizes numerous factors that affect bioavailability and stability of nutrients.

Edible nonocapsules are now created not only to enhance flavors or colors but also to provide more healthy foods. Polymer nanocapsules such as chitosan are considered to be ideal for the encapsulation of bioactive substances (for example, flavonoids and vitamins) to secure and distribute bioactive substances for health-related purposes [2].

Vitamins Vitamins are important elements of human nutrition as they control a variety of biochemical processes inside the human body. Specifically, 13 recognized

Table 5.3 Chitosan-based nanoparticles for nanoencapsulation of vitamins.

Core	Nanoshell	EE (%)	References
• *Hydrophilic vitamins*			
B_2	Alginate/chitosan	56	[79]
C	Chitosan	70	[80]
C	Chitosan/cyclodextrin	>15	[81]
C	Chitosan/zein	51.9%	[82]
B_9, B_{12}, C	Chitosan/*N*,*N*,*N*-trimethyl chitosan	40%	[83]
• *Lipophilic vitamins*			
A (retinol)	Chitosan	60	[84]
E (α-tocopherol)	Chitosan/zein	76.5–86.5	[85]
D_3	Carboxymethyl chitosan/zein	87.9%	[86]
D_3	Carboxymethyl chitosan/soy protein isolate	96.8%	[87]

vitamins are classified into two categories: (i) *lipophilic* vitamins such as vitamins A, E, K, and D and (ii) *hydrophilic* vitamins including vitamin C and B family, including, thiamin (B_1), riboflavin (B_2), nicotinic acid (B_3), pyridoxine (B_6), pantothenic acid (B_5), folate (B_9), and cyanocobalamin (B_{12}). *Hydrophilic* vitamins are usually destroyed throughout cooking operations [77]. For example, during refining and cooking operations, vitamin C is readily oxidized. Light degrades vitamin B_2, while niacin and vitamin B_6 are robust forms under conditions of processing and storage. Considering *lipophilic* ones, A and E vitamins are prone to oxidation due to heat, light, metal ions, etc. The vitamin K structure is significantly changed by visible light and fluorescent. So as to conquer these unwanted circumstances and preserve vitamins throughout food production and storage, nanoencapsulation practices have revealed excessive ability in the preservation and selective distribution of vitamins [78].

Specifically, several chitosan-based nanoparticles have been used for nanoencapsulation of vitamin (Table 5.3). Regarding *hydrophilic* vitamins, alginate/chitosan nanoparticles were produced by Azevedo et al. [79] to encapsulate vitamin B_2 using the ionotropic polyelectrolyte pre-gelation process. The polydispersity index (PDI) of 0.319 ± 0.068 and the mean size of 104.0 ± 67.2 nm of the nanoparticles were recorded, with LC and EE of $2.2 \pm 0.6\%$ and $55.9 \pm 5.6\%$, respectively. In addition, vitamin C was encapsulated by ionotropic gelation using the chitosan nanostructure.

Alishahi et al. [80] employed chitosan/sodium tripolyphosphate (STPP) to prepare nanocarriers designed for vitamin C. In comparison, the vitamin C release rate was the highest at pH = 7.4 and the slowest at 0.1 M HCl, verified by *in vivo* assay in the digestive tract of rainbow trout. Jiménez-Fernández et al. [81] similarly presented chitosan nanoparticles as perfect carriers for transporting vitamin C with EE > 15%. In another study, *N,N,N*-trimethyl chitosan (TMC) nanoparticles have been developed of mean size approx. 196 ± 8 nm and EE of 95% for transporting B_9, B_{12}, and C vitamins [83]. In another sample, a zein/chitosan double-layer nanostructure was applied to encapsulate vitamin C, which was greatly protected from degradation, and just 5% of vitamin C was oxidized at ambient temperature after 10 days of storage [80, 82].

Considering *lipophilic* vitamins, a carboxymethyl chitosan (CMC) and soy protein isolate-complex nanoparticles successfully encapsulated vitamin D_3, which make them an attractive candidate in nanoencapsulation of hydrophobic nutraceuticals [87]. Luo et al. [85] evaluated EE of vitamin D_3 in zein nanoparticles coated with CMC. They found that EE was significantly enhanced to 87.9% after CMC coating, in comparison with 52.2% for that employing plain zein. Moreover, a zein-coated chitosan nanoparticles successfully encapsulated vitamin E (α-tocopherol) and gave improved fortification of its release against GIT environment [85]. Retinol was efficiently encapsulated into a spherical-shaped chitosan nanoparticles by an EE of more than 60% in different formulations [84].

Antioxidants: Phenolic Compounds The strong antioxidant content of phenolic compounds can be used as an antiallergenic, anti-inflammatory, anti-thrombotic, and antimicrobial agent in food items. These are phytochemicals often present in plant sources. They can remove free radicals and can be used as additives in food formulations to extend shelf life of foods without any conflicting effects on their sensory or nutritive value. Mainly, antioxidants can indeed be divided into two enzymatic and nonenzymatic groups [88].

Enzymatic antioxidants, like glutathione peroxidase (GSHpx), can destroy and eliminate free radicals. Nonenzymatic antioxidants include phenolic compounds. Bioavailability of phenolics as well as other antioxidants may be improved via increased solubility, absorption and permeation of the body, and defense of food formulations by nanoencapsulation procedures [89].

Catechins and anthocyanins have also been nanoencapsulated using altered chitosan by ways of an ionic gel strategy. Quercetin-encapsulated chitosan/lecithin nanoparticles were prepared via electrostatic self-assembly technique having 168.58 ± 20.94 nm average diameter where encapsulated quercetin demonstrated better antioxidant properties relative to free quercetin. These findings indicate that quercetin-charged lecithin/chitosan nanoparticles could be used in the production of functional foods [90]. Anthocyanin-charged chitosan nanoparticles were also developed by CMC and chitosan hydrochloride with an EE of 63.15%. Moreover in contrast to the free anthocyanin solution, which proved the efficiency of chitosan nanoencapsulation, the frequency and stability of nanoparticles were investigated in the simulated GIT fluid and beverage model method.

Curcumin as a recognized phenolic compound has been nanoencapsulated by multiple encapsulation methods with chitosan derivatives [91]. Hernandez-Patlan et al. [91] achieved 90% EE of curcumin by chitosan/alginate nanocapsules, and

Table 5.4 Chitosan-based nanoparticles for nanoencapsulation of antioxidants.

Core	Nanoshell	EE (%)	References
Curcumin	Chitosan/alginate	90	[91]
Catechins	Chitosan/TTP	—	[95]
Epigallocatechin gallate	Ferritin–chitosan double shells	9.69%	[96]
Quercetin	Chitosan/lecithin	71.14%	[90]
Resveratrol	Chitosan/γ-poly (glutamic acid)	35%	[97]
Rutin	Chitosan	84.9 ± 4.18	[94]
Chlorogenic acid	Chitosan	59	[93]
Naringenin	Chitosan	80	[92]

they have been found to be stable for three months under storage conditions. Naringenin, a potent flavonoid, was encapsulated by chitosan nanoparticles, which were synthesized by ionic gelation method. Naringenin exhibited a higher free radical scavenging activity in the nanocapsule form [92].

Chlorogenic acid (CGA), as a phenolic substance, was confirmed to be sustainably released from chitosan nanoparticles, which, therefore, by successful distribution of CGA in the body, the fortification of food supplements directed at health benefits may be encouraged [93].

Chitosan nanoparticles efficiently encapsulated rutin, a potent antioxidant flavonoid, with an EE of 84.9 ± 4.18% [94]. Hence, all these studies more significantly suggest that chitosan-based nanoparticles can achieve a better nutritional value of processed functional foods (Table 5.4).

Essential Oils (EOs) Essential oils (EOs) are potential organic volatile liquids or semiliquids derived from plants including certain flowers, seeds, and leaves that are antimicrobial. The application of natural antimicrobials to food has many problems, including poor solubility, instability, oxidation during the process, and their detrimental effects on the sensory characteristics of the eventual food items. Nanoencapsulation may solve these restrictions to some degree instead of plainly adding them to the food [98].

Paula et al. [99] suggested *Lippia sidoides* EO nanoencapsulation was achieved by spray drying using angico gum/chitosan as a wall material varying from 10 to 60 nm in diameter and 77.8% EE [99]. TPP/STPP-chitosan nanoparticles were broadly used to encapsulate *Carum copticum* EO, eugenol, oregano EO, carvacrol, and *Satureja hortensis* L. EO, via emulsion–ionic gelation technique [76, 100–103]. These studies cover a spectrum of particle size between 30 and 200 nm with a wide EE range from 14% to 47%.

For maintaining quality characteristics of food, *Satureja* EO, *Paulownia tomentosa* EO, and thyme EO were nanoencapsulated by chitosan nanoparticles to control undesirable changes and maintain quality of lamb meat, ready-to-cook pork chops,

Table 5.5 Chitosan-based nanoparticles for nanoencapsulation of essential oils.

Core	Nanoshell	EE (%)	References
Mint, thyme, and cinnamon EOs	Chitosan	75.4, 69.7, and 65.2	[106]
Clove EO	Chitosan	45.77	[107]
Mentha piperita EO	Chitosan/cinnamic acid	—	[108]
Coriandrum sativum EO	Chitosan	75.99	[109]
Peppermint and green tea EOs	Chitosan	82 and 81	[110]
Paulownia tomentosa EO	Chitosan	40.6	[111]
Satureja EO	Chitosan/liposomes	46–69	[112]
Thyme EO	Chitosan	30.67 ± 1.62	[113]
Zataria multiflora EO	Chitosan	45.24	[114]
Lippia sidoides EO	Chitosan/cashew gum	70	[115]
Cardamom EO	Chitosan/TTP	> 90	[104]
Oregano EO	Chitosan/TTP	21–47	[102]
Carum copticum EO	Chitosan/STTP	36	[100]
Rosemary EO	Chitosan	>90	[105]
Lippia sidoides EO	Angico gum/chitosan	78	[99]
Lime EO	Chitosan	—	[116]

and beef burgers, respectively. The highest EE in such zone was acquired by Jamil et al. [104] and Abdollahi et al. [105] to encapsulate cardamom EO and rosemary EO, respectively (Table 5.5) [104, 105].

Essential Fatty Acids (EFAs) and Fish-Derived Oils (FOs) FAs are the building units of lipids, which can generally be saturated FAs and unsaturated FAs. The most important omega-3 FAs that are essential for humans are alpha-linolenic acid, eicosapentaenoic acid (EPA), and docosahexaenoic acid (DHA), because our body is unable to generate these FAs due to a lack of unique desaturase enzymes [117]. The richest source of EPA and DHA is fish-derived oils (FOs). Linolenic acids are indeed usually found in the oils of grains, nuts, and plants. In FOs and essential fatty acids (EFAs), oxidation is a big problem because it worsens the flavor and fragrance of the oils and allows dangerous compounds to form [77]. Throughout oxidation reactions, FAs degrade into tiny unstable molecules in a process known as rancidity, which produce off-flavors. In addition, free radicals produced through oxidation within the human body can cause toxic effects. Oxidation in FOs must be either prevented or impaired. Nanoencapsulation is one of the procedures that is used to protect them.

Raeisi et al. [118] prepared a chitosan/Persian gum nanocapsules and investigated their EE for FO. The produced nanocapsules showed a good thermal stability of FOs justifying their suitability for food processing [118]. Moreover, krill oil has been nanoencapsulated in chitosan nanoparticles where EE of the obtained particles were about 33–59%, which will enhance their usage in food processing. Glycol chitosan nanoparticles encapsulated EPA, which were internalized into murine cells by endocytosis. The results showed effective delivery of EPA-glycol chitosan nanoparticles far better than those seen for plain EPA. Thus, it is expected that glycol chitosan nanoparticles may have a wide potential for EPA delivery in food processing fields [119].

5.4.2 Opportunities in Food Packaging

One-third of the world's food is lost, according to the United Nations Food and Agriculture Organization, and this has a significant effect on commercial, political, and environmental problems [120]. Food packaging is a solution to enhance food quality and protection, since it allows biological, physical, and chemical stability to be enhanced. The reactions within a food product cannot be actively regulated by conventional packaging materials. The use of active ingredients such as antimicrobials in sachets, films, and coatings indicates greater market acceptance than those directly added to food [121]. Thus, a new consumer niche in the packaging industry has thus been created [122].

In order to maintain the consistency or prolong the shelf life of food items, active packaging can be characterized as the integration of active systems into packaging films or containers. In fact, antimicrobial packaging is among the most innovative active packaging systems produced in the last decade and is capable of suppressing microbial development while maintaining the quality of food [123].

Post-consumption waste is a big issue with using plastics, including polyvinyl chloride (PVC), while packaging are without doubt the greatest contributor of plastic waste by nearly 63% [124]. Eco-friendly polymers from green materials can be used to address this environmental threat. There is some inconvenience with the use of biopolymers, such as mechanical fatigue, poor barriers, and thermal properties, as measured against nonbiodegradable plastic materials [125]. Due to these boundaries, masses of R&D was done to improve barrier and mechanical characteristics of these biopolymers by using the nanotechnologies perceptions [126].

Nanotechnologies have occupied a serious role in the R&D of advanced food packages that can be segmented into (i) smart packages, (ii) active packages, and (iii) improved packages (Figure 5.12).

5.4.2.1 Chitosan Nanoforms Functionality as Food Packaging Materials

Because of their comprehensive properties, such as biodegradability, biocompatibility, and nontoxicity, chitosan nanoparticles have been found to be outstanding options to be used as food packaging materials [127]. Generally, the first functionality of chitosan nanoforms is being an antimicrobial material. Against several microorganisms, including yeasts, filamentous fungi, and bacteria, the antimicrobial activity of chitosan and its derivatives has been noted. The exact mechanism of the antimicrobial action of chitosan, however, has not been fully explained [3].

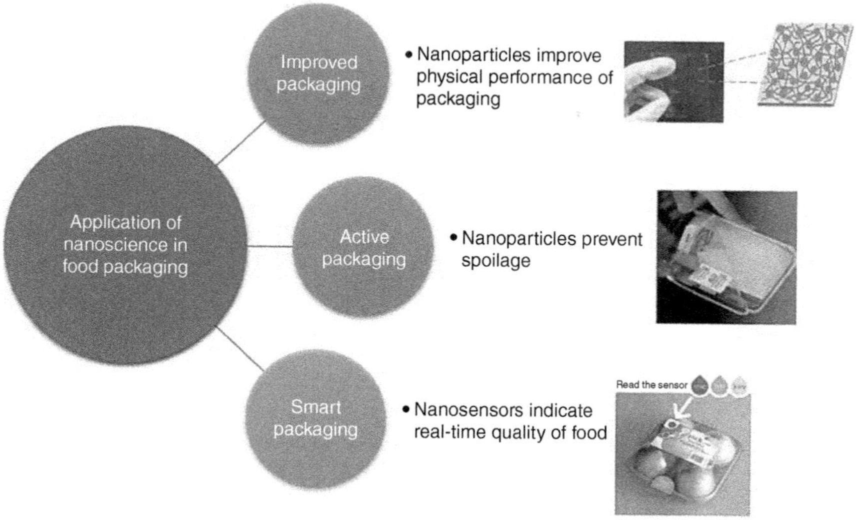

Figure 5.12 Application of nanoscience in food packaging.

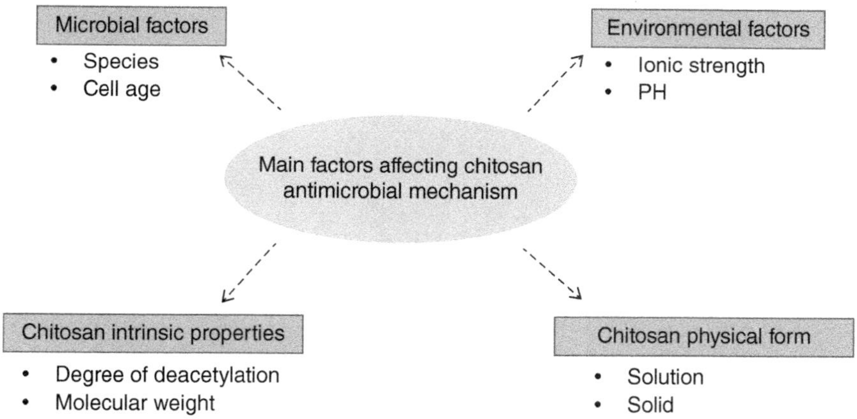

Figure 5.13 Factors controlling the chitosan antimicrobial pathway.

Many factors have been reported to be crucial in the explanation of the chitosan antimicrobial mechanism (Figure 5.13). The literature indicates three potential pathways for the antimicrobial action of chitosan: (i) chitosan adhesion through electrostatic interactions to the microbial cell wall, (ii) metal chelation affecting nutrients transportation, and (iii) nuclei penetration blocking transcription from DNA and protein synthesis (Figure 5.14) [128].

In addition to all mechanisms, chitosan exhibits more efficiency when prepared in nano-sized range (Figure 5.15). In 2019, Pan et al. reported that, relative to plain chitosan, chitosan nanoparticles (<200 nm) have stronger antimicrobial activity due to the high interfacial effect of nanoparticles, which improves their ability to stick to microbial cell walls, infiltrate cells, and kill cell membranes. Divya and Jisha [130]

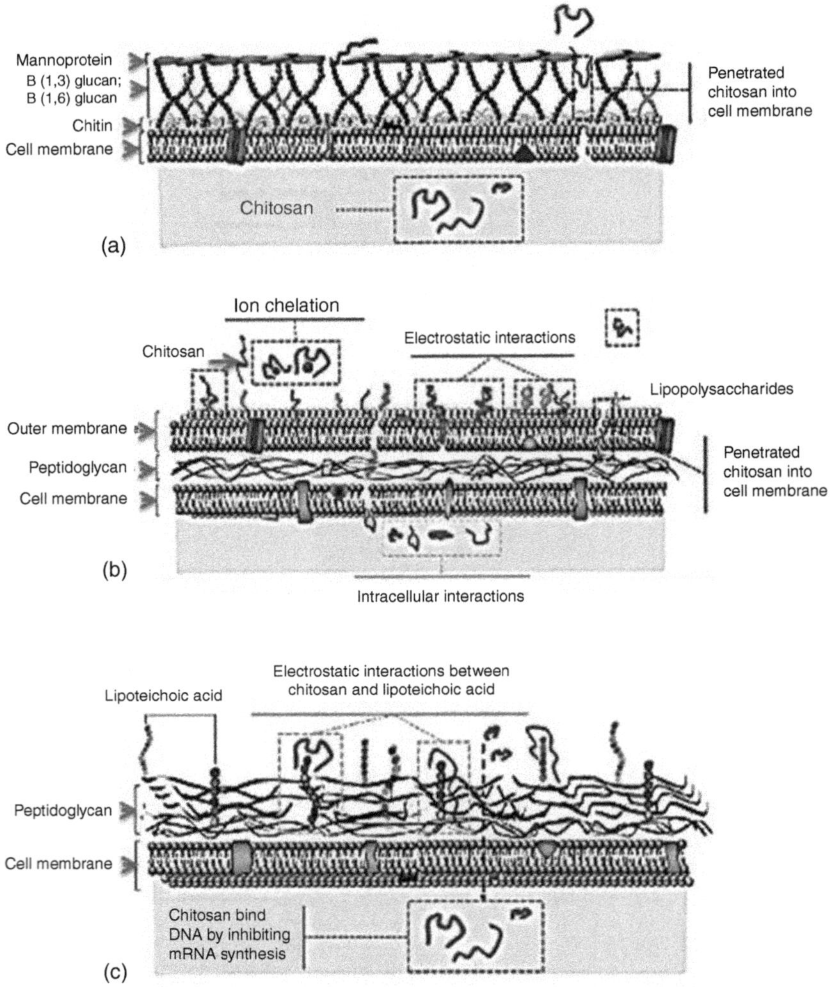

Figure 5.14 Possible chitosan antimicrobial mechanisms through (a) electrostatic interactions, (b) metal chelation, and (c) nuclei penetration.

compared chitosan nanoparticles with Gram+ and Gram− bacterial strains for their antimicrobial activity. The large surface of the nanoparticles allows it to accumulate the bacterial surface, resulting in membrane interruption and leakage of intracellular components and eventual cell mortality.

Another functionality of chitosan nanoforms is being water and gas barriers, and they enhance the barrier, mechanical, and thermal characteristics of food packaging materials. Most of these nanoparticles' incorporation are in packaging films matrices and mainly manufactured through the casting method. According to Abd-Elsalam [131], the association between the polymer matrix and nanoparticles of chitosan focused on enhancing the mechanical, barrier, and antimicrobial performance of such materials.

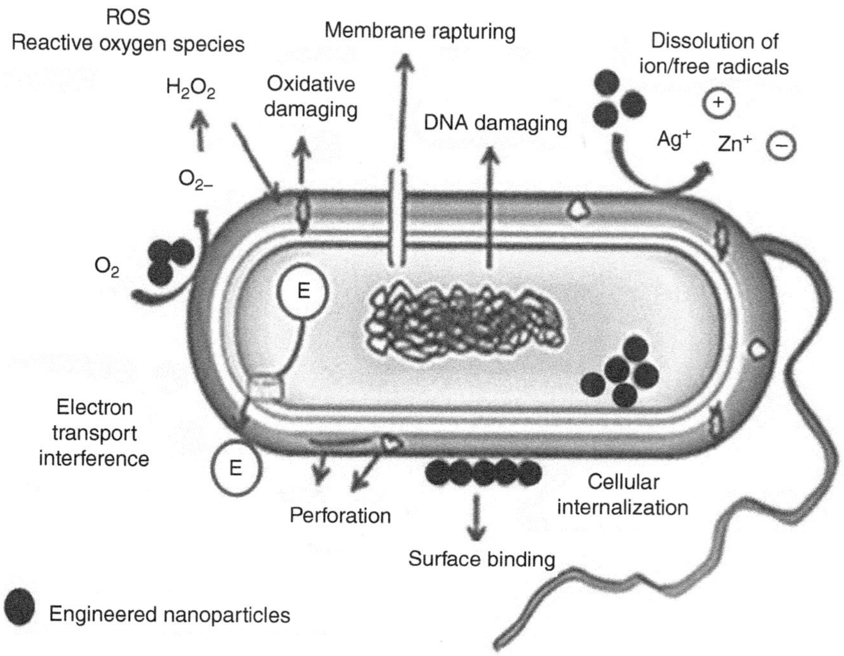

Figure 5.15 Antimicrobial activity mechanism of nanoparticles. Source: Adapted from Suresh et al. [129].

Fruit and Vegetables Preservation Protecting fruits and vegetables after harvesting is very important. Fresh products (e.g. fruits and vegetables) experience consistency degradation during postharvest transportation and storage due to various physiological reactions, such as postharvest oxygenation, ripening, development of ethylene, and senescence. Microbiological spoilage results in losses of around 15–50% of fruits and vegetables produced worldwide after harvesting [132]. It is very difficult to choose a packaging material because different foodstuffs have different degrees of transpiration and oxygenation. Owing to the fermentation mechanism inside packaging, a major harm to fruits and vegetables is found [132].

Antimicrobial nanocomposite films made of chitosan and other incorporated active ingredients are the most common forms of the nanocomposites that have been used to achieve controlled release of antimicrobials, inhibition of foodborne pathogens, controlled postharvest decay of food (e.g. fruits and vegetables), increased shelf life, and reduced weight loss [19].

Antoniou et al. [133] complexed tara gum and chitosan nanoparticles to yield palatable nanocomposite film that owned improved tensile strength and barrier to moisture characteristics, which render it a respectable candidate for food packaging [133] (Figure 5.16). Chitosan/TiO_{2}/SiO_2, chitosan/SiO_2, and CS/gelatin/Ag nanocomposite films have been investigated to preserve various agricultural commodities such as *Ginkgo biloba* seeds, jujube, and red grapes, respectively. They all found to delay ripening and prolong shelf life [134, 135]. Table 5.6 shows numerous studies on the manufacturing of chitosan-based nanoparticles and their

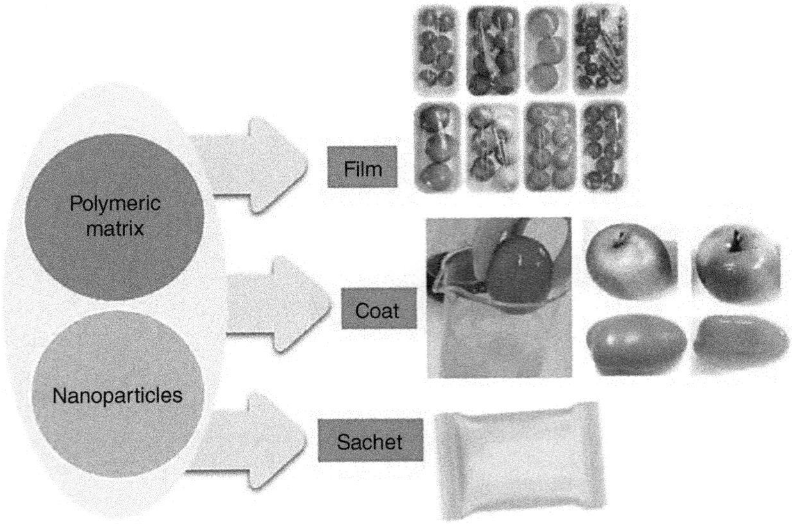

Figure 5.16 Application of chitosan nanocomposites in fruit and vegetables preservation.

incorporation to fruit and vegetables preservation including tomatoes, mangoes, grapes, etc.

Meat and Fish Products Preservation Foodborne infections are scarcely recorded, and the precise effect of foodborne illnesses and outbreaks on customers is often difficult to reliably determine. Meat food is responsible for a comparatively large number of foodborne illness (approximately 500–1500 incidents) recorded annually around 1998 and 2015 in the US. Meats and fish are chemically comminuted and extremely perishable foods and can introduce surface pollutants into the inner layers of muscle foods. Once incorporated in the inner membranes, it is relatively more difficult for foodborne pathogens to be inactivated or eliminated by cooking. Table 5.7 shows numerous studies on the manufacturing of chitosan-based films and their incorporation to fish and meat products including pork, poultry, and meat [146].

In meat packaging, Arkoun et al. [147] found that electrospun chitosan-based nanofibers act as bioactive nanobarriers to meat infection by increasing the shelf life of raw red meat up to seven days. The findings of Lin et al. in 2018 show that ε-polylysine/chitosan nanofibers have been effective in inhibiting the *salmonella* and preserving chicken color and flavor, indicating that they could significantly be used as a potent antibacterial product to improve its shelf life and quality in packaged food. Furthermore, poly(butylene adipate-*co*-terephthalate) film (PBAT) with various levels of chitosan nanofibers demonstrated anti-foodborne pathogens activity – for example, Gram+ *Staphylococcus aureus* and Gram- *Salmonella enteritidis*; these antibacterial films satisfy the increasing demand in the food and beverage industry for green packaging [148].

In order to prevent microbial spoilage of fish fillets, Ceylan et al. developed electrospun chitosan/thymol/liquid smoke nanofibers in 2018 [149]. In order to reduce almost 60% of the total mesophilic bacteria, the electrospun chitosan/thymol/liquid

Table 5.6 Chitosan-based nanoparticles for fruit and vegetables preservation.

Nanocomposite materials	Form	Fruit or vegetable	Main effects	References
Chitosan/TiO_{2}/SiO_2	Film	*Ginkgo biloba* seeds	Mechanical strength, antioxidant, and barrier	[134]
Chitosan/cellulose/acetate phthalate/ZnO	Film	Grape	Thermal stability and antibacterial	[136]
Chitosan/TiO_2	Film	Tomato	Delay ripening, ethylene, photodegradation ability, and mechanical strength	[137]
Chitosan/SiO_2/clay	Film	Strawberry	Prolonged shelf life	[138]
Chitosan/SiO_2	Coat	Jujube	Lessened loss of weight and flavonoid content protection	[139]
Chitosan/SiO_2	Coat	Longan fruit (*Dimocarpus longan* Lour. cv Shijia)	Prolonged shelf life and reduced weight loss	[140]
CS/gelatin/Ag	Film	Red grapes	Extended shelf life by 14 d	[135]
CS/ZnO/polyethylene	Film	Okra	Extended shelf life from 5 d to 12 d	[141]
Chitosan/TiO_2	Film	Mango	Enhanced firmness and quality of mangoes	[8]

Table 5.7 Chitosan-based nanoparticles for meat and fish products preservation.

Nanocomposite materials	Form	Meat or fish product	Main effects	References
Chitosan/cellulose	Film	Ground meat	Antibacterial activity and prolonged shelf life	[142]
Chitosan/ZnO	Pouches	Raw meat	Antibacterial activity and prolonged shelf life	[143]
Chitosan/ montmorillonite	Film	Minced camel's meat	Antibacterial activity and prolonged shelf life	[144]
Chitosan/montmorillonite/ rosemary and ginger essential oil	Film	Poultry	Antibacterial and antioxidation activity and prolonged shelf life	[145]
Chitosan/clay/rosemary essential oil	Coat	Silver carp fillets	Retarded lipid oxidation	[105]

smoke nanofibers were successful. In addition, the authors have suggested that nanofibers are thermostable up to 150 °C, which is within the range of temperatures used in traditional food preservation techniques.

Chitosan/poly(ethylene oxide) nanofibers filled with tea tree oil and added to certain food bacterial pathogens were developed by Cui et al. in 2018 and displayed a stable anti-biofilm function in chicken meat samples over four days [150].

Dairy Products Preservation Microbial infection of dairy foodstuffs is a worldwide food safety and quality concern. Pathogens of concern in dairy products include both Gram+ bacteria including *Listeria monocytogenes* and *S. aureus* and Gram- bacteria including *Salmonella* spp. and *Cronobacter sakazakii*. Preservation using natural antimicrobials is desirable due to increased consumer demand for clean label foods. Cheese packaging is among the most key steps in the long road from the producer to the customer, in which most cheese factories are far away from consumption. The packaging of cheese must, by handling and shipping, protect the food from mechanical loss and bad environmental conditions [151]. Table 5.8 shows various types of cheese and butter that have been preserved using chitosan nanocomposites, which improved the antibacterial activity as well as prolonged shelf life.

5.4.2.2 Chitosan-Based Nanoparticles Toxicity and Fate in Human Body

Notwithstanding the possible advantages of nanotechnology in the food industry, little is understood about the safety implications of the use of nanotechnology in

Table 5.8 Chitosan-based nanoparticles for dairy products preservation.

Nanocomposite materials	Form	Meat or fish product	Main effects	References
Chitosan/carboxymethyl cellulose/ZnO	Film	Soft white cheese	Antibacterial activity and prolonged shelf life	[151]
Chitosan/polyvinyl alcohol (PVA)/TiO_2	Film	Soft white cheese	Antibacterial activity and prolonged shelf life	[151]
Chitosan/gelatin/ZnO	Film	White cheese	Antibacterial activity and prolonged shelf life	[152]
Chitosan/cellulose/Ag/TiO_2	Paper coat	Clarified butter	Antibacterial activity and prolonged shelf life	[153]
Chitosan/monolaurin/cellulose	Film	Ultrafiltered white cheese	*Listeria monocytogenes* inhibition	[154]
Chitosan/cellulose/nisin/ZnO	Paper coat	Ultrafiltered white cheese	*Listeria monocytogenes* inhibition	[155]
Chitosan/polyvinyl alcohol (PVA)/glycerol/TiO_2	Coat	Karish cheese	Mechanical and barrier properties and prolonged shelf life	[156]
Chitosan/polyvinyl alcohol (PVA)/TiO_2	Coat	Ras cheese	Mechanical and barrier properties and prolonged shelf life	[156]

food processing and the inclusion of NPs in food. The chemical composition, size, shape, solubility, electrical charge, and surface structure of the chitosan-based NPs cause them to easily enter membranes or attach to organic compounds like proteins. Consequently, relative to other particles in size, NPs are advantageous in their toxic action. Since chitosan NPs have induced malformations in zebra fish embryos at a size of 200 nm, it is important to research the biosecurity connected with the use of chitosan NPs in food applications [9].

Digestion of Chitosan-Based Nanoparticles Through going through the GIT, nanostructures will alter both structurally and physiochemically. Chitosan-based nanostructures can decay partially due to salivary gland amylase in the mouth at the first point of digestion. Afterward, owing to the bad circumstances of mechanical, acidic, and ionic pressures, almost all of these nanoforms can be strongly changed in the intestine. If the nanoforms migrate into the small intestine, their digestion continues by raising the pH and absorbing the encapsulated ingredients. Nonetheless, based on their particle diameter, stability, composition, and surface properties, the nanoforms may be absorbed in different GIT areas. Via epithelium cells, cationic and hydrophobic nanoparticles are absorbed effectively. This might lead to inflammation, cytotoxicity, fibrosis, immunological responses, and oxidative stress, such as the deposition or excretion of nanoparticles into the blood, lymph, or adjacent organs [157].

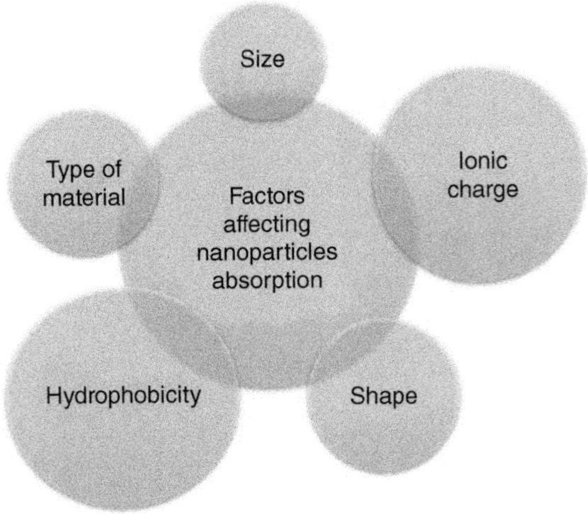

Figure 5.17 Main factors affecting the absorption of nanoparticles in human body.

In view of the previously mentioned, it is important to discover the behavior of all chitosan nanostructures, including bioactive transmission elements, throughout the GIT, which has so far been restrictively carried out.

Absorption of Chitosan-Based Nanoparticles Generally known, there are five factors that mainly affect the absorption of nanoparticles (Figure 5.17). The center of its association with positively charged chitosan nanoparticles is a negatively charged characteristic of the mucus layer and encourages cellular absorption that will improve the bioavailability of nanoencapsulated food components [44].

Complicated species, such as rats, rabbits, dogs, and even volunteers, are used in *in vivo* experiments to estimate body reaction against nanoparticles [111]. *In vivo* studies have shown that hypodermic administrations of chitosan from 5 to 50 mg/kg/day have caused symptoms of cytotoxicity in dogs and rabbits, while oral administration of chitosan equivalent to 6.75 g/day has shown no health effects for human volunteers.

5.5 Updated Regulations in Application of Chitosan-Based Nanoparticles in Food

The primary issue for the use of nanomaterials is their potential migration from nanocoatings and nanofilms to processed foods, which has not yet been clarified due to restricted research on the migration of nanoparticles and their probable toxicity [156]. It should be remembered that the complication of the food matrix makes it impossible to examine the migration and toxicity of nanoparticles. Apart from the physiochemical characteristics, the applied dose may have an effect on the ultimate fate and toxicity of nanostructures in the food processing industries. Taking into account all of the mentioned difficulties with the use of nanoparticles in food such as cytotoxicity, migration to food, complication of food matrixes, and unexpected interactions, description of applied dose and long-term studies are required to ensure the safe use of chitosan-based nanoparticles [158].

Some administrations have established regulations so as to prevent or reduce the risky aspects associated to nanomaterials. Nevertheless, there is no distinct universally acknowledged guidelines for manufacturing, classification, toxicity, and environmental efficacy of nanoparticles. The European Union (EU) is continually reviewing the list of products that may be used to produce packaged foods. Any nano-sized materials are also included in this category even though chitosan is not yet on the list [2].

Related to the list of ingredients that can be used in the production of food packages, there is indeed a list of food additives in Legislation (EC) No 1333/2008, wherein chitosan is also not part of the list. Even so, it has recently been accepted that nanomaterials are not intrinsically dangerous and harmful, but risk assessment is certainly required in the future to expand nanochitosan-based food production [69, 159].

References

1 Singh, T., Shukla, S., Kumar, P. et al. (2017). Application of nanotechnology in food science: perception and overview. *Front. Microbiol.* 8: 1501.

2 Akbari-Alavijeh, S., Shaddel, R., and Jafari, S.M. (2020). Encapsulation of food bioactives and nutraceuticals by various chitosan-based nanocarriers. *Food Hydrocolloids* 105: 105774, 105774.

3 Ahmed, S. and Ikram, S. (2017). *Chitosan: Derivatives, Composites and Applications.* Wiley.

4 May, K.L., Tangso, K.J., Hawley, A. et al. (2020). Interaction of chitosan-based dietary supplements with fats during lipid digestion. *Food Hydrocolloids*: 105965.

5 Navarro-Suarez, S., Flores-Palma, A., Flores-Ruiz, R. et al. (2018). Nanobiomaterials in dentistry. In: *Nanobiomaterials: Nanostructured Materials for Biomedical Applications* (ed. R. Narayan), 297–318. New York: Elsevier.

6 Arunkumar, R., Prashanth, K.V.H., and Baskaran, V. (2013). Promising interaction between nanoencapsulated lutein with low molecular weight chitosan: characterization and bioavailability of lutein in vitro and in vivo. *Food Chem.* 141 (1): 327–337.

7 Kaiser, M., Pereira, S., Pohl, L. et al. (2015). Chitosan encapsulation modulates the effect of capsaicin on the tight junctions of MDCK cells. *Sci. Rep.* 5: 10048.

8 Xing, Y., Yang, H., Guo, X. et al. (2020). Effect of chitosan/nano-TiO_2 composite coatings on the postharvest quality and physicochemical characteristics of mango fruits. *Sci. Horticul.* 263: 109135.

9 Bautista-Baños, S., Romanazzi, G., and Jiménez-Aparicio, A. (eds.) (2016). *Chitosan in the Preservation of Agricultural Commodities.* Academic Press.

10 Kim, S.K. (2010). *Chitin, Chitosan, Oligosaccharides and their Derivatives: Biological Activities and Applications.* CRC Press.

11 Stevens, C.V. (2019). *Chitin and Chitosan: Properties and Applications*. Wiley.
12 Tayel, A.A. (2016). Microbial chitosan as a biopreservative for fish sausages. *Int. J. Biol. Macromol.* 93: 41–46.
13 Berezina, N. (2016). Production and application of chitin. *Phys. Sci. Rev.* 1 (9): 20160048.
14 Crini, G. and Lichtfouse, E. (eds.) (2019). *Sustainable Agriculture Reviews 35: Chitin and Chitosan: History, Fundamentals and Innovations*, vol. 35. Springer.
15 Islam, S., Bhuiyan, M.R., and Islam, M.N. (2017). Chitin and chitosan: structure, properties and applications in biomedical engineering. *J. Polym. Environ.* 25 (3): 854–866.
16 Tokatlı, K. and Demirdöven, A. (2018). Optimization of chitin and chitosan production fromshrimp wastes and characterization. *J. Food Process. Preserv.* 42 (2): e13494.
17 Chang, S.H., Chen, C.H., and Tsai, G.J. (2020). Effects of chitosan on *Clostridium perfringens* and application in the preservation of pork sausage. *Mar. Drugs* 18 (2): 70.
18 Mukherjee, D., Azamthulla, M., Santhosh, S. et al. (2018). Development and characterization of chitosan-based hydrogel as wound dressing materials. *J. Drug Delivery Sci. Technol.* 46: 498–510.
19 Fernandez, J.G. and Ingber, D.E. (2014). Manufacturing of large-scale functional objects using biodegradable chitosan bioplastic. *Macromol. Mater. Eng.* 299 (8): 932–938.
20 He, T., Wang, W., Chen, B. et al. (2020). 5-Fluorouracil monodispersed chitosan microspheres: microfluidic chip fabrication with crosslinking, characterization, drug release and anticancer activity. *Carbohydr. Polym.*: 116094.
21 Su, Z., Han, Q., Zhang, F. et al. (2020). Preparation, characterization and antibacterial properties of 6-deoxy-6-arginine modified chitosan. *Carbohydr. Polym.* 230: 115635.
22 Peniche, C., Argüelles-Monal, W., and Goycoolea, F.M. (2008). Chitin and chitosan: major sources, properties and applications. In: *Monomers, Polymers and Composites from Renewable Resources* (eds. M.N. Belgacem and A. Gandini), 517–542. Elsevier.
23 Philibert, T., Lee, B.H., and Fabien, N. (2017). Current status and new perspectives on chitin and chitosan as functional biopolymers. *Appl. Biochem. Biotechnol.* 181 (4): 1314–1337.
24 Negm, N.A., Hefni, H.H., Abd-Elaal, A.A. et al. (2020). Advancement on modification of chitosan biopolymer and its potential applications. *Int. J. Biol. Macromol.*
25 Chen, W., Li, Y., Yang, S. et al. (2015). Synthesis and antioxidant properties of chitosan and carboxymethyl chitosan-stabilized selenium nanoparticles. *Carbohydr. Polym.* 132: 574–581.
26 Lin, S., Mao, S., Guan, Y. et al. (2012). Effects of dietary chitosan oligosaccharides and *Bacillus coagulans* on the growth, innate immunity and resistance of koi (*Cyprinus carpio koi*). *Aquaculture* 342: 36–41.

27 Wang, Y., Cen, C., Chen, J., and Fu, L. (2020). MgO/carboxymethyl chitosan nanocomposite improves thermal stability, waterproof and antibacterial performance for food packaging. *Carbohydr. Polym.* 236: 116078.

28 Mourya, V.K. and Inamdar, N.N. (2008). Chitosan-modifications and applications: opportunities galore. *React. Funct. Polym.* 68 (6): 1013–1051.

29 Huang, J., Cheng, Z.H., Xie, H.H. et al. (2014). Effect of quaternization degree on physiochemical and biological activities of chitosan from squid pens. *Int. J. Biol. Macromol.* 70: 545–550.

30 Logith Kumar, R., Keshav Narayan, A., Dhivya, S. et al. (2016). A review of chitosan and its derivatives in bone tissue engineering. *Carbohydr. Polym.* 151: 172–188.

31 Pardeshi, C.V. and Belgamwar, V.S. (2016). Controlled synthesis of *N,N,N*-trimethyl chitosan for modulated bioadhesion and nasal membrane permeability. *Int. J. Biol. Macromol.* 82: 933–944.

32 Sahariah, P. and Masson, M. (2017). Antimicrobial chitosan and chitosan derivatives: a review of the structure–activity relationship. *Biomacromolecules* 18 (11): 3846–3868.

33 Rocha, M.A.M., Coimbra, M.A., and Nunes, C. (2017). Applications of chitosan and their derivatives in beverages: a critical review. *Curr. Opin. Food Sci.* 15: 61–69.

34 Verlee, A., Mincke, S., and Stevens, C.V. (2017). Recent developments in antibacterial and antifungal chitosan and its derivatives. *Carbohydr. Polym.* 164: 268–283.

35 Farion, I.A., Burdukovskii, V.F., Kholkhoev, B.C. et al. (2018). Functionalization of chitosan with carboxylic acids and derivatives of them: synthesis issues and prospects of practical use: a review. *Express Polym. Lett.* 12: 1081–1105.

36 Yu, C., Kecen, X., and Xiaosai, Q. (2018). Grafting modification of chitosan. In: *Biopolymer Grafting* (ed. V.K. Thakur), 295–364. Elsevier.

37 Prabaharan, M. and Tiwari, A. (2011). Chemical modifications of chitosan intended for biomedical applications, Chapter 13. In: *Chitin, Chitosan, Oligosaccharides and their Derivatives: Biological Activities and Applications* (ed. S.K. Kim), 173–184. Boca Raton: CRC Press/Taylor & Francis Group LLC.

38 Ramos, V.M., Rodrıguez, N.M., Rodrıguez, M.S. et al. (2003). Modified chitosan carrying phosphonic and alkyl groups. *Carbohydr. Polym.* 51 (4): 425–429.

39 Bornet, A. and Teissedre, P.L. (2011). Chitin, chitosan, and their derivatives in beverage industry. In: *Chitin, Chitosan, Oligosaccharides and their Derivatives: Biological Activities and Applications* (ed. S.K. Kim), 519–530. Boca Raton: CRC Press.

40 Alishahi, A. and Aïder, M. (2012). Applications of chitosan in the seafood industry and aquaculture: a review. *Food Bioprocess Technol.* 5 (3): 817–830.

41 Gassara, F., Antzak, C., Ajila, C.M. et al. (2015). Chitin and chitosan as natural flocculants for beer clarification. *J. Food Eng.* 166: 80–85.

42 Ali, A. and Ahmed, S. (2018). A review on chitosan and its nanocomposites in drug delivery. *Int. J. Biol. Macromol.* 109: 273–286.

43 Lima, E.L., Muñoz, L.C., Harris, R.E., and Caballero, Á.M.H. (2011). 24 Potential applications of chitosan as a marine cosmeceutical. In: *Marine Cosmeceuticals: Trends and Prospects* (ed. S.K. Kim), 319–335. Boca Raton: CRC Press.

44 Liang, J., Yan, H., Puligundla, P. et al. (2017). Applications of chitosan nanoparticles to enhance absorption and bioavailability of tea polyphenols: a review. *Food Hydrocolloids* 69: 286–292.

45 Zia, Q., Tabassum, M., Lu, Z. et al. (2020). Porous poly (L–lactic acid)/chitosan nanofibres for copper ion adsorption. *Carbohydr. Polym.* 227: 115343.

46 Rezaei, S.J.T., Nabid, M.R., Niknejad, H., and Entezami, A.A. (2012). Multifunctional and thermoresponsive unimolecular micelles for tumor-targeted delivery and site-specifically release of anticancer drugs. *Polymer* 53 (16): 3485–3497.

47 Bender, E.A., Adorne, M.D., Colomé, L.M. et al. (2012). Hemocompatibility of poly (ε-caprolactone) lipid-core nanocapsules stabilized with polysorbate 80-lecithin and uncoated or coated with chitosan. *Int. J. Pharm.* 426 (1–2): 271–279.

48 Fachel, F.N.S., Dal Prá, M., Azambuja, J.H. et al. (2020). Glioprotective effect of chitosan-coated rosmarinic acid nanoemulsions against lipopolysaccharide-induced inflammation and oxidative stress in rat astrocyte primary cultures. *Cell. Mol. Neurobiol.* 40 (1): 123–139.

49 Kafshgari, M.H., Khorram, M., Mansouri, M. et al. (2012). Preparation of alginate and chitosan nanoparticles using a new reverse micellar system. *Iran. Polym. J.* 21: 99–107.

50 Fan, W., Yan, W., Xu, Z., and Ni, H. (2012). Formation mechanism of monodisperse, low molecular weight chitosan nanoparticles by ionic gelation technique. *Colloids Surf. B: Biointerfaces* 90: 21–27.

51 Ahmed, S. (ed.) (2018). *Bio-based Materials for Food Packaging: Green and Sustainable Advanced Packaging Materials.* Springer.

52 Garud, N. and Garud, A. (2012). Preparation and in-vitro evaluation of metformin microspheres using non-aqueous solvent evaporation technique. *Trop. J. Pharm. Res.* 11 (4): 577–583.

53 Alshamsan, A. (2014). Nanoprecipitation is more efficient than emulsion solvent evaporation method to encapsulate cucurbitacin I in PLGA nanoparticles. *Saudi Pharm. J.* 22 (3): 219–222.

54 Bhattarai, N., Ramay, H.R., Chou, S.H., and Zhang, M. (2006). Chitosan and lactic acid-grafted chitosan nanoparticles as carriers for prolonged drug delivery. *Int. J. Nanomed.* 1 (2): 181.

55 Devi, C., Kalita, P., Choudhury, D., and Barthakur, M. (2020). *Preparation and Characterization of Gold Nanoparticles Conjugated Insulin*, 29. Smart Healthcare for Disease Diagnosis and Prevention.

56 Ramesh, V., Samuel, S.M., and Selvarajan, E. (2020). Methods for characterizing nanoparticles. In: *Biological Synthesis of Nanoparticles and Their Applications* (eds. L. Karthik, A. Vishnu Kirthi, S. Ranjan and V.M. Srinivasan), 7, 298. Boca Raton: CRC Press/Taylor & Francis Group LLC.

57 Duraccio, D., Silvestre, C., Pezzuto, M. et al. (2013). Polypropylene and polyethylene-based nanocomposites for food packaging applications. In: *Ecosustainable Polymer Nanomaterials for Food Packaging* (eds. C. Silvestre and S. Cimmino), 143–168. Boca Raton: CRC Press/Taylor & Francis Group LLC.

58 Malmiri, H.J., Jahanian, M.A.G., and Berenjian, A. (2012). Potential applications of chitosan nanoparticles as novel support in enzyme immobilization. *Am. J. Biochem. Biotechnol.* 8 (4): 203–219.

59 Jayakumar, R., Reis, R.L., and Mano, J.F. (2007). Synthesis and characterization of pH-sensitive thiol-containing chitosan beads for controlled drug delivery applications. *Drug Delivery* 14 (1): 9–17.

60 Aminabhavi, T.M., Dharupaneedi, S.P., and More, U.A. (2017). The role of nanotechnology and chitosan-based biomaterials for tissue engineering and therapeutic delivery. In: *Chitosan Based Biomaterials Volume 2* (eds. J.A. Jennings and J.D. Bumgardner), 1–29. Woodhead Publishing.

61 Peter, M., Ganesh, N., Selvamurugan, N. et al. (2010). Preparation and characterization of chitosan–gelatin/nanohydroxyapatite composite scaffolds for tissue engineering applications. *Carbohydr. Polym.* 80: 687–694.

62 Seyedi, S.M., Anvaripour, B., Motavassel, M., and Jadidi, N. (2013). Comparative cadmium adsorption from water by nanochitosan and chitosan. *Int. J. Eng. Innov. Technol.* 2 (9): 145–148.

63 Saavedra, G.M., Figueroa, N.E., Poblete, L.A. et al. (2016). Effects of preharvest applications of methyl jasmonate and chitosan on postharvest decay, quality and chemical attributes of *Fragaria chiloensis* fruit. *Food Chem.* 190: 448–453.

64 Dzung, N.A., Khanh, V.T.P., and Dzung, T.T. (2011). Research on impact of chitosan oligomers on biophysical characteristics, growth, development and drought resistance of coffee. *Carbohydr. Polym.* 84 (2): 751–755.

65 Ghadi, A., Mahjoub, S., Tabandeh, F., and Talebnia, F. (2014). Synthesis and optimization of chitosan nanoparticles: potential applications in nanomedicine and biomedical engineering. *Casp. J. Int. Med.* 5 (3): 156.

66 Sudhakar, S., Chandran, S.V., Selvamurugan, N., and Nazeer, R.A. (2020). Biodistribution and pharmacokinetics of thiolated chitosan nanoparticles for oral delivery of insulin in vivo. *Int. J. Biol. Macromol.* 150: 281–288.

67 Huang, H.W., Hsu, C.P., and Wang, C.Y. (2020). Healthy expectations of high hydrostatic pressure treatment in food processing industry. *J. Food Drug Anal.* 28 (1): 1–13.

68 Luo, Y., Wang, Q., and Zhang, Y. (2020). Biopolymer-based nanotechnology approaches to deliver bioactive compounds for food applications: a perspective on the past, present, and future. *J. Agric. Food. Chem.* 68 (46): 12993–13000.

69 Anandharamakrishnan, C. (2014). *Techniques for Nanoencapsulation of Food Ingredients*. New York: Springer.

70 Tachaprutinun, A., Udomsup, T., Luadthong, C., and Wanichwecharungruang, S. (2009). Preventing the thermal degradation of astaxanthin through nanoencapsulation. *Int. J. Pharm.* 374 (1–2): 119–124.

71 Ge, W., Li, D., Chen, M. et al. (2015). Characterization and antioxidant activity of β-carotene loaded chitosan-graft-poly (lactide) nanomicelles. *Carbohydr. Polym.* 117: 169–176.

72 Rahaiee, S., Shojaosadati, S.A., Hashemi, M. et al. (2015). Improvement of crocin stability by biodegradable nanoparticles of chitosan-alginate. *Int. J. Biol. Macromol.* 79: 423–432.

73 Dhiman, A. and Bhalla, D. (2019). Development and evaluation of lycopene loaded chitosan nanoparticles. *Curr. Nanomed.* (Formerly: Recent Patents on Nanomedicine) 9 (1): 61–75.

74 Ko, A., Lee, J.S., Sop Nam, H., and Gyu Lee, H. (2017). Stabilization of black soybean anthocyanin by chitosan nanoencapsulation and copigmentation. *J. Food Biochem.* 41 (2): e12316.

75 Fan, Q., Ma, J., Xu, Q. et al. (2018). Facile synthesis of chitosan-coated silica nanocapsules via interfacial condensation approach for sustained release of vanillin. *Ind. Eng. Chem. Res.* 57 (18): 6171–6179.

76 Woranuch, S. and Yoksan, R. (2013). Eugenol-loaded chitosan nanoparticles: II. Application in bio-based plastics for active packaging. *Carbohydr. Polym.* 96 (2): 586–592.

77 Jafari, S.M. (2017). An overview of nanoencapsulation techniques and their classification. In: *Nanoencapsulation Technologies for the Food and Nutraceutical Industries* (ed. S.M. Jafari), 1–34. Academic Press.

78 Grumezescu, A. (ed.) (2016). *Novel Approaches of Nanotechnology in Food.* Academic Press.

79 Azevedo, M.A., Bourbon, A.I., Vicente, A.A., and Cerqueira, M.A. (2014). Alginate/chitosan nanoparticles for encapsulation and controlled release of vitamin B2. *Int. J. Biol. Macromol.* 71: 141–146.

80 Alishahi, A., Mirvaghefi, A., Tehrani, M.R. et al. (2011). Chitosan nanoparticle to carry vitamin C through the gastrointestinal tract and induce the non-specific immunity system of rainbow trout (*Oncorhynchus mykiss*). *Carbohydr. Polym.* 86 (1): 142–146.

81 Jiménez-Fernández, E., Ruyra, A., Roher, N. et al. (2014). Nanoparticles as a novel delivery system for vitamin C administration in aquaculture. *Aquaculture* 432: 426–433.

82 Dong, F. and Wang, Y. (2016). Encapsulation of vitamin C by a double-layer zein/chitosan structure with improved stability and controlled release. *Int. Nanomed. Nanosurg.* 2 (1).

83 de Britto, D., de Moura, M.R., Aouada, F.A. et al. (2014). Entrapment characteristics of hydrosoluble vitamins loaded into chitosan and *N,N,N*-trimethyl chitosan nanoparticles. *Macromol. Res.* 22 (12): 1261–1267.

84 Kim, D.G., Jeong, Y.I., Choi, C. et al. (2006). Retinol-encapsulated low molecular water-soluble chitosan nanoparticles. *Int. J. Pharm.* 319 (1–2): 130–138.

85 Luo, Y., Zhang, B., Whent, M. et al. (2011). Preparation and characterization of zein/chitosan complex for encapsulation of α-tocopherol, and its in vitro controlled release study. *Colloids Surf., B* 85 (2): 145–152.

86 Luo, Y., Teng, Z., and Wang, Q. (2012). Development of zein nanoparticles coated with carboxymethyl chitosan for encapsulation and controlled release of vitamin D3. *J. Agric. Food. Chem.* 60 (3): 836–843.

87 Teng, Z., Luo, Y., and Wang, Q. (2013). Carboxymethyl chitosan–soy protein complex nanoparticles for the encapsulation and controlled release of vitamin D3. *Food Chem.* 141 (1): 524–532.

88 Jafari, A., Jafarpour, A., and Safari, R. (2017). Influence of chitosan nanocomposite and rosemary (*Rosmarinus officinalis* L.) extract coating on quality of Huso huso fillet inoculated with *Listeria monocytogenes* during refrigerated storage. *J. Aquat. Food Prod. Technol.* 26 (6): 675–685.

89 Assadpour, E. and Mahdi Jafari, S. (2019). A systematic review on nanoencapsulation of food bioactive ingredients and nutraceuticals by various nanocarriers. *Crit. Rev. Food Sci. Nutr.* 59 (19): 3129–3151.

90 Souza, M.P., Vaz, A.F., Correia, M.T. et al. (2014). Quercetin-loaded lecithin/chitosan nanoparticles for functional food applications. *Food Bioprocess Technol.* 7 (4): 1149–1159.

91 Hernandez-Patlan, D., Solis-Cruz, B., Cano-Vega, M.A. et al. (2019). Development of chitosan and alginate nanocapsules to increase the solubility, permeability and stability of curcumin. *J. Pharm. Innov.* 14 (2): 132–140.

92 Kumar, S.P., Birundha, K., Kaveri, K., and Devi, K.R. (2015). Antioxidant studies of chitosan nanoparticles containing naringenin and their cytotoxicity effects in lung cancer cells. *Int. J. Biol. Macromol.* 78: 87–95.

93 Nallamuthu, I., Devi, A., and Khanum, F. (2015). Chlorogenic acid loaded chitosan nanoparticles with sustained release property, retained antioxidant activity and enhanced bioavailability. *Asian J. Pharm. Sci.* 10 (3): 203–211.

94 Ahmad, N., Ahmad, R., Naqvi, A.A. et al. (2016). Rutin-encapsulated chitosan nanoparticles targeted to the brain in the treatment of Cerebral Ischemia. *Int. J. Biol. Macromol.* 91: 640–655.

95 Dube, A., Nicolazzo, J.A., and Larson, I. (2010). Chitosan nanoparticles enhance the intestinal absorption of the green tea catechins (+)-catechin and (−)-epigallocatechin gallate. *Eur. J. Pharm. Sci.* 41 (2): 219–225.

96 Yang, R., Liu, Y., Gao, Y. et al. (2018). Nano-encapsulation of epigallocatechin gallate in the ferritin-chitosan double shells: simulated digestion and absorption evaluation. *Food Res. Int.* 108: 1–7.

97 Jeon, Y.O., Lee, J.S., and Lee, H.G. (2016). Improving solubility, stability, and cellular uptake of resveratrol by nanoencapsulation with chitosan and γ-poly (glutamic acid). *Colloids Surf., B* 147: 224–233.

98 Prakash, B., Kujur, A., Yadav, A. et al. (2018). Nanoencapsulation: an efficient technology to boost the antimicrobial potential of plant essential oils in food system. *Food Control* 89: 1–11.

99 Paula, H.C., Sombra, F.M., Abreu, F.O., and Paul, R. (2010). *Lippia sidoides* essential oil encapsulation by angico gum/chitosan nanoparticles. *J. Braz. Chem. Soc.* 21 (12): 2359–2366.

100 Esmaeili, A. and Asgari, A. (2015). In vitro release and biological activities of *Carum copticum* essential oil (CEO) loaded chitosan nanoparticles. *Int. J. Biol. Macromol.* 81: 283–290.

101 Feyzioglu, G.C. and Tornuk, F. (2016). Development of chitosan nanoparticles loaded with summer savory (*Satureja hortensis* L.) essential oil for antimicrobial and antioxidant delivery applications. *LWT* 70: 104–110.

102 Hosseini, S.F., Zandi, M., Rezaei, M., and Farahmandghavi, F. (2013). Two-step method for encapsulation of oregano essential oil in chitosan nanoparticles: preparation, characterization and in vitro release study. *Carbohydr. Polym.* 95 (1): 50–56.

103 Keawchaoon, L. and Yoksan, R. (2011). Preparation, characterization and in vitro release study of carvacrol-loaded chitosan nanoparticles. *Colloids Surf., B* 84 (1): 163–171.

104 Jamil, B., Abbasi, R., Abbasi, S. et al. (2016). Encapsulation of cardamom essential oil in chitosan nano-composites: in-vitro efficacy on antibiotic-resistant bacterial pathogens and cytotoxicity studies. *Front. Microbiol.* 7: 1580.

105 Abdollahi, M., Rezaei, M., and Farzi, G. (2014). Influence of chitosan/clay functional bionanocomposite activated with rosemary essential oil on the shelf life of fresh silver carp. *Int. J. Food Sci. Technol.* 49 (3): 811–818.

106 (a)Nouri, A. (2019). Chitosan nano-encapsulation improves the effects of mint, thyme, and cinnamon essential oils in broiler chickens. *Br. Poult. Sci.* 60 (5): 530–538; (b) Pathakoti, K., Manubolu, M., and Hwang, H.M. (2017). Nanostructures: Current uses and future applications in food science. *J. Food Drug Anal.* 25 (2): 245–253.

107 Hasheminejad, N., Khodaiyan, F., and Safari, M. (2019). Improving the antifungal activity of clove essential oil encapsulated by chitosan nanoparticles. *Food Chem.* 275: 113–122.

108 Beyki, M., Zhaveh, S., Khalili, S.T. et al. (2014). Encapsulation of *Mentha piperita* essential oils in chitosan–cinnamic acid nanogel with enhanced antimicrobial activity against *Aspergillus flavus*. *Ind. Crops Prod.* 54: 310–319.

109 Das, S., Singh, V.K., Dwivedy, A.K. et al. (2019). Encapsulation in chitosan-based nanomatrix as an efficient green technology to boost the antimicrobial, antioxidant and in situ efficacy of *Coriandrum sativum* essential oil. *Int. J. Biol. Macromol.* 133: 294–305.

110 Shetta, A., Kegere, J., and Mamdouh, W. (2019). Comparative study of encapsulated peppermint and green tea essential oils in chitosan nanoparticles: encapsulation, thermal stability, in-vitro release, antioxidant and antibacterial activities. *Int. J. Biol. Macromol.* 126: 731–742.

111 Zhang, H., Li, X., and Kang, H. (2019). Chitosan coatings incorporated with free or nano-encapsulated *Paulownia tomentosa* essential oil to improve shelf-life of ready-to-cook pork chops. *LWT* 116: 108580.

112 Pabast, M., Shariatifar, N., Beikzadeh, S., and Jahed, G. (2018). Effects of chitosan coatings incorporating with free or nano-encapsulated Satureja plant essential oil on quality characteristics of lamb meat. *Food Control* 91: 185.

113 Ghaderi-Ghahfarokhi, M., Barzegar, M., Sahari, M.A., and Azizi, M.H. (2016). Nanoencapsulation approach to improve antimicrobial and antioxidant activity of thyme essential oil in beef burgers during refrigerated storage. *Food Bioprocess Technol.* 9 (7): 1187–1201.

114 Mohammadi, A., Hashemi, M., and Hosseini, S.M. (2015). Nanoencapsulation of *Zataria multiflora* essential oil preparation and characterization with enhanced antifungal activity for controlling Botrytis cinerea, the causal agent of gray mould disease. *Innov. Food Sci. Emerg. Technol.* 28: 73–80.

115 Abreu, F.O., Oliveira, E.F., Paula, H.C., and de Paula, R.C. (2012). Chitosan/cashew gum nanogels for essential oil encapsulation. *Carbohydr. Polym.* 89 (4): 1277–1282.

116 Sotelo-Boyás, M.E., Correa-Pacheco, Z.N., Bautista-Baños, S., and Corona-Rangel, M.L. (2017). Physicochemical characterization of chitosan nanoparticles and nanocapsules incorporated with lime essential oil and their antibacterial activity against food-borne pathogens. *LWT* 77: 15–20.

117 Gammone, M.A., Riccioni, G., Parrinello, G., and D'Orazio, N. (2019). Omega-3 polyunsaturated fatty acids: benefits and endpoints in sport. *Nutrients* 11 (1): 46.

118 Raeisi, S., Ojagh, S.M., Quek, S.Y. et al. (2019). Nano-encapsulation of fish oil and garlic essential oil by a novel composition of wall material: Persian gum-chitosan. *LWT* 116: 108494.

119 Kim, J., Lee, C.M., Jeong, H.J. et al. (2012). Elevated anti-inflammatory effects of eicosapentaenoic acid based self-aggregated glycol chitosan nanoparticles. *J. Nanosci. Nanotechnol.* 12 (3): 2672–2678.

120 Gustavsson, J., Cederberg, C., Sonesson, U. et al. (2011). *Global Food Losses and Food Waste*. Rome: Food and Agriculture Organization of the United Nations. FAO.

121 Pelissari, F.M., Ferreira, D.C., Louzada, L.B. et al. (2019). Starch-based edible films and coatings: an eco-friendly alternative for food packaging. In: *Starches for Food Application* (eds. M.T.S. Clerici and M. Schmiele), 359–420. Academic Press.

122 Khaneghah, A.M., Hashemi, S.M.B., and Limbo, S. (2018). Antimicrobial agents and packaging systems in antimicrobial active food packaging: an overview of approaches and interactions. *Food Bioprod. Process.* 111: 1–19.

123 Bahrami, A., Delshadi, R., Assadpour, E. et al. (2020). Antimicrobial-loaded nanocarriers for food packaging applications. *Adv. Colloid Interface Sci.* 278: 102140.

124 Tumwesigye, K.S., Oliveira, J.C., and Sousa-Gallagher, M.J. (2016). New sustainable approach to reduce cassava borne environmental waste and develop biodegradable materials for food packaging applications. *Food Packag. Shelf Life* 7: 8–19.

125 Kabir, E., Kaur, R., Lee, J. et al. (2020). Prospects of biopolymer technology as an alternative option for non-degradable plastics and sustainable management of plastic wastes. *J. Cleaner Prod.* 258: 120536.

126 Fahmy, H.M., Eldin, R.E.S., Serea, E.S.A. et al. (2020). Advances in nanotechnology and antibacterial properties of biodegradable food packaging materials. *RSC Adv.* 10 (35): 20467–20484.

127 Naskar, S., Sharma, S., and Kuotsu, K. (2019). Chitosan-based nanoparticles: an overview of biomedical applications and its preparation. *J. Drug Delivery Sci. Technol.* 49: 66–81.

128 Ferreira, D.C.M., de Souza, A.L., da Silveira, J.V.W. et al. (2020). Chitosan nanocomposites for food packaging applications. In: *Multifunctional Hybrid Nanomaterials for Sustainable Agri-Food and Ecosystems* (ed. K.A. Abd-Elsalam), 393–435. Elsevier.

129 Suresh, A.K., Pelletier, D.A., and Doktycz, M.J. (2013). Relating nanomaterial properties and microbial toxicity. *Nanoscale* 5 (2): 463–474.

130 Divya, K. and Jisha, M.S. (2018). Chitosan nanoparticles preparation and applications. *Environ. Chem. Lett.* 16 (1): 101–112.

131 Abd-Elsalam, K.A. (ed.) (2020). *Multifunctional Hybrid Nanomaterials for Sustainable Agri-Food and Ecosystems*. Elsevier.

132 Kumar, S., Ye, F., Dobretsov, S., and Dutta, J. (2019). Chitosan nanocomposite coatings for food, paints, and water treatment applications. *Appl. Sci.* 9 (12): 2409.

133 Antoniou, J., Liu, F., Majeed, H., and Zhong, F. (2015). Characterization of tara gum edible films incorporated with bulk chitosan and chitosan nanoparticles: a comparative study. *Food Hydrocolloids* 44: 309–319.

134 Tian, F., Chen, W., Cai'E, W. et al. (2019). Preservation of *Ginkgo biloba* seeds by coating with chitosan/nano-TiO_2 and chitosan/nano-SiO_2 films. *Int. J. Biol. Macromol.* 126: 917–925.

135 Kumar, S., Shukla, A., Baul, P.P. et al. (2018). Biodegradable hybrid nanocomposites of chitosan/gelatin and silver nanoparticles for active food packaging applications. *Food Packag. Shelf Life* 16: 178–184.

136 Indumathi, M.P., Sarojini, K.S., and Rajarajeswari, G.R. (2019). Antimicrobial and biodegradable chitosan/cellulose acetate phthalate/ZnO nano composite films with optimal oxygen permeability and hydrophobicity for extending the shelf life of black grape fruits. *Int. J. Biol. Macromol.* 132: 1112–1120.

137 Kaewklin, P., Siripatrawan, U., Suwanagul, A., and Lee, Y.S. (2018). Active packaging from chitosan-titanium dioxide nanocomposite film for prolonging storage life of tomato fruit. *Int. J. Biol. Macromol.* 112: 523–529.

138 Barikloo, H. and Ahmadi, E. (2018). Effect of nanocomposite-based packaging and chitosan coating on the physical, chemical, and mechanical traits of strawberry during storage. *J. Food Meas. Charact.* 12 (3): 1795–1817.

139 Yu, Y., Zhang, S., Ren, Y. et al. (2012). Jujube preservation using chitosan film with nano-silicon dioxide. *J. Food Eng.* 113 (3): 408–414.

140 Shi, S., Wang, W., Liu, L. et al. (2013). Effect of chitosan/nano-silica coating on the physicochemical characteristics of longan fruit under ambient temperature. *J. Food Eng.* 118 (1): 125–131.

141 Al-Naamani, L., Dutta, J., and Dobretsov, S. (2018). Nanocomposite zinc oxide-chitosan coatings on polyethylene films for extending storage life of okra (*Abelmoschus esculentus*). *Nanomaterials* 8 (7): 479.

142 Dehnad, D., Mirzaei, H., Emam-Djomeh, Z. et al. (2014). Thermal and antimicrobial properties of chitosan–nanocellulose films for extending shelf life of ground meat. *Carbohydr. Polym.* 109: 148–154.

143 Rahman, P.M., Mujeeb, V.A., and Muraleedharan, K. (2017). Flexible chitosan-nano ZnO antimicrobial pouches as a new material for extending the shelf life of raw meat. *Int. J. Biol. Macromol.* 97: 382–391.

144 Khezrian, A. and Shahbazi, Y. (2018). Application of nanocomposite chitosan and carboxymethyl cellulose films containing natural preservative compounds in minced camel's meat. *Int. J. Biol. Macromol.* 106: 1146–1158.

145 Pires, J.R.A., de Souza, V.G.L., and Fernando, A.L. (2018). Chitosan/montmorillonite bionanocomposites incorporated with rosemary and ginger essential oil as packaging for fresh poultry meat. *Food Packag. and shelf life* 17: 142–149, 192.

146 Kumar, S., Mukherjee, A., and Dutta, J. (2020). Chitosan based nanocomposite films and coatings: emerging antimicrobial food packaging alternatives. *Trends Food Sci. Technol.*

147 Arkoun, M., Daigle, F., Holley, R.A. et al. (2018). Chitosan-based nanofibers as bioactive meat packaging materials. *Packag. Technol. Sci.* 31 (4): 185–195.

148 Díez-Pascual, A.M. and Díez-Vicente, A.L. (2015). Antimicrobial and sustainable food packaging based on poly (butylene adipate-co-terephthalate) and electrospun chitosan nanofibers. *RSC Adv.* 5 (113): 93095–93107.

149 Ceylan, Z., Sengor, G.F.U., and Yilmaz, M.T. (2018). Nanoencapsulation of liquid smoke/thymol combination in chitosan nanofibers to delay microbiological spoilage of sea bass (*Dicentrarchus labrax*) fillets. *J. Food Eng.* 229: 43–49.

150 Cui, H., Bai, M., Li, C. et al. (2018). Fabrication of chitosan nanofibers containing tea tree oil liposomes against *Salmonella* spp. in chicken. *LWT* 96: 671–678.

151 Youssef, A.M., El-Sayed, S.M., El-Sayed, H.S. et al. (2016). Enhancement of Egyptian soft white cheese shelf life using a novel chitosan/carboxymethyl cellulose/zinc oxide bionanocomposite film. *Carbohydr. Polym.* 151: 9–19.

152 Amjadi, S., Emaminia, S., Nazari, M. et al. (2019). Application of reinforced ZnO nanoparticle-incorporated gelatin bionanocomposite film with chitosan nanofiber for packaging of chicken fillet and cheese as food models. *Food Bioprocess Technol.* 12: 1205–1219.

153 Apjok, R., Mihaly Cozmuta, A., Peter, A. et al. (2019). Active packaging based on cellulose-chitosan-Ag/TiO_2 nanocomposite for storage of clarified butter. *Cellulose* 26: 1923–1946.

154 Lotfi, M., Tajik, H., Moradi, M. et al. (2018). Nanostructured chitosan/monolaurin film: preparation, characterization and antimicrobial activity against *Listeria monocytogenes* on ultrafiltered white cheese. *LWT* 92: 576–583.

155 Divsalar, E., Tajik, H., Moradi, M. et al. (2018). Characterization of cellulosic paper coated with chitosan-zinc oxide nanocomposite containing nisin and its application in packaging of UF cheese. *Int. J. Biol. Macromol.* 109: 1311–1318.

156 Youssef, A.M., El-Sayed, S.M., El-Sayed, H.S. et al. (2018). Novel bionanocomposite materials used for packaging skimmed milk acid coagulated cheese (Karish). *Int. J. Biol. Macromol.* 115: 1002–1011.

157 Kulkarni, A.S., Ghugre, P.S., and Udipi, S.A. (2016). Applications of nanotechnology in nutrition: potential and safety issues. In: *Novel Approaches of Nanotechnology in Food* (ed. A.M. Grumezescu), 509–554. Academic Press.

158 Bhatia, S. (2016). Chitosan based nanomaterials and its applications. In: *Systems for Drug Delivery* (ed. S. Bhatia), 55–117. Cham: Springer.

159 Jeevanandam, J., Barhoum, A., Chan, Y.S. et al. (2018). Review on nanoparticles and nanostructured materials: history, sources, toxicity and regulations. *Beilstein J. Nanotechnol.* 9 (1): 1050–1074.

6

Nutrients-Based Nanocarriers and Its Application in Packaging Systems

Leidy T. Sanchez[1], *N. David Rodriguez-Marin*[1], *Magda I. Pinzon*[1], *and Cristian C. Villa*[2]

[1] *Universidad del Quindío, Programa de Ingenieria de Alimentos, Facultad de Ciencias Agroindustriales, Carrera 15 Calle 12N, Armenia 630004, Colombia*
[2] *Universidad del Quindío, Programa de Quimica, Facultad de Ciencias Básicas y Tecnologías, Carrera 15 Calle 12N, Armenia 630004, Colombia*

Encapsulation is a technique used in the food industry for the protection and preservation of bioactive, volatile, and easily degradable compounds from biochemical and thermal deterioration. It is based in the creation of an external membrane or wall of one material over another material [1]. Over the last decades, the rise of nanotechnology applications on the food industry has allowed the development of nanoencapsulation techniques, which can improve the water solubility, thermal stability, and oral bioavailability of bioactive compounds [2]. It is believed that nanocarriers or nanovehicles could provide increasing encapsulation efficiencies and more bioavailability through increased surface-to-volume ratios. Besides, due to their small particle size, nanocarriers have shown higher mucoadhesive within the small intestine, as well as higher interactions with the cell membranes, allowing them to penetrate into the target cells and release their cargos [3]. Although there are several types of nanocarriers that have been developed for applications in different fields, nanocarriers for food applications must be produced from nontoxic and biocompatible food grade materials. In general, nanocarriers for food applications can be classified according to the nature of their material and synthesis methods; this includes lipid-based, protein-based, and carbohydrate-based nanocarriers.

6.1 Lipid-Based Nanocarrier

Lipid-based nanocarriers have shown great potentials to encapsulate and release various nutrients and nutraceutical molecules (hydrophilic, lipophilic, and amphiphilic compounds) in a sustained and controlled manner. They have also shown the capability to improve solubility and encapsulation efficiency of hydrophobic bioactive compounds, decrease their volatility, and enhance their

Nanotechnology-Enhanced Food Packaging, First Edition. Edited by Jyotishkumar Parameswaranpillai, Radhakrishnan Edayileveettil Krishnankutty, Aswathy Jayakumar, Sanjay Mavinkere Rangappa, and Suchart Siengchin.

target specificity [4–6]). Among lipid-based nanocarriers four subgroups can be identified. Figure 6.1 shows schematic representations of the different types of lipid-based nanocarriers.

6.1.1 Nanoemulsions

Nanoemulsions are colloidal dispersion with droplet size in the order of 100 nm that are formulated by combination and stabilization of two immiscible phases using a surfactant as a stabilizing agent. Due to their small droplet size, nanoemulsions are kinetically and thermodynamically stable; furthermore, they tend to be transparent and present high encapsulation efficiency [7, 8]. Likewise, nanoemulsions are metastable and can be diluted with water without change in the droplet size distribution [4]. Nanoemulsion mainly consists of three components – oil, water, and surfactant. The correct mixing of these components determines the stability and property of emulsion. Several kinds of surfactants can be used to form nanoemulsions and, by changing their molecular characteristics a surfactant, can be used to give the desired effect to the nanoemulsion, such as controlling droplet size [9]. Furthermore, there are several methods for nanoemulsion formation; these include high energy methods such as high-pressure homogenization and sonication, while low energy methods include phase inversion and spontaneous emulsification [7].

Over the last decade, there have been several studies about the potential of nanoemulsions for delivery of poorly water-soluble food ingredients, such as fish oil and lipophilic vitamins and antioxidants. Furthermore, nanoemulsions have shown the capability to improve bioactive molecules solubilization and potential for enhancing bioavailability and absorption in the gastrointestinal tract [4]. Nanoemulsions have been specially used for nanoencapsulation of lipophilic vitamins; examples include the encapsulation of carotenoids and precursors of vitamin A [10], vitamin E [11], vitamin D [12], and vitamin K [13]. Likewise, they have been used for encapsulation of antioxidants and other bioactive molecules such as curcumin and quercetin [5, 14–17].

6.1.2 Nanoliposomes

Liposomes can be considered as one of the most studied lipid-based carriers, as interest in them started in the latter half of the twenty-first century. In general, they can be described as spherical or almost spherical closed, continuous bilayered structures made mainly of lipid and/or phospholipid molecules [18]. Over the last decades, the term nanoliposome has been used to exclusively refer to liposomes with diameters in the nanoscale. Both liposomes and nanoliposomes share the same structural features; however, due to their smaller particle size, nanoliposomes provide more surface area and have shown to increase solubility, enhance bioavailability, and improve controlled lipophilic molecules [19]. There are several methods for nanoliposomes formation; however, they are all based on the same principles. When exposed to an aqueous phase, lipid or phospholipid molecules tend to organize themselves in order to diminish their hydrophobic surface in

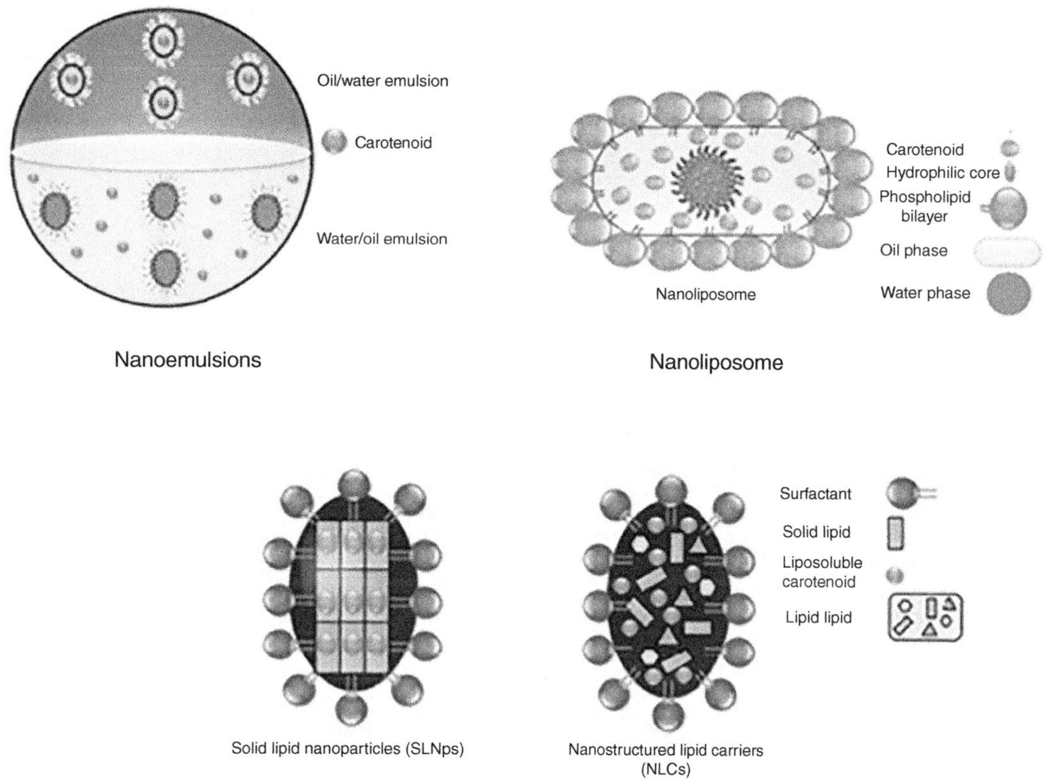

Figure 6.1 Schematic representation of the different lipid-based nanocarriers.

contact with aqueous phase and ultimately by forming a continuous monolayer. Once a sufficient energy input is added to the systems, in form of homogenization, agitation, heat, and/or sonication, they create a continuous core–shell structure in which the aqueous compartments of the media completely entrap into a phospholipid bilayer [20]. One of the most interesting aspects of nanoliposomes is that their physicochemical properties can be changed through several factors including ionic strength, temperature, and pH. Furthermore, they heavily depend on their composition, so properties such as permeability, size, and superficial charge can be modified by changing the composition of lipid mixtures [18].

6.1.3 Solid Lipid Nanoparticles (SLNps)

Solid lipid nanoparticles (SLNps) are similar to nanoemulsions, as they are spherical lipophilic droplets stabilized by a surfactant and dispersed in an aqueous phase; however, their lipid phase is solid at room temperature [21, 22]. SLNps have shown great versatility for encapsulation and special release of bioactive molecules as the movement of a molecule in a solid lipid is lower than in liquid oil. They have also shown several interesting properties for nutrient delivery, such as bioavailability, better physical stability, controlled release, and lack of biotoxicity [10]. The oil phase of SLNps can be produced from a single solid lipid species or by using a mixture of lipid classes; thus, encapsulation and mobility of incorporated compounds can be controlled by the composition of the lipid matrix state [23]. Production of SNLps is relatively easy and can be rapidly escalated to industrial levels. Most of the synthesis methods include emulsification of the lipid phase at a relative high temperature, followed by a high energy method (sonication, high shear/pressure homogenization, and high-speed stirring [24].

6.1.4 Nanostructured Lipid Carriers (NLCs)

Nanostructured lipid carriers (NLCs) can be described as SLNps that are formed by lipid droplets that are partially crystallized, thus presenting a less-ordered crystalline structure or anamorphous solid structure [25]. The recrystallization of the lipid phase during SLNps formation can reduce the bioactive molecule solubility, leading to its uncontrolled expulsion from the SLNps. In order to overcome this problem, NLCs are formed by incorporating oil into the core of a solid lipid, allowing higher loading capacity and a more controlled release [26, 27]. Production methods of NLCs are similar to that of SLNPs, including hot/cold homogenization, solvent emulsification–evaporation, and phase inversion techniques [28].

6.2 Carbohydrate-Based Nanocarriers

Carbohydrate-based nanocarriers are mostly formed by polysaccharides found in nature (plants, algae, microbes, and animals). They are mostly formed by monosaccharide units linked through glycoside bonds. The characteristics of the different carbohydrate-based nanocarriers depend on the type of monosaccharide and glycoside bonds that form each polysaccharide. Research on carbohydrate-based nanocarriers has bloomed over the last decades, due to their biocompatibility, relatively easy formation, and high encapsulation capacity.

6.2.1 Starch Nanoparticles (SNPs) and Nanocrystals (SNCs)

Starch is a polysaccharide used as an energy reserve in plants and is composed of glucose units, linked in two different types of molecules. Amylose is a linear molecule formed by glucose units linked through α-1,4 bonds, while amylopectin is a branched molecule formed by linear units linked through branched point with α-1,6 bonds [29–31]. Physicochemical properties of the different starches are heavily dependent of the amylose/amylopectin ratios on the starch granule; furthermore, this ratio depends on its botanical source. Starch can be easily modified through chemical, physical, and enzymatic methods. Through these modifications it is easy to tailor the physicochemical properties of starch to different applications.

Two types of starch-based nanomaterials can be used as carriers of nutrients and other bioactive molecules. Starch nanoparticles (SNPs) are mostly amorphous particles obtained by nanoprecipitation of gelatinized starch, using either antisolvent methods or by high energy methods [31–33]. On the other hand, starch nanocrystals (SNCs) are highly crystalline structures obtained by acid or enzymatic hydrolysis of the amorphous regions of the starch granule [34–36]. Both types of starch-based nanocarriers can be modified to increase encapsulation efficiency either through esterification reactions or cross-linking [37–39].

6.2.2 Chitosan Nanoparticles

Chitosan is polysaccharide derivate from chitin. It is composed of deacetylated units of D-glucosamine and acetylated units of *N*-acetyl-D-glucosamine connected through β-(1,4) glycosidic linkages [40]. Due to the presence of positive-charge amino groups, chitosan is a cationic biopolymer, that is only soluble at pH under 6; and it is insoluble at neutral pH. Chitosan nanoparticles can be obtained through several methods, the most common being ionic gelation. In this method an opposite-charged macromolecule is carefully added to a chitosan suspension. The most common molecule used is tripolyphosphate (TPP), a nontoxic molecule commonly used in the food industry. Concentration, ratios of ingredients, way of mixing, and pH are the affective factors in gelation process. Other methods include reverse microemulsions templates, emulsification, and coacervation [40, 41].

6.2.3 Alginate Nanoparticles

Alginates are polysaccharides obtained from algae and bacterial sources. They are conformed by (1–4)-linked units of β-D-mannuronate (M) and α-L-guluronic (G) acid residues in pyranosic conformation [10, 42]. Physicochemical properties of the different alginates are derived from the M/G ratio on the polymer sequence. The

M/G ratio and sequence depend on the source of the used algae, and as with other polysaccharides alginate can also be chemically modified to alter its properties [43]. Alginate-based nanosystems can either be nanocapsule, nanospheres, or nanosized systems with diameters generally ranging from 10 to 1000 nm in size [44]. The method used to form the nanoparticles determines the type of nanosystems obtained. Alginate nanocapsules can be described as a vesicular type system, in which the bioactive molecule is confined to an oily or aqueous liquid core, surrounded by the polymeric membrane. Furthermore, nanospheres are spherical nanoparticles with a gelled interior in which the entrapped component is physically dispersed [45–48].

6.3 Protein-Based Nanocarriers

Protein-based nanocarriers have attracted much attention in the food industry due to their nutritional and functional attributes, which include the ability to stabilize colloidal systems through surface absorption; they can form different structures such as spheres, tubes, or fibers and have the ability to inhibit oxidation through chelation or free radical scavenging [49]. Likewise, proteins are complex systems with a variety of functional groups that can interact with different types of molecules, allowing a wide range of bioactive molecules that can be encapsulated. Protein-based nanocarriers can be made from different types of proteins from either animal or plant sources.

Examples of protein-based nanocarriers include casein nanomicelles, which are obtained from milk. They are biocompatible, biodegradable, GRAS, and have excellent emulsification, solubilizing, gel forming properties, gelation, heat and acid stability, water holding, and binding of small molecules and ions. They have been used for nanoencapsulation of vitamins and other nutraceutical molecules [10, 50, 51]. Other examples include nanoparticles made from gelatin, a protein extracted from animal sources that have shown great biocompatibility and high encapsulation efficiency [52, 53]. Nanoparticles made from zein, a protein present in corn and other cereals, have also shown potential as carriers of bioactive molecules [54, 55].

6.4 Applications of Nanocarriers in Active and Bioactive Food Packaging

One of the most important threads in food science and technology over the last couple of decades has been the development of active and bioactive biodegradable food packaging. Active food packaging incorporates active molecules, such as antioxidants and antimicrobials that are released during food storage, increasing shelf life of fruits, vegetables, meat, and fish products [56–58]. On the other hand, bioactive packaging incorporate bioactive molecules into the edible polymeric matrix in order to contribute to health benefits for consumers. Incorporation of vitamins, antioxidants, probiotics, and prebiotics is common in bioactive food packaging design [2, 59, 60]. In both cases, nanocarriers have played an important role, as they can be used to protect the bioactive molecules during storage while providing a controlled or targeted release.

Vitamins have shown great versatility in both active and bioactive food packaging, as most of them present antioxidant activity that can be used during food storage, while once consumed they offer several health and metabolic benefits to the consumer. Starch-based nanomaterials have been used to incorporate β-carotene (a vitamin A precursor) into biodegradable and edible food packaging. Hari et al. [61] used SNCs to encapsulate β-carotene and incorporated the nanocarrier system into chitosan–gelatin composite films. Results showed that incorporation of the nanocarrier improved some of the physicochemical properties of the films, such as a decreased swelling index, solubility in water, degradation, and moisture content when compared to that of control chitosan–gelatin film, due to the lipophilic nature of the bioactive molecule. Likewise, antioxidant activity assays (DPPH) showed a 92.3% of radical scavenging activity, while sustained release around 52% of β-carotene was observed during 12 days [61]. On the other hand, Assis et al. [62] used SLNps to incorporate and protect β-carotene molecules into cassava starch film. They observed that nanoencapsulation increased mechanical and barrier properties of the films, when compared with films with free β-carotene molecules incorporated. Likewise, the addition of the antioxidant in the nanometric scale results in greater stability and solubility in comparison with the free antioxidant. They used the films to control the oxidation rate of sunflower oil, observing that films with the nanoencapsulated molecule decreased the oxidation rate significantly [62]. A similar approach was used by Assis et al. [63] to encapsulate lycopene, another vitamin A precursor, using SLNps as carrier into cassava starch films; results showed an enhanced antioxidant activity compared to the free lycopene [63].

Another vitamin that has been incorporated into composite films is α-tocopherol, a precursor to vitamin E, known for its high antioxidant activity and lipophilic nature. Noronha et al. [64] used SLNps to encapsulate α-tocopherol. Incorporation into carboxymethyl cellulose (CMC) films leads to changes in their mechanical properties, such as decrease in tensile strength and an increase in elongation at break values. Furthermore, the hydrophobic nature of the nanocarriers decreased water vapor permeability, while lightness and yellowish color were intensified. Finally, films with the nanoencapsulated vitamin showed high antioxidant activity and excellent barrier properties against UV and visible light [64]. Similar behavior was observed using vitamin E SLNps in CMC and guar gum films [65]. Besides vitamins, nanocarriers have allowed the incorporation of other bioactive molecules, such as essential oils with antimicrobial activity [56, 66, 67]. Likewise, peptides and polyphenols have also been nanoencapsulated for incorporation into food packages [68–71].

6.5 Outlooks and Perspectives

Nanotechnology continues to have an impact in nutrient delivery, as many studies continue to show its effectiveness in stabilization and controlled release of different nutrient and nutraceutical molecules that otherwise can be lost during food storage and digestion. However, there are several hurdles that must be overcome before its full impact can be achieved; among them, large-scale production of some types of nanocarriers can be very difficult, increasing prices and limiting their industrial application. On the other hand combinations of different types of materials for nanocarriers formation can be used as a way to improve not only stability during storage but also through passage in the digestive track.

References

1 Saifullah, M., Shishir, M.R.I., Ferdowsi, R. et al. (2019). Micro and nano encapsulation, retention and controlled release of flavor and aroma compounds: a critical review. *Trends Food Sci. Technol.* 86: 230–251. https://doi.org/10.1016/j.tifs.2019.02.030.

2 Ezhilarasi, P.N., Karthik, P., Chhanwal, N., and Anandharamakrishnan, C. (2013). Nanoencapsulation techniques for food bioactive components: a review. *Food Bioprocess Technol.* 6 (3): 628–647. https://doi.org/10.1007/s11947-012-0944-0.

3 Assadpour, E. and Mahdi Jafari, S. (2019). A systematic review on nanoencapsulation of food bioactive ingredients and nutraceuticals by various nanocarriers. *Crit. Rev. Food Sci. Nutr.* 59 (19): 3129–3151. https://doi.org/10.1080/10408398.2018.1484687.

4 Fathi, M., Mozafari, M.R., and Mohebbi, M. (2012). Nanoencapsulation of food ingredients using lipid based delivery systems. *Trends Food Sci. Technol.* 23 (1): 13–27. https://doi.org/10.1016/j.tifs.2011.08.003.

5 Yousefi, M., Ehsani, A., and Jafari, S.M. (2019). Lipid-based nano delivery of antimicrobials to control food-borne bacteria. *Adv. Colloid Interface Sci.* 270: 263–277. https://doi.org/10.1016/j.cis.2019.07.005.

6 Assadpour, E. and Jafari, S.M. (2019). Advances in spray-drying encapsulation of food bioactive ingredients: from microcapsules to nanocapsules. *Annu. Rev. Food Sci. Technol.* 10 (1): 103–131.

7 Safaya, M. and Rotliwala, Y.C. (2020). Nanoemulsions: a review on low energy formulation methods, characterization, applications and optimization technique. *Mater. Today: Proc.* 27: 454–459. https://doi.org/10.1016/j.matpr.2019.11.267.

8 Salvia-Trujillo, L., Soliva-Fortuny, R., Rojas-Graü, M.A. et al. (2017). Edible Nanoemulsions as carriers of active ingredients: a review. *Annu. Rev. Food Sci. Technol.* 8 (1): 439–466. https://doi.org/10.1146/annurev-food-030216-025908.

9 Dasgupta, N., Ranjan, S., and Gandhi, M. (2019). Nanoemulsion ingredients and components. *Environ. Chem. Lett.* 17 (2): 917–928. https://doi.org/10.1007/s10311-018-00849-7.

10 Rehman, A., Tong, Q., Jafari, S.M. et al. (2020). Carotenoid-loaded nanocarriers: a comprehensive review. *Adv. Colloid Interface Sci.* 275: 102048. https://doi.org/10.1016/j.cis.2019.102048.

11 Saberi, A.H., Fang, Y., and McClements, D.J. (2013). Fabrication of vitamin E-enriched nanoemulsions: factors affecting particle size using spontaneous emulsification. *J. Colloid Interface Sci.* 391: 95–102. https://doi.org/10.1016/j.jcis.2012.08.069.

12 Guttoff, M., Saberi, A.H., and McClements, D.J. (2015). Formation of vitamin D nanoemulsion-based delivery systems by spontaneous emulsification: factors affecting particle size and stability. *Food Chem.* 171: 117–122. https://doi.org/10.1016/j.foodchem.2014.08.087.

13 Campani, V., Biondi, M., Mayol, L. et al. (2016). Development of nanoemulsions for topical delivery of vitamin K1. *Int. J. Pharm.* 511 (1): 170–177. https://doi.org/10.1016/j.ijpharm.2016.07.004.

14 Joung, H.J., Choi, M.-J., Kim, J.T. et al. (2016). Development of food-grade curcumin nanoemulsion and its potential application to food beverage system: antioxidant property and in vitro digestion. *J. Food Sci.* 81 (3): N745–N753. https://doi.org/10.1111/1750-3841.13224.

15 Kim, D.M., Hyun, S.S., Yun, P. et al. (2012). Identification of an emulsifier and conditions for preparing stable nanoemulsions containing the antioxidant astaxanthin. *Int. J. Cosmet. Sci.* 34 (1): 64–73. https://doi.org/10.1111/j.1468-2494.2011.00682.x.

16 Schwingel, T.E., Klein, C.P., Nicoletti, N.F. et al. (2014). Effects of the compounds resveratrol, rutin, quercetin, and quercetin nanoemulsion on oxaliplatin-induced hepatotoxicity and neurotoxicity in mice. *Naunyn-Schmiedeberg's Arch. Pharmacol.* 387 (9): 837–848. https://doi.org/10.1007/s00210-014-0994-0.

17 Zorzi, G.K., Caregnato, F., Moreira, J.C.F. et al. (2016). Antioxidant effect of nanoemulsions containing extract of *Achyrocline satureioides* (Lam) D.C.—Asteraceae. *AAPS PharmSciTech* 17 (4): 844–850. https://doi.org/10.1208/s12249-015-0408-8.

18 Khorasani, S., Danaei, M., and Mozafari, M.R. (2018). Nanoliposome technology for the food and nutraceutical industries. *Trends Food Sci. Technol.* 79: 106–115. https://doi.org/10.1016/j.tifs.2018.07.009.

19 Demirci, M., Caglar, M.Y., Cakir, B., and Gülseren, İ. (2017). Encapsulation by nanoliposomes, Chapter 3. In: *Nanoencapsulation Technologies for the Food and Nutraceutical Industries* (ed. S.M. Jafari), 74–113. Academic Press.

20 Rostamabadi, H., Falsafi, S.R., and Jafari, S.M. (2019). Nanoencapsulation of carotenoids within lipid-based nanocarriers. *J. Controlled Release* 298: 38–67. https://doi.org/10.1016/j.jconrel.2019.02.005.

21 Müller, R.H., Radtke, M., and Wissing, S.A. (2002). Solid lipid nanoparticles (SLN) and nanostructured lipid carriers (NLC) in cosmetic and dermatological preparations. *Adv. Drug Delivery Rev.* 54: S131–S155. https://doi.org/10.1016/S0169-409X(02)00118-7.

22 Shah, M.R., Imran, M., and Ullah, S. (2017). Solid lipid nanoparticles, Chapter 1. In: *Lipid-Based Nanocarriers for Drug Delivery and Diagnosis* (eds. M.R. Shah, M. Imran and S. Ullah), 1–35. William Andrew Publishing.

23 da Silva Santos, V., Badan Ribeiro, A.P., and Andrade Santana, M.H. (2019). Solid lipid nanoparticles as carriers for lipophilic compounds for applications in foods. *Food Res. Int.* 122: 610–626. https://doi.org/10.1016/j.foodres.2019.01.032.

24 Rabinarayan, P. and Padilama, S. (2012). Preparation and characterization of solid lipid nanoparticles – a review. *Curr. Drug Discovery Technol.* 9 (1): 2–16. https://doi.org/10.2174/157016312799304552.

25 Tamjidi, F., Shahedi, M., Varshosaz, J., and Nasirpour, A. (2013). Nanostructured lipid carriers (NLC): a potential delivery system for bioactive food molecules. *Innovative Food Sci. Emerg. Technol.* 19: 29–43. https://doi.org/10.1016/j.ifset.2013.03.002.

26 Babazadeh, A., Ghanbarzadeh, B., and Hamishehkar, H. (2016). Novel nanostructured lipid carriers as a promising food grade delivery system for rutin. *J. Funct. Foods* 26: 167–175. https://doi.org/10.1016/j.jff.2016.07.017.

27 Jaiswal, P., Gidwani, B., and Vyas, A. (2016). Nanostructured lipid carriers and their current application in targeted drug delivery. *Artif. Cells Nanomed. Biotechnol.* 44 (1): 27–40. https://doi.org/10.3109/21691401.2014.909822.

28 Aditya, N.P., Macedo, A.S., Doktorovova, S. et al. (2014). Development and evaluation of lipid nanocarriers for quercetin delivery: A comparative study of solid lipid nanoparticles (SLN), nanostructured lipid carriers (NLC), and lipid nanoemulsions (LNE). *LWT Food Sci. Technol.* 59 (1): 115–121. https://doi.org/10.1016/j.lwt.2014.04.058.

29 Hsieh, C.-F., Liu, W., Whaley, J.K., and Shi, Y.-C. (2019). Structure, properties, and potential applications of waxy tapioca starches – a review. *Trends Food Sci. Technol.* 83: 225–234. https://doi.org/10.1016/j.tifs.2018.11.022.

30 Magallanes-Cruz, P.A., Flores-Silva, P.C., and Bello-Perez, L.A. (2017). Starch structure influences its digestibility: a review. *J. Food Sci.* 82 (9): 2016–2023. https://doi.org/10.1111/1750-3841.13809.

31 Rostamabadi, H., Falsafi, S.R., and Jafari, S.M. (2019b). Starch-based nanocarriers as cutting-edge natural cargos for nutraceutical delivery. *Trends Food Sci. Technol.* 88: 397–415. https://doi.org/10.1016/j.tifs.2019.04.004.

32 Ahmad, M., Gani, A., Masoodi, F.A., and Rizvi, S.H. (2020). Influence of ball milling on the production of starch nanoparticles and its effect on structural, thermal and functional properties. *Int. J. Biol. Macromol.* 151: 85–91. https://doi.org/10.1016/j.ijbiomac.2020.02.139.

33 Qiu, C., Hu, Y., Jin, Z. et al. (2019). A review of green techniques for the synthesis of size-controlled starch-based nanoparticles and their applications as nanodelivery systems. *Trends Food Sci. Technol.* 92: 138–151. https://doi.org/10.1016/j.tifs.2019.08.007.

34 Dai, L., Li, C., Zhang, J., and Cheng, F. (2018). Preparation and characterization of starch nanocrystals combining ball milling with acid hydrolysis. *Carbohydr. Polym.* 180: 122–127. https://doi.org/10.1016/j.carbpol.2017.10.015.

35 Hao, Y., Chen, Y., Li, Q., and Gao, Q. (2018). Preparation of starch nanocrystals through enzymatic pretreatment from waxy potato starch. *Carbohydr. Polym.* 184: 171–177. https://doi.org/10.1016/j.carbpol.2017.12.042.

36 Sessini, V., Arrieta, M.P., Kenny, J.M., and Peponi, L. (2016). Processing of edible films based on nanoreinforced gelatinized starch. *Polym. Degrad. Stab.* 132: 157–168. https://doi.org/10.1016/j.polymdegradstab. 2016.02.026.

37 Acevedo-Guevara, L., Nieto-Suaza, L., Sanchez, L.T. et al. (2018). Development of native and modified banana starch nanoparticles as vehicles for curcumin. *Int. J. Biol. Macromol.* 111: 498–504. https://doi.org/10.1016/j.ijbiomac.2018.01.063.

38 Nieto-Suaza, L., Acevedo-Guevara, L., Sánchez, L.T. et al. (2019). Characterization of *aloe vera*-banana starch composite films reinforced with curcumin-loaded starch nanoparticles. *Food Struct.*: 100131. https://doi.org/10.1016/j.foostr.2019.100131.

39 Santoyo-Aleman, D., Sanchez, L.T., and Villa, C.C. (2019). Citric-acid modified banana starch nanoparticles as a novel vehicle for β-carotene delivery. *J. Sci. Food Agric.* https://doi.org/10.1002/jsfa.9918.

40 Akbari-Alavijeh, S., Shaddel, R., and Jafari, S.M. (2020). Encapsulation of food bioactives and nutraceuticals by various chitosan-based nanocarriers. *Food Hydrocolloids* 105: 105774. https://doi.org/10.1016/j.foodhyd.2020.105774.

41 Orellano, M.S., Porporatto, C., Silber, J.J. et al. (2017). AOT reverse micelles as versatile reaction media for chitosan nanoparticles synthesis. *Carbohydr. Polym.* 171: 85–93. https://doi.org/10.1016/j.carbpol.2017.04.074.

42 Aristizabal-Gil, M.V., Santiago-Toro, S., Sanchez, L.T. et al. (2019). ZnO and ZnO/CaO nanoparticles in alginate films. Synthesis, mechanical characterization, barrier properties and release kinetics. *LWT* 112: 108217. https://doi.org/10.1016/j.lwt.2019.05.115.

43 Paques, J.P., van der Linden, E., van Rijn, C.J.M., and Sagis, L.M.C. (2014). Preparation methods of alginate nanoparticles. *Adv. Colloid Interface Sci.* 209: 163–171. https://doi.org/10.1016/j.cis.2014.03.009.

44 Hamidi, M., Azadi, A., and Rafiei, P. (2008). Hydrogel nanoparticles in drug delivery. *Adv. Drug Delivery Rev.* 60 (15): 1638–1649. https://doi.org/10.1016/j.addr.2008.08.002.

45 Ghayempour, S. and Mortazavi, S.M. (2015). Preparation and investigation of sodium alginate nanocapsules by different microemulsification devices. *J. Appl. Polym. Sci.* 132 (17) https://doi.org/10.1002/app.41904.

46 Lertsutthiwong, P., Noomun, K., Jongaroonngamsang, N. et al. (2008). Preparation of alginate nanocapsules containing turmeric oil. *Carbohydr. Polym.* 74 (2): 209–214. https://doi.org/10.1016/j.carbpol.2008.02.009.

47 Urzedo, A.L., Gonçalves, M.C., Nascimento, M.H.M. et al. (2020). Multifunctional alginate nanoparticles containing nitric oxide donor and silver nanoparticles for biomedical applications. *Mater. Sci. Eng., C* 112: 110933. https://doi.org/10.1016/j.msec.2020.110933.

48 Zhang, C., Shi, G., Zhang, J. et al. (2017). Targeted antigen delivery to dendritic cell via functionalized alginate nanoparticles for cancer immunotherapy. *J. Controlled Release* 256: 170–181. https://doi.org/10.1016/j.jconrel.2017.04.020.

49 Fathi, M., Donsi, F., and McClements, D.J. (2018). Protein-based delivery systems for the nanoencapsulation of food ingredients. *Compr. Rev. Food Sci. Food Saf.* 17 (4): 920–936. https://doi.org/10.1111/1541-4337.12360.

50 Chen, H., Wooten, H., Thompson, L., and Pan, K. (2019). Nanoparticles of casein micelles for encapsulation of food ingredients, Chapter 2. In: *Biopolymer Nanostructures for Food Encapsulation Purposes* (ed. S.M. Jafari), 39–68. Academic Press.

51 Rehan, F., Ahemad, N., and Gupta, M. (2019). Casein nanomicelle as an emerging biomaterial—a comprehensive review. *Colloids Surf., B* 179: 280–292. https://doi.org/10.1016/j.colsurfb.2019.03.051.

52 Geh, K.J., Hubert, M., and Winter, G. (2016). Optimisation of one-step desolvation and scale-up of gelatine nanoparticle production. *J. Microencapsulation* 33 (7): 595–604. https://doi.org/10.1080/02652048.2016.1228706.

53 Tarhini, M., Greige-Gerges, H., and Elaissari, A. (2017). Protein-based nanoparticles: from preparation to encapsulation of active molecules. *Int. J. Pharm.* 522 (1): 172–197. https://doi.org/10.1016/j.ijpharm.2017.01.067.

54 Hu, D., Lin, C., Liu, L. et al. (2012). Preparation, characterization, and in vitro release investigation of lutein/zein nanoparticles via solution enhanced dispersion by supercritical fluids. *J. Food Eng.* 109 (3): 545–552. https://doi.org/10.1016/j.jfoodeng.2011.10.025.

55 Luo, Y., Teng, Z., and Wang, Q. (2012). Development of zein nanoparticles coated with carboxymethyl chitosan for encapsulation and controlled release of vitamin D3. *J. Agric. Food. Chem.* 60 (3): 836–843. https://doi.org/10.1021/jf204194z.

56 Rehman, A., Jafari, S.M., Aadil, R.M. et al. (2020). Development of active food packaging via incorporation of biopolymeric nanocarriers containing essential oils. *Trends Food Sci. Technol.* 101: 106–121. https://doi.org/10.1016/j.tifs.2020.05.001.

57 Topuz, F. and Uyar, T. (2020). Antioxidant, antibacterial and antifungal electrospun nanofibers for food packaging applications. *Food Res. Int.* 130: 108927. https://doi.org/10.1016/j.foodres.2019.108927.

58 Vilela, C., Kurek, M., Hayouka, Z. et al. (2018). A concise guide to active agents for active food packaging. *Trends Food Sci. Technol.* 80: 212–222. https://doi.org/10.1016/j.tifs.2018.08.006.

59 Khan, A., Huq, T., Khan, R.A. et al. (2014). Nanocellulose-based composites and bioactive agents for food packaging. *Crit. Rev. Food Sci. Nutr.* 54 (2): 163–174. https://doi.org/10.1080/10408398.2011.578765.

60 Orozco-Parra, J., Mejía, C.M., and Villa, C.C. (2020). Development of a bioactive synbiotic edible film based on cassava starch, inulin, and *Lactobacillus casei*. *Food Hydrocolloids* 104: 105754. https://doi.org/10.1016/j.foodhyd.2020.105754.

61 Hari, N., Francis, S., Rajendran Nair, A.G., and Nair, A.J. (2018). Synthesis, characterization and biological evaluation of chitosan film incorporated with β-carotene loaded starch nanocrystals. *Food Packag. Shelf Life* 16: 69–76. https://doi.org/10.1016/j.fpsl.2018.02.003.

62 Assis, R.Q., Pagno, C.H., Costa, T.M.H. et al. (2018). Synthesis of biodegradable films based on cassava starch containing free and nanoencapsulated β-carotene. *Packag. Technol. Sci.* 31 (3): 157–166. https://doi.org/10.1002/pts.2364.

63 Assis, R.Q., Lopes, S.M., Costa, T.M.H. et al. (2017). Active biodegradable cassava starch films incorporated lycopene nanocapsules. *Ind. Crops Prod.* 109: 818–827. https://doi.org/10.1016/j.indcrop.2017.09.043.

64 Noronha, C.M., de Carvalho, S.M., Lino, R.C., and Barreto, P.L.M. (2014). Characterization of antioxidant methylcellulose film incorporated with α-tocopherol nanocapsules. *Food Chem.* 159: 529–535. https://doi.org/10.1016/j.foodchem.2014.02.159.

65 Mirzaei-Mohkam, A., Garavand, F., Dehnad, D. et al. (2019). Optimisation, antioxidant attributes, stability and release behaviour of carboxymethyl cellulose films incorporated with nanoencapsulated vitamin E. *Prog. Org. Coat.* 134: 333–341. https://doi.org/10.1016/j.porgcoat.2019.05.026.

66 Espitia, P.J.P., Fuenmayor, C.A., and Otoni, C.G. (2019). Nanoemulsions: synthesis, characterization, and application in bio-based active food packaging. *Compr. Rev. Food Sci. Food Saf.* 18 (1): 264–285. https://doi.org/10.1111/1541-4337.12405.

67 Ju, J., Chen, X., Xie, Y. et al. (2019). Application of essential oil as a sustained release preparation in food packaging. *Trends Food Sci. Technol.* 92: 22–32. https://doi.org/10.1016/j.tifs.2019.08.005.

68 Meira, S.M.M., Zehetmeyer, G., Werner, J.O., and Brandelli, A. (2017). A novel active packaging material based on starch-halloysite nanocomposites incorporating antimicrobial peptides. *Food Hydrocolloids* 63: 561–570. https://doi.org/10.1016/j.foodhyd.2016.10.013.

69 Oliveira Filho, J.G.d., Rodrigues, J.M., Valadares, A.C.F. et al. (2019). Active food packaging: alginate films with cottonseed protein hydrolysates. *Food Hydrocolloids* 92: 267–275. https://doi.org/10.1016/j.foodhyd.2019.01.052.

70 Santos, J.C.P., Sousa, R.C.S., Otoni, C.G. et al. (2018). Nisin and other antimicrobial peptides: production, mechanisms of action, and application in active food packaging. *Innovative Food Sci. Emerg. Technol.* 48: 179–194. https://doi.org/10.1016/j.ifset.2018.06.008.

71 Siripatrawan, U. (2016). Active food packaging from chitosan incorporated with plant polyphenols, Chapter 14. In: *Novel Approaches of Nanotechnology in Food* (ed. A.M. Grumezescu), 465–507. Academic Press.

7

Active Packaging Systems Based on Metal and Metal Oxide Nanoparticles

Lina F. Ballesteros[1,2], *Hafsae Lamsaf*[2,3], *Sebastian Calderon V*[2,3], *Miguel A. Cerqueira*[3], *Lorenzo Pastrana*[3], *and José A. Teixeira*[1]

[1] *University of Minho, CEB - Centre of Biological Engineering, Campos Gualtar, Braga, 4710-057, Portugal*
[2] *University of Minho, CFUM-UP - Department of Physics, Campus Azurém, Guimarães, 4800-058, Portugal*
[3] *INL - International Iberian Nanotechnology Laboratory, Av. Mestre José Veiga s/n, Braga, 4715-330, Portugal*

List of Abbreviations

CMC	carboxymethylcellulose
CMCh	carboxymethyl–chitosan
CS	chitosan
DSC	differential scanning calorimetry
EFSA	European Food Safety Authority
EVOH	ethylene vinyl alcohol
FCC	face-centered cubic
FDA	Food and Drug Administration
FTIR	Fourier-transform infrared spectroscopy
GE	gelatin
HPMC	hydroxypropyl methylcellulose
LCA	life cycle assessment
LDPE	low-density polyethylene
MCP	mechanochemical processing
NPs	nanoparticles
OBG	olive flounder bone gelatin
PBAT	poly(butylene adipate-*co*-terephthalate)
PE	polyethylene
PEG	polyethylene glycol
PET	polyethylene terephthalate
PLA	poly(lactic) acid
PP	polypropylene
PVA	polyvinyl alcohol
PVC	polyvinyl chloride

Nanotechnology-Enhanced Food Packaging, First Edition. Edited by Jyotishkumar Parameswaranpillai, Radhakrishnan Edayileveettil Krishnankutty, Aswathy Jayakumar, Sanjay Mavinkere Rangappa, and Suchart Siengchin.

PVD physical vapor deposition
ROS reactive oxygen species
UNEP United Nations Environment Program
XRD X-ray diffraction

7.1 Introduction

The quality and preservation of food are strongly affected by the packaging materials used for their protection and preservation. Consequently, food packaging manufacturing has strived to develop in recent years innovative and suitable materials that can guarantee product safety, shelf life extension, cost-efficiency, and consumer demands [1]. It is expected the package protects the food from environmental conditions including oxygen, light, mechanical damages, moisture, and microbial contamination, thus ensuring safe transportation and storage of the food and, at the same time, satisfying the environmental issues related to the packaging industry (e.g. being biodegradable or appropriate for recycling or refilling). For supplying these needs, active packaging materials are an emerging technology with a relevant impact on the quality of foods.

Active packaging systems cover several aspects that play a significant role when determining the shelf life of food. Chemical processes (e.g. lipid oxidation), physical processes (e.g. dehydration and respiratory rate of fruit and vegetables), and biological characteristics (e.g. decomposition by microorganisms) are some of them [2, 3]. Thus, an active material with the desired features for a selected food can be obtained through the incorporation of suitable components into the packaging, which allows the release/absorption of substances into or from the packaged food and the environment surrounding it, as defined by the European regulations (EC) No 1935/2004 and (EC) No 450/2009 [4].

Plastics produced from the petrochemical sources have been extensively utilized as packaging materials [5], due to their manufacturing costs and relevant mechanical features, as well as lower permeability moisture [6, 7]. However, the use of petroleum-based packaging is increasingly criticized and limited due to its environmental impact. Thus, bio-based packaging materials have captured the attention of the food packaging industry, becoming potential materials to develop active systems for food [8–10]. The use of bio-based polymers for packaging provides several advantages over traditional polymers, thanks to their biodegradable nature and nontoxic characteristics. However, these materials present, in most cases, low water resistance and weak mechanical properties, limiting their applications in food [1, 11]. These drawbacks can be solved using nanomaterials, which improve the mentioned properties and, in turn, potentiate functional characteristics of the packaging including antimicrobial agents, optical properties related with the color changes, oxygen scavengers, humidity sensors, gases removers/emitters, and mechanical performance, among others.

Metal and metal oxide nanomaterials provide unique chemical, physical, biological, and optical characteristics, due to a large number of surface atoms that offer higher reactivity. Besides, the larger surface area, surface–volume ratio, and surface charges compared to the microscale counterparts provide distinctive properties [12–14]. The interest in pure metal compounds such as zinc, silver, copper, and gold, or a combination of them, as well as metal oxides including zinc oxide, titanium dioxide, and magnesium oxide at the nanometric scale, has increased [14–16]. Therefore, the use of these nanoparticles (NPs) on traditional or biological polymers has allowed the development of active packaging systems for foods with new functionalities, which will be discussed in the present chapter. The different types of metal NPs applied to foods, the main production methods, and the incorporation of them into the packaging are addressed. Additionally, the main features and limitations related to the effect of metal-based NPs on the packaging properties, as well as the possible particle migration to the food and environment when discarded, are discussed.

7.2 Metal and Metal Oxide Nanoparticles Used in Active Food Packaging

Metal and metal oxide NPs have been evaluated in different fields of science and technology because of their low toxicity, high stability, durability, and selectivity compared with organic NPs. They usually exhibit interesting physicochemical properties due to diverse shape, size, and high surface-to-volume ratios, thus increasing their reactivity and functional properties, which result very attractive to the food packaging industry. Some examples of these kinds of NPs used in active food packaging and their respective functional properties are shown in Figure 7.1.

Noble metal NPs, for instance, are usually incorporated in active food packaging in their metallic state, due to the stability of these metals in natural environments. Ag and Au NPs are the most common choices due to their proven functional capabilities, including antimicrobial, optical, and catalytical properties. However, in some cases, especially for Ag the production process affects the oxidation states. For instance, Ag NPs are commonly produced by salts reductions, which lead to different Ag oxidation states, very likely to be in the precursor's states or compounds such as $AgNO_3$ and AgCl [17]. Nonetheless, some authors claim the existence of oxides such as AgO NPs [18]. The chemical stability of the noble metal NPs also may lead to weaker chemical bonding with the surrounding environment and, therefore, more prompt to migration, which requires additional consideration for the design of food packaging materials, including the size of the particles [19] and barrier layers (against migration) [20].

Alkali and alkaline-earth NPs have been less reported when compared with transition metal and noble metal NPs for the functionalization of food packaging materials. Among the most reported, the alkali metal oxide MgO NPs have shown to be highly efficient against Gram-positive and Gram-negative bacteria [21], as well as safe and inexpensive to be used in food packaging when compared with Ag NPs

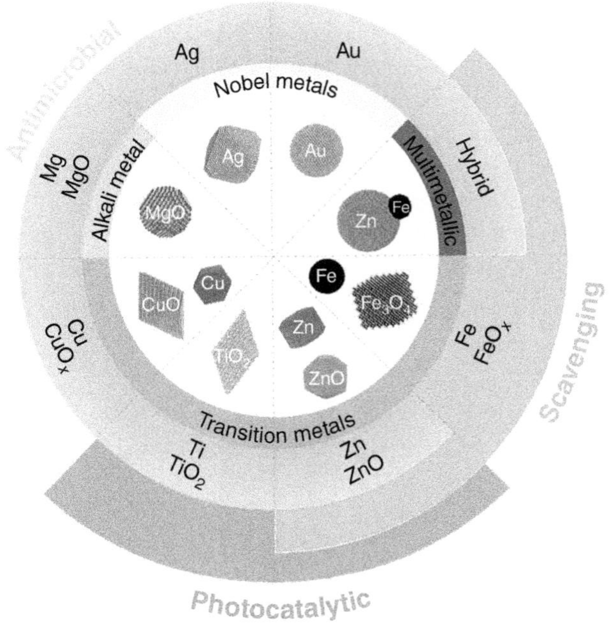

Figure 7.1 Metal and metal oxide nanoparticles used in active food packaging materials and their respective functional properties.

[22]. This metal oxide has been regarded as a potential antibacterial agent with multiple functionalities, including its use as a photocatalytic disinfecting agent, and as a blocker against the oxidizing effects of ultraviolet (UV) radiation [22].

Transition metal NPs are normally incorporated in oxide form [11] due to their spontaneous oxidation mechanisms. These oxides have the advantage of offering chemical stability, as well as several functional characteristics, such as antibacterial (e.g. ZnO, CuO, and FeO_x), optical (e.g. AlO_x, FeO_x, and ZnO), and catalytic properties (e.g. ZnO and TiO_2). The most common transition metal NPs used in food packaging are Zn, Ti, and Fe [23]. Transition metal NPs can be also created with large proportions of metallic state, for instance, in core-shell nanostructures with a metallic core and an oxide shell, which protect the core from oxidation, creating a reservoir for the metal to be used [24].

One of the most used transition metal oxides in active food packaging is ZnO, which is described as a multifunctional inorganic compound with a large range of applications. ZnO NPs could be used as antimicrobial agents to avoid the deterioration of food [25] or as ultraviolet light absorbents to protect food that is sensitive to UV radiations [26]. In addition, Cu and CuO NPs have demonstrated to be a potential antibacterial additive for food packaging applications. Evenly, Cu is preferred over silver because of its lower cost, ease to mix with polymers, and relatively higher physicochemical stability [27].

TiO_2 NPs have also proved multiple capacities in plastic packaging. They can block UV rays and then use their energy to degrade certain microorganisms [28]; they have low cost, are nontoxic, and have photostability [29]. TiO_2 NPs decrease fast food damage including oxidation, odors, bad flavors, quality reductions, and microorganisms growth by eliminating the oxygen present during the packaging of the product [28]. Likewise, TiO_2 is used in different polymers to improve O_2 scavenging potential [30].

Furthermore, **hybrid NPs** have inspired the research community due to their multifunctional character. The production of hybrid NPs not only allows combining different properties provided by their constituents within a single entity but also allows obtaining properties that emerge from the interface, leading to advanced characteristics. A limited number of studies have utilized bimetallic NPs, such as nano-alloys and hybrid NPs, to provide functional characteristics. Such hybrid NPs composed of multimetallic NPs counting two or more metal components presented better properties compared with the pure NPs. Hence, multimetallic NPs including CuPd, PtPd, AuPd, FePd, AuFeAg, and FeAgPt, as well as others, have shown excellent physicochemical, magnetic, optical, and electronic features, when compared with monometallic NPs [12].

Apart from metal and metal oxide NPs, several polymeric NPs such as chitosan [31, 32], poly(methacrylic acid) [31], starch [22], and others [33] have been used in active food packaging materials to provide functional characteristics. However, these particles are out of the scope of this chapter, and the authors refer the readers to other references for this subject.

7.3 Methods of Production of Metal and Metal Oxide Nanoparticles

Different physical, chemical, and biological procedures can be utilized to synthesize metal and metal oxide NPs. The most employed are described in the following section.

7.3.1 Physical Synthesis

NPs production by physical methods is characterized by the absence of chemical solvent contamination during the preparation of samples and the uniform distribution of NPs and their shapes and sizes.

Magnetron sputtering belongs to **physical vapor deposition (PVD)** techniques; it is a vacuum coating process to deposit metals by ejecting material in a plasma environment from a "target" that is a source onto a "substrate." Nanostructures produced by pulsed magnetron sputtering allow good control of the size and distribution of the NPs. Thus, authors have used this method to prepare Zn, Ag, Au, and TiO_2 NPs, among others. The quantity of deposited atoms depends on the deposition conditions, which can reduce or increase the movement of the adatoms or the amount of material placed on the substrate, whereas, the power increases the electrical field so that the Ar ions accelerate and, consequently, raises the sputtering yield in the discharge [24, 34].

Laser ablation is a known method to produce NPs by focusing a pulsed laser beam on a target made of the material to be deposited. It has been used to produce metallic NPs for food packaging applications. In an investigation made by Longano et al., the pulsed laser ablation was used to generate Cu NPs in liquid. For that, a metallic target immersed in acetone at atmospheric pressure and room temperature, and ytterbium-doped fiber laser were used [35]. The same technique was used for the production of <100 nm Ag NPs [36, 37] and also Ag/Cu NPs [38]. The absence of chemical reagents solution is one of the most important advantages of laser ablation [39].

Evaporation–condensation method has been used to prepare different metallic NPs by a tube furnace at atmospheric pressure [39]. There are several drawbacks of this technique because of the large space the tube furnace occupies. Thus, it consumes a great amount of energy to raise the temperature around the material and needs a long time to be thermally stable, in addition to the high power of important kilowatts and the preheating time to achieve the operating temperature [39]. In an article published by Jung et al., Ag NPs were synthesized using aerosol generation equipment in a small ceramic heater with a local heating area, which is connected to AC power. The NPs of 4.6–156.8 nm were produced by this technique [40].

7.3.2 Chemical Synthesis

Different kinds of metal and metal oxide NPs have been synthesized by a reduction chemical reaction using a reducing agent. Chemical methods can have the advantage to produce a considerable amount of NPs during a short time, but some of them are not eco-friendly and hazardous to human health [41].

For instance, in NPs chemical synthesis, the reducing agents are used to reduce Ag ions and form (Ag^0) as a metal, which leads to the production of metallic colloidal Ag particles [39]. Protective agents are used to avoid the agglomeration of NPs and stabilize their formation on the surface [39]. Ag NPs between 20 and 200 nm of different shapes can be produced by a Tollens synthesis method, which can control size in one step, due to a chemical reaction between $Ag(NH_3)^{2+}$ (aq) and a Tollens reagent [42].

Mechanochemical processing (MCP) is another experimental method known by its simplicity, low cost, and environmental safety; since the reaction involved does not include toxic solvent, it is suitable for large-scale NPs production [43]. Various NPs, such as ZnS, CdS, ZnO, $LiMn_2O_4$, SiO_2, and CeO, have been produced using the MCP technique [44]. The principle is about chemical reactions activated at the nanoscale and a physical size reduction process that is occurred after grinding the material in a conventional ball mill. For instance, experimental details of ZnO NPs production were described in [44]. Crystalline ZnO NPs have been prepared by milling $ZnCl_2$ and Na_2CO_3 with NaCl (an inert diluent accompanying the precursors), using the ball–powder collisions in a conventional ball mill. This later acts as a low-temperature chemical reactor, following the chemical reaction: $ZnCl_2 + Na_2CO_3 + 8NaCl \rightarrow ZnCO_3 + 10NaCl$. The nanostructured solution was heated (170–380 °C) to be thermally decomposed from $ZnCO_3$ to ZnO and then washed by deionized water to separate the NaCl from ZnO and dried. ZnO NPs produced are between 20 and 30 nm [44]. Another study that prepared ZnO NPs for packaging with MCP is reported in [43].

On the other hand, there is **sol-gel synthesis**, which is a known method to synthesize inorganic nano-compound. For instance, ZnO NP is one of the known elements that has been synthesized with this method. In a typical procedure, $C_4H_6O_4Zn$ is dissolved in double distilled water and heated to 50 °C while adding absolute alcohol gradually. Afterward, H_2O_2 is added dropwise to the vessel until the solution becomes clear. After 24 hours, the solution is dried at 80°, and ZnO NPs are obtained [45].

7.3.3 Biological Synthesis

Biological synthesis or green synthesis methods of NPs have attracted appreciable attention in developing nontoxic chemicals, renewable, and eco-friendly materials, minimizing the energy consumption. Some authors state that this process generates metal and metal oxide NPs with higher catalytic reactivity and larger specific surface area [46], and, generally, it results in more sustainable materials providing enhancements over the other traditional methods. These biological synthesis methods use **microbes**, **plants**, or **their derivatives** to reduce metallic ions present in a solution. Depending on the method, reducing sugars, antioxidants, flavonoids, among other compounds present in the cell membranes or plant extracts helps to form the NPs using a precursor agent. The medium composition, temperature, pH, salt concentration, and incubation period are some of the most relevant parameters, being responsible for the control of the size and distribution of the NPs. Such as in a work, Ag NPs were synthesized from a mixture of fruit peel extracts (*Vitis vinifera*, *Carica papaya*, and *Citrullus lanatus*) and $AgNO_3$, obtaining NPs of size about 30 nm [47]. Similarly, green synthesis of Ag NPs has been done from pomegranate seed extracts and $AgNO_3$ aqueous solution with different pH values [48]. ZnO NPs of 30–60 nm have been also prepared using a large variety of green materials, including gelatin, starch, plants, algae, bacteria, and fungi [41].

7.4 Incorporation of Metal and Metal Oxide Nanoparticles into Food Packaging Materials

A large diversity of materials and methods can be used to incorporate functional metal and metal oxide NPs into food packaging to provide the packaging with a large variety of functional properties. However, in this chapter, four techniques to produce nanocomposites are addressed, including widely industrial practices to more recent available technologies. There is a wide spectrum of polymers used for food packaging, including both nondegradable and biodegradable polymers. The most common petroleum-based plastics used as food packaging materials incorporated with meta-based NPs are low-density polyethylene (LDPE), polyvinyl chloride (PVC), and ethylene vinyl alcohol (EVOH). The LDPE is transparent, flexible, and resistant to moisture; however, the oxygen barrier property is its main weakness [49]. PVC is transparent and stiff, with an excellent barrier to oil, while EVOH is usually found as a thin film with good resistance to oxygen but very sensitive to humidity [50]. On the other hand, biopolymers incorporated with metallic NPs include chitosan, sodium benzoate, nisin, benzoic acid, sorbic acid, sodium sorbate, lactic acid, citric acid, plant extracts [51], pullulan, cellulose, sodium and calcium alginate, and starch [50], showing advantages over traditional polymers due to their eco-friendly character and low or nontoxicity.

The most common metal-based NPs used in food packaging are Ag, ZnO, and TiO_2, due to their potential to reduce microbial activity. However, there is a large variety of particles composition that can be found in the literature, including iron oxides, manganese oxides, gold, and copper oxide, among others. The incorporation of such NPs into polymeric matrices was shown to be very challenging due to surface interactions of the NPs with the matrix [52], promoting aggregation, and, hence, the selection of the method of incorporation should be carefully studied. Among the different methods, extrusion, casting, PVD, and electrospinning will be addressed.

7.4.1 Extrusion

Extrusion is the most widely used technique to produce packaging materials based on polymers. Briefly, it consists of pressing polymeric masterbatch fed through a hopper and push through a die. The extrusion speed, temperature gradient, the polymer feed rate, and the shape of the die are the most relevant process parameters. In the context of the production of metal-polymer nanocomposites, the incorporation of NPs is usually carried out by the production of masterbatch with dispersed NPs that will be later used for the production of the films. During the masterbatch production, the NPs in powder form are mixed with granule polymer of interest and then extruded at specific temperatures, highly dependent on the matrix. Subsequently, the masterbatch is cooled down and dried (e.g. 80 °C) [19, 53–58].

Despite being the most industrially available methodology, it is known to produce low dispersion of the NPs [53, 59], forming agglomerates of micron sizes [54]. Nonetheless, extrusion methodology is known to produce films with better mechanical, barrier, and microstructural properties, when compared with casting methodologies [60]. Thus, some efforts have been devoted to improving the dispersion of the NPs, for instance, by the inclusion of compatibilizers such as paraffin [53].

7.4.2 Casting

Casting is one of the most common methodologies reported to evaluate the feasibility and functionality of metal-based NPs dispersed into polymeric matrices for food packaging materials [61–69]. Its versatility allows to obtain both biodegradable and conventional packaging materials, as well as evaluate advanced architectures such as multilayers. The process consists of dissolving the polymer into appropriate solvents, creating what is known as the film-forming solution, which contains all the additives and functional compounds required for the food packaging, such as the metallic NPs. This solution is poured onto a substrate where the application speed, film length, and film drying speed can be controlled, also allowing the adjustment of the thickness to micrometric scales. The dispersion of the NPs is attained by ultrasonic vibration and/or stirring procedures.

Among the advantages of this technique is its versatility to create different architectures and shapes of the food packaging, including printability capacities.

7.4.3 Physical Vapor Deposition

PVD system includes a variety of techniques that produced films or, in an early stage, NPs deposited directly on the (polymeric) substrates. The techniques include sputtering, evaporation, and laser ablation, among others. Each of these methodologies differs in the method in which the material is ejected from the surface of the initial raw materials, known as the target, to deposit on the substrate of interest. Nonetheless, all of them are done under a vacuum and rely on physical processes instead of chemicals. For instance, in magnetron sputtering, an electric field is created between two plates and an ionized nonreactive gas bombards the target, ejecting atoms/molecules to the substrates. In laser ablation, on the other hand, the evaporation of the materials is caused by a high power laser-focused on the surface of the target, and in evaporation, the material is evaporated at high temperatures caused by large currents passed through a filament usually of made of tungsten.

These methodologies have the versatility of producing *in situ* NPs very well distributed with narrow size distribution [24], which allows the incorporation of a very large spectrum of composition, morphologies, and combinations of NPs. It has the advantage of being a process much eco-friendlier compared with other methods because it avoids the use of hazardous chemicals and reduces one step into the production process, being the NPs or nanofilms directly produced onto the substrate. However, it also has the disadvantage of using high vacuum environments, high temperatures, and/or water-cooling systems to dissipate the heat. These requirements need the use of expensive machinery.

7.4.4 Electrospinning

Electrospinning technology has received a lot of attention in the last two decades for its versatility in the fabrication of nanofiber mats with high specific surface areas [70]. It consists of a high voltage-driven process where nano- and microfibers/particles are made from a polymer solution that contains all the additives, including the NPs, needed for the packaging material. The typical setup requires the polymeric solution in a syringe, a pump, a high voltage power source, and a collector. The most common parameters for controlling the morphology and structure of the electrospun fibers are the polymer-based solution, its viscosity and surface tension, the evaporation speed of the solvent that allows the fiber integrity, the applied voltage, and the distance between the needle and the collector.

The incorporation of the NPs in this methodology is carried out by creating a colloidal solution of the polymeric solution and the NPs, which will be directly spun from the reservoir. This methodology is proven to obtain better dispersion of NPs, when compared to extrusion [59], and very well dispersed NPs can be observed in several studies [18, 59, 70, 71]. However, special care needs to be taken when considering the size of the fibers and the NPs, since a large NPs/fiber size ratio can lead to agglomeration of the particles [72].

7.5 Effect of Metal and Metal Oxide Nanoparticles on Active Packaging Properties

The incorporation of metal and metal oxide NPs into packaging materials affect a great variety of properties of the package, including the structure, morphology, and mechanical performance, as well as the antibacterial, photocatalytic, and oxygen scavenger characteristics, which are considered active functional properties. Therefore, the effects of these types of NPs on the main properties of food packaging materials, and more specifically on the active systems, are described in the following section.

7.5.1 Structure

The structural properties of the composite depend on the crystallinity of both the polymers and the NPs, which have a slightly different meaning. The metal and metal oxide NPs are solid with well-defined crystalline structures (or lack of it), while the polymers are usually regarded as materials with different degrees of crystallinity, which highly depends on the production methods and even on the inclusion of fillers, such as the type of NPs.

In most cases, it has been demonstrated that the addition of metal and metal oxide NPs to polymeric matrix alters its degree of crystallinity. The increase or reduction of the crystallinity depends on the interaction between the matrix and NPs and the amount of the NPs used. Therefore, a very large variety of possibilities can be encountered. However, it is clear that the effect of the NPs on the nanocomposite crystallinity highly depends on the quantity of NPs introduced. The changes in the crystallinity are more obviously observed when the content of the NPs increases [54, 62, 71], while with low particles amount, no significant changes are observed [55, 57, 64, 67]. For instance, ethyl cellulose/gelatin nanofibers containing ZnO NPs showed a decrease in the peak intensity and broadening of the polymer peak as the NPs are introduced [71]. Bumbudsanpharoke et al. have also shown that increasing the Ag NPs content lead to a loss of crystallinity of LDPE [54]. This amorphization effect is usually attributed to the capacity of the NPs to serve as nucleation points. On the contrary, some authors suggest an increase in the crystallinity of some polymers when incorporating ZnO [73] or Fe NPs [74].

Additional structural changes in the nanocomposite have been reported, where the NPs bond to functional groups in the polymeric matrix. For instance, Tankhiwale and Bajpai showed that Ag NPs bind to the functional groups of chitosan, shifting the peaks in the Fourier-transform infrared spectroscopy (FTIR) measurements [75]. Others have shown that Ag NPs reduced the intensity of polar functional group peaks present in FTIR [76].

7.5.2 Morphology

The morphology of the NPs can be controlled by the production method, which in turn depends on the thermodynamic or kinematical process. Among the most reported shapes are spherical, rods, fibers, tubes, and flakes NPs, which together with the aspect ratio, geometry, size, and distribution of the NPs can be used to tune the packaging characteristics [77, 78], including antibacterial [20, 79], mechanical, and barrier properties [14]. Table 7.1 shows some examples of these nanocomposite systems, detailing the application, shape, and size of the NPs.

The shape of the NPs has been pointed out as a key parameter for antibacterial behavior. Triangular, spherical, rod, and tubular NPs have shown variation in the antimicrobial power of Ag, TiO_2, and ZnO NPs, nanotubular particles being more effective in the case of TiO_2 and ZnO compare with spherical NPs due to a needle effect, while truncated triangular NPs and nanocubes of Ag showed better responses when compared with other shapes such as spherical, nanorods, and wires [20], likely due to the less stable edges that can ionize easier in such NPs.

The morphology together with the directionality of the polymer chains imposed by the production process also defines the degree of anisotropy of the properties in the nanocomposites [77]. For example, Lepot et al. have reported that ZnO nanorods with large aspect ratios can be aligned with the film's extrusion process and thus obtain anisotropic mechanical properties [82].

Another important aspect of the nanocomposite morphology is the size of the NPs and their degree of agglomeration. The particles size not only influences the properties of the matrix (e.g. porosity) and mechanical properties [55] but also determines the behavior of such particles within the nanocomposite, e.g. limiting their mobility [19] and aggregation and, therefore, affecting migration and toxicity. The size of the NPs has been reported to modify the antibacterial mechanism of Ag [20], and some report that the reduction of size can significantly improve the antibacterial effect for ZnO [91]. Besides, the particles size is important when studying the migration of the NPs from packaging materials [87], due to the affinity of the NPs and the matrix and because of the effects that the NPs size has on the dissolution and reactivity. For instance, Hannon et al. reported that the increased size-dependent dissolution of Cu NPs can cause higher migration than Ag NPs [92].

The methodology used to incorporate the NPs affects significantly their degree of agglomeration, as previously discussed; however, several authors have proposed different strategies to reduce the agglomeration of the NPs, including compounds

Table 7.1 Examples of metal and metal oxide NPs incorporated in food packaging materials for different applications.

Metal based	Polymeric matrix	Shape	Application	Size (nm)	References
MgO	PLA	Polyhedral	Gas barrier	<60	[80]
ZnO	No specified	Nanorings, nanohelix	Antifungal, antimicrobial	No specified	[11]
ZnO	PVA	Nanoplates (flowerlike structures)	Antibacterial	18–85	[81]
ZnO	PP	Nanorods	Oxygen barrier	50–200 (diameter)	[82]
ZnO	PBAT	Nanoparticles	Antimicrobial	<60	[83]
ZnO		Nanoparticles	Antibacterial	30–40	[84]
ZnO	CMC	Nanoparticles	Antibacterial biocomposite	20–30	[85]
Ag	CMC	Nanoparticles	Antibacterial biocomposite	<50	[85]
Ag	Silk fibroin	Spherical	Antibacterial	No specified	[18]
Ag	PE	Round	Antibacterial	<70	[86]
Ag	Chitosan–konjac glucomannan–cassava starch	Spherical	Moisture resistant antimicrobial	5	[64]
Ag	PE, PP	Nanoparticles	Antibacterial, antiodor Keep food fresh	20–70	[87]
Ag	No specified	Nanoparticles	Antifungal	7–100	[88]
Au	PVA	Nanoparticles	Shelf life extension	No specified	[67]
TiO_2	LDPE	Nanopowder	Antibacterial	<25	[56]
TiO_2	PBAT	Spherical	Antibacterial barrier biodegradable film	55	[66]
TiO_2	Gelatin/agar	Nanoparticles	UV absorbent	10–15	[89]
TiO_2	Silk fibroin	Nanoparticles	Biodegradable packaging	<20	[29]
CuO	CMC	Nanoparticles	Antibacterial biocomposite	35	[85]
Fe	Silicon	Nanopowder	Oxygen scavenger	<10	[90]

PLA, polylactic acid; PVA, polyvinyl alcohol; PP, polypropylene; PBAT, poly(butylene adipate-*co*-terephthalate); CMC, carboxymethylcellulose; PE, polyethylene; LDPE, low-density polyethylene.

that help to bind the NPs to the polymeric matrix, use different crystalline matrices, or simply by a reaction between the polymer and the NPs. Some authors reported a reduction of agglomeration employing compatibilizers, such as paraffin [53] and maleated polypropylene (PP-g-MA) [74], which bind to the NPs, helping to increase their dispersion. In the case of TiO_2, it is suggested that the mixture of anatase and rutile phases helps to prevent the agglomeration of the NPs in LDPE during extrusion [57]. Díez-Pascual et al. demonstrated that the interactions between the –OH groups of the ZnO surface and the polar moieties of the biopolymeric matrix prevent NPs aggregation [73].

Furthermore, the incorporation of NPs to the polymeric materials is known to provide additional roughness to the nanocomposite due to their agglomeration [54, 59, 61, 85, 93, 94], worsening as the number of NPs increases [81, 83, 85, 95, 96]. These aggregates are typically several times larger than the individual NPs, and their size and structure are highly dependent on the amount, size of the NPs, and the chemical bonding between the polymeric matrix and the NPs [13, 77]. Some studies show that Ag NPs can agglomerate in the interphase for compounds blends, for instance, Stará et al. show that Ag tends to agglomerate between the interface of polyethylene (PE) and a superabsorbent polymer [97].

7.5.3 Mechanical Properties

The mechanical properties of the nanocomposite rely on the interaction between the NPs and the polymeric matrix [54]; thus, when there is a strong interaction between these components, the stress is more effectively transferred from the polymer matrix, and the mechanical properties are enhanced [49]. This interaction is affected by the dispersion of the NPs and in turn, alters the tensile strength of nanocomposite [62]. For example, salt bridges between the carboxylic groups of the polymers and ZnO NPs have been suggested to create favorable interactions for the film structure, consequently improving the rigidity of the gelatin films [98].

The degree of aggregation of the NPs also plays a significant role in the mechanical properties of the packaging materials. It is frequently reported that the high dispersion of the NPs produces an improvement in the mechanical properties of the nanocomposites [53]. However, significant aggregation of the particles can deteriorate the mechanical properties, even below the values of the unloaded polymers [49, 66, 81, 98]. The impact of the NPs on the mechanical properties of the nanocomposite generally depends on the amount of the NPs loaded to the polymers. By determining the proper amount of NPs in the composite, the overall mechanical properties can be improved compared to the "blank"/unfilled polymers [6, 11, 51, 85, 93, 99]. It has been demonstrated that a very low amount of NPs does not offer a significant improvement on the mechanical properties [59, 74], and high amounts may even deteriorate them [99]. Such modification depends on the interfacial interactions between the NPs and the polymer, the dispersion inside the matrix, and the amount [85]. In fact, as the NPs increase, more agglomeration is usually reported, which may explain the deterioration of the mechanical properties. For instance,

Castro Mayorga et al. suggested a negligible impact of CuO on the mechanical properties of the films due to the low CuO loading and the high NPs dispersion [59].

As previously mentioned, the aspect ratio of the NPs and the production process of the matrix is another critical feature in the mechanical properties. For example, ZnO/PP showed a marked isotropic improvement in the mechanical properties, when spherical NPs are used, while ZnO nanorods offered a significant improvement in the direction of the large aspect ratio direction [82].

Generally, the **tensile strength** of the nanocomposite follows the same trends previously mentioned; it is not significantly affected by small amounts of particles, and it can be significantly improved by increasing the amount of the NPs [57, 63, 64, 83]. For instance, some authors suggest that to obtain a significant modification in the tensile strength, Ag NPs content should range between 5 and 20 wt% for spherical fillers and 2 and 7 wt% for higher aspect ratio NPs [54]. It has been suggested that although the crystallinity of PE is reduced by incorporating Ag NPs, the tensile strength is enhanced [54]. Nonetheless, some studies demonstrate that an overload of the NPs into the polymeric matrix can deteriorate the tensile strength of the films. For instance, Venkatesan et al. demonstrate that the tensile strength and modulus of poly(butylene adipate-*co*-terephthalate) (PBAT) films increased with TiO_2 content up to 7 wt% and decreased thereafter [66].

On the other hand, although some studies have reported that the **elongation at break** of nanocomposites increases upon increasing NPs content due to the loss of crystallinity [54], it is commonly reported that the introduction of the NPs reduces the flexibility of the polymers and, therefore, decreases the elongation at break [62, 63, 67, 83].

7.5.4 Barrier Properties

Polymeric materials are more prompt to gas and water vapor permeability than other conventional food packaging materials such as glass, metal cans, and paper. Thus, researchers have focused a lot of attention on different approaches to reduce such permeability and consequently extend the shelf life of food without the need for preservatives. One of these strategies is to create nanocomposite; thanks to the formation of two-phase materials between the metallic/metal-based NPs and the polymeric matrix, the diffusion coefficient of gases is reduced. This is usually explained by the reduction of the available area for diffusion due to the impermeable NPs and the increase in distance that a gas needs travel across the film [82]. However, the changes in the permeability are highly dependent on structural parameters, including the volume fraction of filler, aspect ratio, and the dispersion of fillers in the polymer matrix [100], as demonstrated by Bhunia et al. who studied the diffusion behavior in nanocomposite using finite element methods for different volume fractions, aspect ratio, intercalation width, exfoliation, and orientation angle. Besides, Lepot et al. suggest that the higher the aspect ratio of the filler, the greater the decrease in permeability due to the longer diffusion path that the gas needs to go through; no significant differences were found between spherical and nanorod NPs [82]. On the other hand, Zhao et al. discussed a significant preservation quality of green tea, using based packaging material, prepared by adding NPs of Ag, TiO_2, and attapulgite to PE. During one-year ambient storage, the material showed an improvement in the barrier properties. Meanwhile, the contents of amino acid, tea polyphenols, chlorophyll, and ascorbic acid were increased compared to the normal packaging material [101].

It has been extensively demonstrated with a different types of NPs, including TiO_2, ZnO, $CaCO_3$, Ag, and MgO, among others that the barrier effect of films is enhanced against a variety of gases oxygen, carbon dioxide, and water vapor [57, 82, 102–104]. However, it is also shown that overfilling the polymer with NPs could have a detrimental effect on the permeability [102].

Some authors suggest a more complex behavior of the permeability of the nanocomposite. As previously exposed, the incorporation of some NPs promotes changes in the crystallinity of the polymeric matrix, which in turn modify the gas diffusion coefficient [103]. This together with the effect of impermeable particles leads to an overall increase or reduction of permeability.

7.5.5 Antimicrobial Activity

The principal function of antimicrobial agents is to decrease food spoilage and subsequently avoid the growth of microorganisms. As a result, the incorporation of metal (e.g. Ag, Zn, Ti, and Cu) and metal oxide (e.g. ZnO and TiO_2) NPs into the food packaging to generate active compounds with antimicrobial properties has brought significant progress in the food materials field. Antimicrobial food packaging materials containing these NPs have presented significant effects against an extensive range of pathogens such as viruses, fungi, yeasts, and bacteria [11, 69], thus increasing their biocidal spectrum in comparison to the antibiotics [105].

On the other hand, the antimicrobial metal agents possess benefits over organic agents related to lower toxicity, selectivity, and specificity [11]. For example, ZnO, CaO, and MgO have proved that high antimicrobial capacity relates to the production of reactive oxygen species (ROS), and, in turn, they offer a relevant advantage by transporting essential minerals to the body [106].

In general, metal NPs possess not only relevant antimicrobial properties but also stability at high temperatures and pressures when compared to organic antimicrobial compounds. This characteristic is appreciated by the food packaging industry since these nanomaterials can tolerate very demanding processing conditions [68, 107].

Therefore, this section summarizes the mechanisms of action and potential applications of the most used metal particles as antimicrobial compounds in active food packaging. Some examples are shown in Table 7.2.

7.5.5.1 Silver Nanoparticles

Ag NPs have been extensively studied and incorporated into different food polymer matrices as antimicrobial agents. Their mechanism is related to the cell wall disruption through the generation of ROS [122], interaction with thiol groups of enzymes [123], and changes in mitochondrial membrane permeability [124].

Table 7.2 Examples of metal and metal oxide NPs incorporated in food packaging materials for antimicrobial applications.

NPs	Polymeric matrix	Microorganism	NP size (nm)	Nanoparticle concentration	References
ZnO	Bacterial cellulose	*S. aureus* *E. coli*	30	5% wt	[108]
ZnO	Polycarbonate	*S. aureus* *E. coli*	2–30	0.05–1% wt	[109]
ZnO	LDPE	*B. subtilis* *E. aerogenes*	50–400	2 and 4% wt	[110]
ZnO	PVA–GE	*B. subtilis* *E. coli*	~300	Film	[111]
ZnO	OBG	*L. monocytogenes*	<50	1%, 3%, and 5% of the gelatin	[98]
ZnO–CS	PE	*S. enterica* *E. coli* *S. aureus*	55	0.1% solution	[112]
Ag	Starch	*S. aureus* *E. coli* *C. albicans*	No specified	0.3, 0.5, 0.8, and 1.0 mM of silver nitrate	[113]
Ag	HPMC	*E. coli* *S. aureus*	41 and 100	No mentioned	[114]
Ag	CMCh	*S. aureus* *S. faecalis* *P. aeruginosa* *C. albicans*	9–16	1–3% silver nitrate	[115]
Ag	LDPE	*E. coli* *S. aureus* *C. albicans*	5.5	0.1, 0.3, 0.5, 3, and 5% wt of polymer	[116]
TiO_2	CS	*E. coli* *S. aureus* *C. albican* *A. niger*	50–80	0.05 g of NPs/0.5 g of CS	[117]
TiO_2	PBAT	*E. coli* *S. aureus*	55	1, 3, 5, 7, and 10% w/w	[66]
TiO_2	PVA–CS	*E. coli* *S. aureus* *L. monocytogenes* *S. enterica*	15–30	0.05–0.3% w/v	[118]
CuO	Alginate CMC K-carrageenan	*E. coli* *L. monocytogenes*	No specified	80 mg NPs/4 g of biopolymer	[119]

Table 7.2 (Continued)

NPs	Polymeric matrix	Microorganism	NP size (nm)	Nanoparticle concentration	References
CuO	Starch	*E. coli* *S. aureus*	No specified	Up to 3% wt	[120]
CuO	Cotton fibers	*E. coli* *L. monocytogenes* *S. aureus* *S. epidermis*	100–200	7.4 mg of NPs/100 mg cotton	[121]

NPs, nanoparticles; LDPE, low-density polyethylene; PVA, polyvinyl alcohol; GE, gelatin; OBG, Olive flounder bone gelatin; CS, chitosan; PE, polyethylene; HPMC, hydroxypropyl methylcellulose; CMCh, carboxymethyl–chitosan; PBAT, poly(butylene adipate-*co*-terephthalate); PVA–CS, polyvinyl alcohol–chitosan; CMC, carboxymethylcellulose.

Ag NPs activity seems to depend on the release of Ag^+ ions, binding to electron donor groups in biological molecules including phosphates, hydroxyls, thiols, and amines [15, 20]. Although Ag NPs have demonstrated notable antimicrobial activity against Gram-positive and Gram-negative bacteria, their effect has been stronger against Gram-negative [20, 108–111, 125, 126]. Bactericide impacts have been reported against *Escherichia coli, Staphylococcus aureus, Salmonella choleraesuis, Bacillus cereus, Pseudomonas aeruginosa,* and *Listeria innocua* bacteria, owing to increased toxic influence on the cells of bacteria [11].

Siqueira et al., for example, studied the antimicrobial activity of Ag NPs incorporated into carboxymethylcellulose (CMC) films for food packaging applications. The results demonstrated that Ag NPs inhibited the growth of the microorganisms tested [127]. Additionally, LDPE antimicrobial films coated with Ag NPs exhibited great potential to develop packaging materials with antimicrobial properties [128]. As a final point, Ag NPs have been incorporated in sodium alginate coatings on carrots and pears [129], in LDPE films to coat fresh-cut sliced carrots [130], and in cellulose to produce an absorbent pad to meats [130]. As a result, Ag NPs into these polymeric matrices reduced microbial growth in all the food tested.

7.5.5.2 Zinc Oxide Nanoparticles

ZnO is one of the most utilized metal oxides, because of its notable antimicrobial characteristics and safety (since is listed as Generally Recognized as Safe (GRAS) by the Food and Drug Administration (FDA)). ZnO NPs performance has been associated with different mechanisms of action and chemical species, including the generation of ROS owing to the semiconductor properties of metal oxides [131], allowing ZnO NPs to participate in light-catalyzed redox reactions [15] and the disruption of the cell membrane upon interaction with the NPs, especially with the intrinsic features of Zn^{2+} ions [11]. It seems that in the humidity medium, the ZnO NPs can produce oxidative stress, thus affecting the cell wall microorganism by the generation of H_2O_2 active species on its surface [112, 132].

ZnO NPs exhibited a more marked effect against Gram-positive (e.g. *S. aureus*) than Gram-negative (e.g. *E. coli*) bacteria, and the bactericidal efficiency can rise when the particle size decreases [91]. Moreover, He et al. demonstrated the antifungal capacity of ZnO NPs against *Penicillium expansum* and *Botrytis cinerea* [133].

ZnO NPs have been identified as potential antimicrobials when added into several polymeric matrices including pullulan [134], bacterial cellulose [108], and LDPE films [110], among others. For example, PLA–ZnO NPs composite films demonstrated strong antibacterial activity against foodborne pathogenic bacteria, *E. coli*, and *Listeria monocytogenes*, besides the increase of minced fish cake shelf life [135]. On the other hand, black grapefruits [132] and meat and poultry products [134] were also coated with films containing ZnO NPs, which contribute to the decrease of bacterial growth.

7.5.5.3 Copper and Copper Oxide Nanoparticles

Cu and CuO NPs possess relevant biological, chemical, and physical characteristics [11]. Additionally, Cu presents a broader spectrum of activity, when compared with other metals due to its dual nature to donate or receive electrons [20]. The mechanism of antimicrobial activity from CuO NPs is not known. It seems to be an integration of various actions including emancipating of Cu^{2+} ions, their biochemical pathway, and disruption cell membrane by DNA damage and cellular enzymes [136].

The efficiency of microorganism growth inhibition using CuO depends on the concentration and NPs size [11]. They have proved activity into polymer matrices against bacteria (Gram-negative and Gram-positive) and fungi [27, 120, 137]. Yoon et al., for instance, have defined Cu NPs as an antimicrobial agent against *E. coli* and *Bacillus subtilis*. As a result, the reaction of Cu NPs showed higher susceptibility than Ag NPs against *E. coli* [138]. The effectivity of CuO NPs was also demonstrated when incorporated into LDPE films to avoid coliforms in cheese [139].

7.5.5.4 Titanium Dioxide Nanoparticles

TiO_2 NPs have been widely studied as an antimicrobial agent. In fact, FDA has suggested and accepted their use in cosmetics and healthcare, as well as in food materials, due to their low toxicity. The photocatalytic mechanism of these NPs endorses ROS production, which is related to bacterial inhibition [140, 141], phospholipids peroxidation, and loss of membrane integrity [20].

TiO_2 NPs present the ability to react with O_2 and –OH, which are captivated on the surface for acquiring hydroxyl and oxygen-free radicals [142]. Thus, it has evaluated the linear correlation between the OH concentration and the bacteria mortality as well as with the inhibition percentage [11].

TiO_2 NPs have been embedded into chitosan films, presenting inactivation against Gram-negative and Gram-positive bacteria, such as *E. coli* and *S. aureus,* and fungi including *Aspergillus niger* and *Candida albicans* [117]. Chitosan film containing TiO_2 was also used in the films to coat grapes, achieving to extend the fruit stability and reduce surface microbial contamination [20].

7.5.5.5 Gold Nanoparticles

Au metal NPs are extensively utilized in the medical field for different biological applications including nanoscale devices, because of their chemical inertia and resistance to oxidation [14]. However, Au NPs incorporated into films for food packaging applications have been considered as promising materials for food safety and the shelf life extension of food [143].

The suggested mechanism of antimicrobial activity from Au NPs is ascribed to the production of ROS, which causes a rise of the oxidative stress of microbial cells and release of intracellular lactate dehydrogenase enzyme into the extracellular medium [144]. It has also been shown that Au NPs join the DNA of bacteria and inactivate the transcription of DNA [145].

As an example, stabilized Au NPs (by an ionic silsesquioxane) have been incorporated into bio-based films of quinoa (*Chenopodium quinoa* W.) starch. The results showed an active nanomaterial with high antibacterial activity against foodborne pathogens (99% and 98% inhibition of bacterial growth against *E. coli* and *S. aureus*, respectively). Additionally, an improvement in the mechanical, optical, and morphological properties were demonstrated with the presence of Au NPs, besides maintaining the thermal and barrier characteristics when compared to the standard bio-based films [143].

7.5.5.6 Other Nanoparticle Systems

The antimicrobial characteristics of mixtures of metal and/or metal oxide NPs and systems of hybrid NPs (bimetallic and trimetallic) have been also studied to potentialize the characteristics of individual NPs or even achieve synergistic effects. As a result, Ag, ZnO, and CuO NPs were tested in the preparation of CMC nanocomposite films. The results confirmed the antibacterial activity of the films against *S. aureus* and *E. coli*. Moreover, the addition of NPs increased the tensile strength of the materials. The nanocomposite films also exhibited a higher resistance compared with the pure CMC film. Therefore, this new material could be considered a promissory active film, thus raising the shelf life of the food [48].

Likewise, the combination of Fe and Zn oxides NPs has produced a compound with antibacterial magnetic and antibacterial properties. Their NPs showed that the higher ratio of Zn/Fe has a relevant antibacterial potential against *S. aureus* and *E. coli* [146].

7.5.6 Scavenger Properties

The presence of oxygen during packaging, as well as the permeation of oxygen during storage, can be evaded by using oxygen scavenging compounds, which helps to increase package security and preserve the food quality by reducing, for example, microbial spoilage, nutrient loss, discoloration, rancidity, off-flavor, and organoleptic deterioration. One of the advantages of incorporating oxygen scavenger, as an integral part of the polymer, in form of a nanocomposite is to avoid the necessity of additional components, including self-adhesive labels, cards, and sachets, which can lead to an inefficient activity of the functional compound.

One of the most effective and known oxygen scavengers is an iron-based powder, which can remove oxygen confined in the packaging headspace or permeating through the packaging material during storage by an electrochemical reaction [147]. Khalaj et al. have shown, for instance, that nanocomposite formed with Fe NPs can reduce the oxygen and water vapor permeability up to 55% and 77% [74]. Moreover, Foltynowicz et al. studied the nanopowder of Fe in low and high humidity, showing that the nano-oxygen scavengers absorbs oxygen the wet and dry environment. The rate of scavenging is higher at 100% relative humidity three times than micro-scavengers [90].

Some studies have also shown the scavenging properties of TiO_2 nanocrystalline films through a photocatalytic process. Although these films were not directly applied to food packaging nanocomposite, they showed the potential to be used as oxygen scavengers in different applications [30]. Pd NPs have also displayed significant capacity to remove residual oxygen in the presence of H_2 [148], which has been successfully applied for the removal of 2 vol% oxygen in the headspace of modified atmosphere packed ham within 35 minutes after packaging [148].

In general, zerovalent metals such as Cu, Zn, and Mg NPs have not been extensively explored in the field of oxygen scavenging. However, these metals have great potential due to their spontaneous oxidation in air and their activation in humid environments. In fact, several of these metals could display multifunctional properties, due also to their antimicrobial properties. Nonetheless, special care needs to be taken in the production and protection of the zerovalent metal NPs due to their easy oxidative behavior. Only recently, the authors have also demonstrated O_2 scavenging properties of Zn/ZnO core-shell NPs; the results indicated an acceleration in the oxidation process at high humidity that allowed it to be used in food packaging applications [24].

7.5.7 Photocatalytic Properties

One of the most known photocatalytic agents for a wide range of application is TiO_2 due to its electronic configuration [91]. However, a large variety of metal-based NPs possess catalytic properties, including ZnO, Cu_2O, and Fe_2O_3, among others. As such, these NPs can be incorporated in food packaging materials to promote oxidation processes accelerated by photocatalysis, e.g. to avoid microbial growth without the need of releasing metallic ions [91], to disintegrate pesticides after harvesting [149], and to help in the cleaning operations in the food industry due to their self-cleaning properties [150].

Bodaghi et al., for instance, demonstrate the inactivation of bacteria under UVA illumination of the composite films with TiO_2, which displayed excellent antimicrobial activity and ethylene photodegradation [57]. Some authors have also demonstrated that the extent of bacteria inactivation caused by a photo-disinfectant such as TiO_2 depends on the kind of artificial light sources, stating that under black-light bulb illumination, the photocatalytic disinfection activity was eight times higher than under the fluorescent-light bulb illumination [151].

Furthermore, it is worth considering that the use of different additives, binders, and components in the production of the nanocomposite may affect the photocatalytic properties of the final product. Yemmireddy et al. suggested that the increase of binding agents concentration, such as polyurethane, polycyclic, and shellac may decrease the exposed NPs on the surface, which may reduce the bactericidal property of TiO_2 [152].

7.5.8 Optical Properties

The color of the food packaging materials not only influences the aesthetic of the product but also is essential when limited interaction between light radiation and the foodstuff is required. In terms of aesthetics, color has been reported as one of the most critical aspects for consumers to select the products and make subconscious judgments in a couple of minutes over a series of branches for the same product. There are even psychological studies on the effect colors have on human behavior and emotions [153]. Such elements of design are important when selecting additives to the polymers such as metallic-based NPs because they can significantly alter the base color of the polymers or interact with the colorant additives. However, the studies of metallic or metal-based NPs have been focused on how the NPs influence the light absorption and in some cases the transparency of the films.

Ebrahimi et al. have shown that the absorption of light in the visible and UV range can be tuned by using different NPs compositions since each type of NPs will absorb at its own wavelength range [85]. The changes in the color of the films are highly dependent on the chemistry of the NPs, since each metal or metal oxide will absorb light at a different position in the spectrum. Ag NPs, for instance, have been reported to gradually alter the color of LDPE from transparent to dark yellow [54]. A combination of TiO_2/Ag, on the other hand, can change the same LDPE polymer to a more reddish color [53].

Additionally, metal-based NPs have been reported to significantly reduce the transparency of the nanocomposites, which depends on the amount of the NPs introduced [98]. Ebrahimi et al. have shown that Ag, ZnO, and CuO NPs reduced the transparency of CMC films to a large extent [85]. These characteristics have potential application for foods sensitive to light radiation. For example, Vejdan et al. suggested that TiO_2 NPs at a concentration of around 2% can considerably prevent the transmission of UV light, showing potential to avoid photooxidation [89].

Finally, some authors have correlated the change of color with other functional properties. It has been shown that the color change in nanocomposite incorporated with core/shell Au/Ag nanorods is related to the growth of *E. coli* bacteria, thus potentially using the NPs as sensors in the packaging [154].

7.6 Migration of Nanoparticles

The increasing interest to incorporate metal NPs in food packaging has led to concerns related to the NPs migration to the food and the environment. Migration studies have confirmed that only a tiny quantity of NPs or ions migrate from the packaging to the foods or food simulants and that the toxicological risks for the consumers when exposed to the ingestion of these NPs are low [16]. Nevertheless, the regulation related to the use of metal NPs in food packaging is not completely developed yet, even in the major economies. Therefore, research is still needed to determine the effects of the ingestion of these NPs in the gastrointestinal system, and if the absorption could affect other organs.

On the other hand, the release of metal NPs from the food packaging to the environment is perturbing since they could generate enormous pollution. Migration of this type of NPs to nonfood matrices results very complexly due to the affinity that some particles present to organic matter and others are dissolving into ions or forming aggregates [155]. Thus, similar to the NPs migration from packaging to foods, their migration to the environment is still a subject of investigation, in order to understand the evolution and proper treatment when packaging is discarded.

Considering these issues and the advantages that lead to the use of metal and metal oxide NPs in the development of active packaging, some aspects related to food safety and NPs regulation are addressed in the following section.

7.6.1 Food Safety and Regulations

Active packaging materials containing metal and metal oxides NPs provide several advantages including the inhibition of microbial growth and the reduction of chemical preservatives, among others already discussed throughout this chapter. However, the safety issues related to the migration of the NPs from the packaging materials to the foods remain [156]. This migration can be related to the dissolution, diffusion, and abrasion of the packaging surface, allowing the transfer of the NPs to the food when both in contact, thus causing a negative effect on human health [16]. On the other hand, the toxicology of the NPs is influenced by their physicochemical properties such as composition, size, surface-area-to-volume ratio morphology, and solubility [6, 157]. Additionally, roughness, charge, hydrophilic nature, and the chemistry of the NPs are factors that significantly affect the cytotoxicity of the NPs since biological systems are mainly affected by the surface of the NPs [12].

The majority of studies about NPs migration into foodstuffs have been focused on Ag NPs, and only a few are referring to the migration of metal oxides such as ZnO and TiN [7]. According to the results, the migration of NPs tends to rise as the temperature and time increase, though the quantities reported have been low. Song et al., for example, studied the migration of Ag NPs from PE using ethanol at 95% (v/v) and acetic acid at 3% (w/v) as food simulants, at different times and temperatures [158]. The results showed that the quantification of Ag in ethanol rises with time, while temperature did not show any significant effect. Besides, the migration of Ag slightly increased with the time and temperature when acetic acid was used, before reaching a steady state. On the other hand, the migration of Ag and TiO_2 NPs (incorporation of 1% and 0.1%, respectively, in polymeric materials formed into food containers) from the packaging to real foodstuffs such as apples, soft cheese, powdered milk, bread, and butter, among others, when storage at 40 °C for 10 days, was evaluated. The results of migration determined insignificance levels of Ag and TiO_2 NPs in all the foods [159].

Concerning human health, some authors have reported that Ag NPs may damage the cells by modifying the function of mitochondria, increasing membrane permeability, and generating ROS [158]. Moreover, other studies have determined that Ag and TiO_2 NPs after entering through the gastrointestinal tract can pass into the blood and be deposited in some organs [160, 161], including the liver and spleen [162]. Thus, it is appropriate to know how the body metabolizes and eliminates these types of NPs and identify possible adverse effects to the health [12].

As a final point, and although the existing information to understand the migration phenomenon is limited, the food packaging industry must develop safe materials, controlling parameters such as the particle size, concentration, molecular weight, and solubility of the metal NPs, as well as the structure and properties of the polymer used (e.g. viscosity), composition of food, and time and temperature of storage.

7.6.2 Regulation

Regulations concerning the packaging materials in contact with foods have been ratified for a long time ago. Nevertheless, these protocols, which differ from country to country, are continually revised and modified due to the listed components that can be added or removed. Each country or group of countries has stipulated their regulations about what kind of materials are satisfactory as components of the packaging and what are their limits according to the incorporation into packaging materials and migration into foodstuffs.

Information regarding the use of metal and metal oxide NPs in food packaging is restricted. However, some countries recently have started to incorporate nanomaterials into packaging materials. The United States, for example, is one the most advanced in the development of these nanomaterials together with the European Union; nevertheless, the protocols concerning the particle size limits and what is considered as a safe material are not harmonized yet.

Although FDA has not defined yet a legal approach to regulating the nanotechnology in food contact substances [87], FDA has proportioned some standards to the industry, considering it as a nanomaterial with a particle size between 1 nm and 1 μm. Additionally, FDA recommends that the GRAS status does not essentially indicate that the material is safe if it is used at nanoscale level [16]. Therefore, a case-by-case evaluation should be needed considering the food package and the additive in the packaging. According to metal oxide NPs, FDA has included TiO_2 as a food contact substance that has demonstrated to be safe for specific applications [163, 164] and ZnO as a GRAS substance [165].

On the other hand, the European Union through the European Food Safety Authority (EFSA) has enacted rules about nanotechnology for food applications. They defined nanomaterial to that with a particle size in the range of approximately 1–100 nm. Moreover, the document "Guidance on the risk assessment of the application of nanoscience and nanotechnologies in the food and feed chain" [166] provides information to the industry concerning the physicochemical characteristics of nanomaterials and possible procedures to determine the risks. However, the lack of knowledge to properly validate all the applications in food materials limits their safety and marketing. Additionally, European Commission published in 2016, the Regulation (EU) No 2016/1416 [167] improving and correcting the Regulation (EU) No 10/2011 on plastic materials and articles, intended to come into contact with food; however, the condition that the nanomaterials must be assessed on a case-by-case basis to evaluate their risk was maintained. Currently, TiN and ZnO NPs have been approved for plastic-based food packaging applications. TiN was authorized as an additive up to 20 mg/kg in the manufacture of polyethylene terephthalate (PET) bottles [168], and the specific migration limit for ZnO NPs decreased to 5 mg/kg food or food simulant express as zinc, according to the EFSA Regulation (EU) No 2016/1416 [167]. Materials such as aluminum oxide and TiO_2 have been included in the list of approved additives for food contact materials, but not in NPs form [16]. Additionally, the specific migration limit of Ag into foods from a material containing Ag is 0.05 mg/kg food or food stimulant [169]. It has been estimated that adults may ingest between 20 and 80 μg/day of Ag, but only a fraction of this being in NPs form [13].

7.7 Environmental Impact of Active Food Packaging Materials

7.7.1 Biodegradability

The biodegradability issue is a big challenge in food packages applications. Thus, researchers are still working on a combination of polymers of biological origin and metallic NPs, in order to improve their properties while remaining biodegradable.

Biodegradation is a reversible process that converts the polymer material into CO_2, H_2O, methane, biomass, and an inorganic compound. It is fragmentation and chemical reactions through microorganisms' activity that cause a loss of mechanical properties [170]. In the last years, biodegradable polymers have attracted huge intention due to their environmentally friendly alternative character. In particular, poly(lactic) acid (PLA) is one of the most known as the basis of many investigations because of its mechanical, biodegradability, and biocompatibility properties [171]. Ramos et al. made a study on the evaluation of the influence of Ag NPs on the degradation of PLA-based nanocomposites. The results demonstrated that the addition of Ag NPs accelerates the rate of degradation, which makes this nanocomposite material environmentally friendly for active food packaging with good antioxidant and antimicrobial properties. The disintegration of the material has been explained by the increase of the diffusion process of water molecules through its nanostructure that promoted hydrolysis [171]. In a study on the biodegradability of the hybrid nanocomposites of starch/lysine and ZnO NPs, it has been demonstrated that its solubility was decreased with increasing ZnO NPs content [172]. The biodegradability of PLA/$CaCO_3$ NPs was enhanced with increasing NPs and demonstrated a high loss of weight [102]. Besides, TiO_2 NPs incorporated into the synthetic plastic matrix are used to increase biodegradability [173].

On the other hand, the environmental fate and effect of metal and metal oxide NPs in the ecosystem are slight identified. Part et al., for example, defined the probable destiny of this kind of NPs (primarily Ag and TiO_2) in waste, suggesting that the degradation procedures are not affected by NPs, but the communities of microbes found there could be affected by their presence [174], since metallic NPs are not biodegradable at the end of their life.

7.7.2 Recyclability

Recycling must have real environmental benefits. Therefore, some packaging waste could be recycled and reused as secondary raw material in order to eliminate a part of the waste and to save raw materials. That is to say, the transformation of recovered packaging into a secondary material must not pollute or consume more energy than it provides. In an investigation paper, the authors have made a study on the extrusion process of nanomaterials including $CaCO_3$, Ag, and ZnO NPs as part of different pure polymer matrices of PE, PP, and PET. They concluded that the recycled materials exhibited a visible change of color in all samples along with slight changes in brightness in nano-reinforced PET and yellowness in nano-reinforced PE, PP, and in PET–Ag. It is interesting to notice that after the recycling process the quality and the mechanical properties of the final recycled nanomaterials have not been affected. Moreover, PET results demonstrated that the increase of additional nanomaterial into the recycled PET matrix (especially PET–Ag) could have a significant influence on the final material, because of a slight degradation of the polymer. The authors suggested that this change on the recycled nanomaterial could be required for specific applications [175].

7.7.3 Life Cycle Assessment

The environmental performance of a product can be measured by its life cycle assessment (LCA) through a study on its environmental impact during its lifetime, using a methodology regulated by ISO 14040 International Series of Standards [176]. This later allows quantification of the amounts of gas emissions, energy consumed, and hazardous substances emitted by a product during its life cycle. Defining the complete life cycle of food packaging is one of the most important aims that the researchers work on. Silvestre et al. described the steps of the life cycle from the extraction as raw material, going by safety and eco-friendly through the production, the use and reuse/recycle, and ending by disposal [177]. The United Nations Environment Program (UNEP, 2004) believes that the life cycle approach covers the opportunities and risks associated with a product or technology along the entire life chain. Meanwhile, it is mandatory to take into account the reduction of food losses during the LCA [178]. In an article published by Nowack et al., it is demonstrated that nano-Ag incorporated into plastics are preferable to the organic counterparts as antibacterial agents, since the organic matters may not survive the plastic because of high temperatures used in the manufacturing process [179]. Zhang et al. carried out an LCA on metallic NPs as an active agent for food packaging film. Ag and TiO_2 NPs and a combination of them were used with PLA as a host material. The results indicated that the packaging film that utilized TiO_2 NPs had lower environmental impacts in global warming potential and nonrenewable fossil energy consumption, which was around 65% of the film with Ag NPs. It was seen in the case of both NPs that the environmental impacts have decreased by 2–5% compared with TiO_2 NPs films [180]. Likewise, the authors have mentioned the lack of transparency in the data inventory of nanomaterials given by the producers, which produces many obstacles to make more LCA studies [180].

7.8 Conclusions and Future Trends

Recently, the use of metal and metal oxide NPs has gained significant attention in food packaging applications, thanks to their interesting physicochemical and functional properties. It has been shown that pure metal NPs such as zinc, silver, copper, and gold, or a combination of them, as well as metal oxides including zinc oxide, copper oxide, and titanium dioxide, provide to packaging materials multifunctional properties among which can highlight the antimicrobial, photocatalyst, and oxygen scavenging characteristics (typical features of active systems). These types of NPs also possess the capability to enhance the mechanical, optical, and barrier characteristics of the packaging. Therefore, these novel materials demonstrated to be suitable for preserving and increasing the shelf life of food.

However, the use of nanotechnology in the food packaging industry requires much more researcher to overcome the barriers related to the regulations, food safety, and environmental impact. Besides, the standardization and technology development to produce these nanomaterials at the industry level and, thus, boost their use in the global market is another issue that needs to be worked on. These studies should be performed based on collaborations between government regulatory agencies, food and food packaging industries, consumers, and multidisciplinary experts in order to achieve the desired safety and effectiveness to the active packaging systems based on metal and metal oxide NPs.

Acknowledgments

This work was supported by the Portuguese Foundation for Science and Technology (FCT) in the framework of the project NANOXYPACK cofinanced via FEDER (PT2020) POCI-01-0145-FEDER-030789.

References

1 Ballesteros, L.F., Michelin, M., Vicente, A.A. et al. (2018). *Lignocellulosic Materials and Their Use in Bio- Based Packaging*. Springer.

2 Ahvenainen, R. (2003). *Novel Food Packaging Techniques* (ed. R. Ahvenainen), 23–39. Woodhead Publishing Limited/CRC Press LLC.

3 Souza, V.G.L. and Fernando, A.L. (2016). Nanoparticles in food packaging: biodegradability and potential migration to food - a review. *Food Packag. Shelf Life* 8: 63–70. https://doi.org/10.1016/j.fpsl.2016.04.001.

4 Consultation, A.P. (2009). Guidelines on submission of a dossier for safety evaluation by the EFSA of active or intelligent substances present in active and intelligent materials and articles intended to come into contact with food. *EFSA J.* 7 https://doi.org/10.2903/j.efsa.2009.1208.

5 Andrady, A.L. and Neal, M.A. (2009). Applications and societal benefits of plastics. *Philos. Trans. R. Soc. London, Ser. B* 364 (1526): 1977–1984.

6 Wyser, Y., Adams, M., Avella, M. et al. (2016). Outlook and challenges of nanotechnologies for food Packaging. *Packag. Technol. Sci.* 29: 615–648.

7 Störmer, A., Bott, J., Kemmer, D., and Franz, R. (2017). Critical review of the migration potential of nanoparticles in food contact plastics. *Trends Food Sci. Technol.* 63: 39–50.

8 Sharma, R., Jafari, S.M., and Sharma, S. (2020). Antimicrobial bio-nanocomposites and their potential applications in food packaging. *Food Control* 112 (January): 107086. https://doi.org/10.1016/j.foodcont.2020.107086.

9 Zare, M., Namratha, K., Ilyas, S. et al. (2019). Smart fortified PHBV-CS biopolymer with ZnO-Ag nanocomposites for enhanced shelf life of food Packaging. *ACS Appl. Mater. Interfaces* 11 (51): 48309–48320.

10 Li, W., Li, L., Zhang, H. et al. (2018). Evaluation of PLA nanocomposite films on physicochemical and microbiological properties of refrigerated cottage cheese. *J. Food Process. Preserv.* 42 (1): 1–9.

11 Al-Tayyar, N.A., Youssef, A.M., and Al-hindi, R. (2020). Antimicrobial food packaging based on sustainable bio-based materials for reducing foodborne pathogens: a review. *Food Chem.* 310 (November 2019): 125915. https://doi.org/10.1016/j.foodchem.2019.125915.

12 Basavegowda, N., Mandal, T.K., and Baek, K.H. (2020). Bimetallic and trimetallic nanoparticles for active food Packaging applications: a review. *Food Bioprocess Technol.* 13 (1): 30–44.

13 Enescu, D., Cerqueira, M.A., Fucinos, P., and Pastrana, L.M. (2019). Recent advances and challenges on applications of nanotechnology in food packaging. A literature review. *Food Chem. Toxicol.* 134 (May): 110814. https://doi.org/10.1016/j.fct.2019.110814.

14 Costa, C., Conte, A., Alessandro, M., and Nobile, D. (2016). Use of metal nanoparticles for active Packaging applications. *Antimicrob. Food Packag.*: 399–406. https://doi.org/10.1016/B978-0-12-800723-5.00031-0.

15 de Azeredo, H.M.C., Otoni, C.G., Assis, O.B.G. et al. (2018). Nanoparticles and antimicrobial food Packaging. *Ref. Module Food Sci.*: 1–13.

16 Garcia, C.V., Shin, G.H., and Kim, J.T. (2018). Metal oxide-based nanocomposites in food packaging: applications, migration, and regulations. *Trends Food Sci. Technol.* 82 (September): 21–31. https://doi.org/10.1016/j.tifs.2018.09.021.

17 Bi, Y., Westerband, E.I., Alum, A. et al. (2018). Antimicrobial efficacy and life cycle impact of silver-containing food containers. *ACS Sustainable Chem. Eng.* 6 (10): 13086–13095.

18 Li, W., Wang, J., Chi, H. et al. (2012). Preparation and antibacterial activity of polyvinyl alcohol/regenerated silk fibroin composite fibers containing Ag nanoparticles. *J. Appl. Polym. Sci.* 123 (1): 20–25.

19 Bott, J., Störmer, A., and Franz, R. (2014). A comprehensive study into the migration potential of nano silver particles from food contact polyolefins. *ACS Symp. Ser.* 1159: 51–70.

20 Azeredo, H.M.C., Otoni, C.G., Corrêa, D.S. et al. (2019). Nanostructured antimicrobials in food Packaging—recent advances. *Biotechnol. J.* 14 (12): 1–9.

21 Stoimenov, P.K., Klinger, R.L., Marchin, G.L., and Klabunde, K.J. (2002). Metal oxide nanoparticles as bactericidal agents. *Langmuir* 18 (17): 6679–6686.

22 Reig, C.S., Lopez, A.D., Ramos, M.H., and Ballester, V.A.C. (2014). Nanomaterials: a map for their selection in food Packaging applications. *Packag. Technol. Sci.* 27: 839–866.

23 Pietroiusti, A., Magrini, A., and Campagnolo, L. (2016). New frontiers in nanotoxicology: gut microbiota/microbiome-mediated effects of engineered nanomaterials. *Toxicol. Appl. Pharmacol.* 299: 90–95.

24 Calderon, V.S., Gomes, B., Ferreira, P.J., and Carvalho, S. (2017). Zinc nanostructures for oxygen scavenging. *Nanoscale* 9 (16): 5254–5262.

25 Sirelkhatim, A., Mahmud, S., Seeni, A. et al. (2015). Review on zinc oxide nanoparticles: antibacterial activity and toxicity mechanism. *Nano-Micro Lett.* 7 (3): 219–242.

26 EFSA – European Food Safety Authority (2017). Safety assessment of the substance zinc oxide, nanoparticles, for use in food contact materials. *EFSA J.* 14 (3): 4408.

27 Shankar, S. and Rhim, J.W. (2014). Effect of copper salts and reducing agents on characteristics and antimicrobial activity of copper nanoparticles. *Mater. Lett.* 132: 307–311.

28 Guès-Bergeron, A. (2014). Nanotechnologies, alimentation et cosmétiques: le cas du dioxyde de titane et du dioxyde de titane nanométrique et la part du Québec dans cette industrie.

29 Feng, X.X., Zhang, L.L., Chen, J.Y. et al. (2007). Preparation and characterization of novel nanocomposite films formed from silk fibroin and nano-TiO_2. *Int. J. Biol. Macromol.* 40 (2): 105–111.

30 Peiró, A.M., Colombo, C., Doyle, G. et al. (2006). Photochemical reduction of oxygen adsorbed to nanocrystalline TiO_2 films: a transient absorption and oxygen scavenging study of different TiO_2 preparations. *J. Phys. Chem. B* 110 (46): 23255–23263.

31 De Lima, R., Feitosa, L., do Pereira, A., E.S. et al. (2010). Evaluation of the genotoxicity of chitosan nanoparticles for use in food packaging films. *J. Food Sci.* 75 (6): 89–96.

32 Melo, P.T.S., Nunes, J.C., Otoni, C.G. et al. (2019). Combining Cupuassu (*Theobroma grandiflorum*) puree, pectin, and chitosan nanoparticles into novel edible films for food packaging applications. *J. Food Sci.* 84 (8): 2228–2233.

33 Nakazato, G., Kobayashi, R.K.T., Seabra, A.B., and Duran, N. (2017). Use of nanoparticles as a potential antimicrobial for food packaging. In: *Food Preservation* (ed. A.M. Grumezescu), 413–447. Elsevier Inc.

34 Zawadzka, K., Kisielewska, A., Piwonski, I. et al. (2016). Mechanisms of antibacterial activity and stability of silver nanoparticles grow non magnetron sputtered TiO_2 coatings. *Bull. Mater. Sci.* 39 (1): 57–68.

35 Longano, D., Ditaranto, N., Cioffi, N. et al. (2012). Analytical characterization of laser-generated copper nanoparticles for antibacterial composite food packaging. *Anal. Bioanal.Chem.* 403 (4): 1179–1186.

36 Rhim, J.W., Wang, L.F., Lee, Y., and Hong, S.I. (2014). Preparation and characterization of bio-nanocomposite films of agar and silver nanoparticles: laser ablation method. *Carbohydr. Polym.* 103 (1): 456–465.

37 Ratti, M., Naddeo, J.J., Tan, Y. et al. (2016). Irradiation with visible light enhances the antibacterial toxicity of silver nanoparticles produced by laser ablation. *Appl. Phys. A Mater. Sci. Process.* 122 (4): 1–7.

38 Picca, R.A., Di Maria, A., Riháková, L. et al. (2016). Laser ablation synthesis of hybrid copper/silver Nanocolloids for prospective application as Nanoantimicrobial agents for food packaging. *MRS Adv.* 1 (56): 3735–3740.

39 Abbasi, E., Milani, M., Aval, S.F. et al. (2016). Critical reviews in microbiology silver nanoparticles: synthesis methods , bio- applications and properties. *Crit. Rev. Microbiol.* 42 (2): 173–180.

40 Jung, J.H., Oh, H.C., Noh, H.S. et al. (2006). Metal nanoparticle generation using a small ceramic heater with a local heating area. *J. Aerosol Sci.* 37 (12): 1662–1670.

41 Mirzaei, H. and Darroudi, M. (2017). Zinc oxide nanoparticles: biological synthesis and biomedical applications. *Ceram. Int.* 43 (1): 907–914.

42 Sergeev, G.B. (2003). Cryochemistry of metal nanoparticles. *J. Nanopart. Res.* 5 (5–6): 529–537.

43 Espitia, P.J.P., de Soares, N., F.F., dos Coimbra, J.S., R. et al. (2012). Zinc oxide nanoparticles: synthesis, antimicrobial activity and food packaging applications. *Food Bioprocess Technol.* 5 (5): 1447–1464.

44 Aghababazadeh, R., Mazinani, B., Mirhabibi, A., and Tamizifar, M. (2006). ZnO nanoparticles synthesised by mechanochemical processing. *J. Phys. Conf. Ser.* 26: 312.

45 Alwan, R.M., Kadhim, Q.A., Sahan, K.M. et al. (2015). Synthesis of zinc oxide nanoparticles via sol – gel route and their characterization. *Nanosci. Nanotechnol.* 5 (1): 1–6.

46 Raliya, R. and Tarafdar, J.C. (2014). Biosynthesis and characterization of zinc, magnesium and titanium nanoparticles: an eco-friendly approach. *Int. Nano Lett.* 4 (1): 3–10.

47 Kowsalya, E., Mosachristas, K., Balashanmugam, P. et al. (2019). Biocompatible silver nanoparticles/poly(vinyl alcohol) electrospun nano fi bers for potential antimicrobial food packaging applications. *Food Packag. Shelf Life* 21 (August): 100379.

48 Mohseni, M.S., Khalilzadeh, M.A., Mohseni, M. et al. (2020). Green synthesis of Ag nanoparticles from pomegranate seeds extract and synthesis of Ag-starch nanocomposite and characterization of mechanical properties of the films. *Biocatal. Agric. Biotechnol.* 25 (March): 101569. https://doi.org/10.1016/j.bcab.2020.101569.

49 Reščeк, A., Katančić, Z., Kratofil Krehula, L. et al. (2018). Development of double-layered PE/PCL films for food Packaging modified with zeolite and magnetite nanoparticles. *Adv. Polym. Tech.* 37 (3): 837–842.

50 Carbone, M., Donia, D.T., Sabbatella, G., and Antiochia, R. (2016). Silver nanoparticles in polymeric matrices for fresh food packaging. *J. King Saud Univ. Eng. Sci.* 28 (4): 273–279. https://doi.org/10.1016/j.jksus.2016.05.004.

51 Dobrucka, R. and Ankiel, M. (2019). Possible applications of metal nanoparticles in antimicrobial food packaging. *J. Food Saf.* 39 (2): 1–7.

52 Ng, L.Y., Mohammad, A.W., Leo, C.P., and Hilal, N. (2013). Polymeric membranes incorporated with metal/metal oxide nanoparticles: a comprehensive review. *Desalination* 308: 15–33. https://doi.org/10.1016/j.desal.2010.11.033.

53 Hosseini Nasab, N., Jalili, M.M., and Farrokhpay, S. (2018). Application of paraffin and silver coated titania nanoparticles in polyethylene nanocomposite food packaging films. *J. Appl. Polym. Sci.* 135 (8): 1–7.

54 Bumbudsanpharoke, N., Lee, W., and Ko, S. (2018). A comprehensive feasibility study on the properties of LDPE-Ag nanocomposites for food packaging applications. *Polym. Compos.* 39 (9): 3178–3186.

55 Ahmed, J., Arfat, Y.A., Bher, A. et al. (2018). Active chicken meat packaging based on polylactide films and bimetallic Ag–Cu nanoparticles and essential oil. *J. Food Sci.* 83 (5): 1299–1310.

56 Marcous, A., Rasouli, S., and Ardestani, F. (2017). Low-density polyethylene films loaded by titanium dioxide and zinc oxide nanoparticles as a new active Packaging system against *Escherichia coli* O157:H7 in fresh calf minced meat. *Packag. Technol. Sci.* 30: 693–701.

57 Bodaghi, H., Mostofi, Y., Oromiehie, A. et al. (2015). Synthesis of clay-TiO_2 nanocomposite thin films with barrier and photocatalytic properties for food packaging application. *J. Appl. Polym. Sci.* 132 (14): 4–11.

58 Cushen, M., Kerry, J., Morris, M. et al. (2014). Evaluation and simulation of silver and copper nanoparticle migration from polyethylene nanocomposites to food and an associated exposure assessment. *J. Agric. Food. Chem.* 62 (6): 1403–1411.

59 Castro Mayorga, J.L., Fabra Rovira, M.J., Cabedo Mas, L. et al. (2018). Antimicrobial nanocomposites and electrospun coatings based on poly(3-hydroxybutyrate-*co*-3-hydroxyvalerate) and copper oxide nanoparticles for active packaging and coating applications. *J. Appl. Polym. Sci.* 135 (2): 1–11.

60 Cerqueira, M.A.P.R., Pereira, R.N.C., da Silva Ramos, O.L. et al. (2017). *Edible Food Packaging: Materials and Processing Technologies*, Contemporary Food Engineering. CRC Press.

61 Keshk, S.M.A.S., El-Zahhar, A.A., Haija, M.A., and Bondock, S. (2019). Synthesis of a magnetic nanoparticles/dialdehyde starch-based composite film for food Packaging. *Starch/Staerke* 71 (1–2): 1–7.

62 Mohammadpour Velni, S., Baniasadi, H., and Vaziri, A. (2018). Fabrication and characterization of antimicrobial starch-based nanocomposite films and modeling the process parameters via the RSM. *Polym. Compos.* 39 (Ccd): E584–E591.

63 Batra, M., Malik, G.K., and Mitra, J. (2020). Enhancing the properties of gelatin–chitosan bionanocomposite films by incorporation of silica nanoparticles. *J. Food Process Eng* 43 (2): 1–12.

64 Nair, S.B., Alummoottil, N.J., and Moothandasserry, S. (2017). Chitosan-konjac glucomannan-cassava starch-nanosilver composite films with moisture resistant and antimicrobial properties for food-packaging applications. *Starch/Staerke* 69 (1–2): 1–12.

65 Sahraee, S., Milani, J.M., Ghanbarzadeh, B., and Hamishehkar, H. (2020). Development of emulsion films based on bovine gelatin-nano chitin-nano ZnO for cake packaging. *Food Sci. Nutr.* 8 (2): 1303–1312.

66 Venkatesan, R. and Rajeswari, N. (2017). TiO2 nanoparticles/poly(butylene adipate-co-terephthalate) bionanocomposite films for packaging applications. *Polym. Adv. Technol.* 28 (12): 1699–1706.

67 Chowdhury, S., Teoh, Y.L., Ong, K.M. et al. (2020). Poly(vinyl) alcohol crosslinked composite packaging film containing gold nanoparticles on shelf life extension of banana. *Food Packag. Shelf Life* 24 (November 2019): 100463. https://doi.org/10.1016/j.fpsl.2020.100463.

68 Eskandarabadi, S.M., Mahmoudian, M., Farah, K.R. et al. (2019). Active intelligent packaging film based on ethylene vinyl acetate nanocomposite containing extracted anthocyanin, rosemary extract and ZnO/Fe-MMT nanoparticles. *Food Packag. Shelf Life* 22 (September 2018): 100389. https://doi.org/10.1016/j.fpsl.2019.100389.

69 Martínez-Abad, A., Lagaron, J.M., and Ocio, M.J. (2012). Development and characterization of silver-based antimicrobial ethylene-vinyl alcohol copolymer (EVOH) films for food-packaging applications. *J. Agric. Food. Chem.* 60 (21): 5350–5359.

70 Kowsalya, E., MosaChristas, K., Balashanmugam, P. et al. (2019). Biocompatible silver nanoparticles/poly(vinyl alcohol) electrospun nanofibers for potential antimicrobial food packaging applications. *Food Packag. Shelf Life* 21 (September 2018): 100379. https://doi.org/10.1016/j.fpsl.2019.100379.

71 Liu, Y., Li, Y., Deng, L. et al. (2018). Hydrophobic ethylcellulose/gelatin nanofibers containing zinc oxide nanoparticles for antimicrobial packaging. *J. Agric. Food. Chem.* 66: 9498–9506.

72 Pirsa, S., Shamusi, T., and Kia, E.M. (2018). Smart films based on bacterial cellulose nanofibers modified by conductive polypyrrole and zinc oxide nanoparticles. *J. Appl. Polym. Sci.* 135 (34): 1–10.

73 Díez-Pascual, A.M. and Díez-Vicente, A.L. (2014). ZnO-reinforced poly(3-hydroxybutyrate-co-3-hydroxyvalerate) bionanocomposites with antimicrobial function for food packaging. *ACS Appl. Mater. Interfaces* 6 (12): 9822–9834.

74 Khalaj, M.J., Ahmadi, H., Lesankhosh, R., and Khalaj, G. (2016). Study of physical and mechanical properties of polypropylene nanocomposites for food packaging application: nano-clay modified with iron nanoparticles. *Trends Food Sci. Technol.* 51: 41–48. https://doi.org/10.1016/j.tifs.2016.03.007.

75 Tankhiwale, R. and Bajpai, S.K. (2010). Silver-nanoparticle-loaded chitosan lactate films with fair antibacterial properties. *J. Appl. Polym. Sci.* 115 (5): 1894–1900.

76 Bayani Bandpey, N., Aroujalian, A., Raisi, A., and Fazel, S. (2017). Surface coating of silver nanoparticles on polyethylene for fabrication of antimicrobial milk packaging films. *Int. J. Dairy Technol.* 70 (2): 204–211.

77 Šupová, M., Martynková, G.S., and Barabaszová, K. (2011). Effect of Nanofillers dispersion in polymer matrices: a review. *Sci. Adv. Mater.* 3 (1): 1–25.

78 Han, J.W., Ruiz-Garcia, L., Qian, J.P., and Yang, X.T. (2018). Food packaging: a comprehensive review and future trends. *Compr. Rev. Food Sci. Food Saf.* 17 (4): 860–877.

79 Hoseinnejad, M., Jafari, S.M., and Katouzian, I. (2018). Inorganic and metal nanoparticles and their antimicrobial activity in food packaging applications. *Crit. Rev. Microbiol.* 44 (2): 161–181.

80 Swaroop, C. and Shukla, M. (2018). Nano-magnesium oxide reinforced polylactic acid bio films for food packaging applications. *Int. J. Biol. Macromol.* 113: 729–736.

81 Gharoy Ahangar, E., Abbaspour-Fard, M.H., Shahtahmassebi, N. et al. (2015). Preparation and characterization of PVA/ZnO nanocomposite. *J. Food Process. Preserv.* 39 (6): 1442–1451.

82 Lepot, N., Van Bael, M.K., Van den Rul, H. et al. (2010). Influence of incorporation of ZnO nanoparticles and biaxial orientation on mechanical and oxygen barrier properties of polypropylene films for food Packaging applications. *J. Appl. Polym. Sci.* 116 (5): 2658–2667.

83 Venkatesan, R. and Rajeswari, N. (2017). ZnO/PBAT nanocomposite films: investigation on the mechanical and biological activity for food packaging. *Polym. Adv. Technol.* 28 (1): 20–27.

84 Liu, Q., Zhang, M., Fang, Z. xiang, and Rong, X. hong (2014). Effects of ZnO nanoparticles and microwave heating on the sterilization and product quality of vacuum-packaged Caixin. *J. Sci. Food Agric.* 94 (12): 2547–2554.

85 Ebrahimi, Y., Peighambardoust, S.J., Peighambardoust, S.H., and Karkaj, S.Z. (2019). Development of antibacterial Carboxymethyl cellulose-based nanobiocomposite films containing various metallic nanoparticles for food Packaging applications. *J. Food Sci.* 84 (9): 2537–2548.

86 Dehnavi, A.S., Aroujalian, A., Raisi, A., and Fazel, S. (2013). Preparation and characterization of polyethylene/silver nanocomposite films with antibacterial activity. *J. Appl. Polym. Sci.* 127 (2): 1180–1190.

87 Bumbudsanpharoke, N. and Ko, S. (2015). Nano-food packaging: an overview of market, migration research, and safety regulations. *J. Food Sci.* 80 (5): R910–R923.

88 Sardella, D., Gatt, R., and Valdramidis, V.P. (2019). Metal nanoparticles for controlling fungal proliferation: quantitative analysis and applications. *Curr. Opin. Food Sci.* 30: 49–59. https://doi.org/10.1016/j.cofs.2018.12.001.

89 Vejdan, A., Ojagh, S.M., and Abdollahi, M. (2017). Effect of gelatin/agar bilayer film incorporated with TiO_2 nanoparticles as a UV absorbent on fish oil photooxidation. *Int. J. Food Sci. Technol.* 52 (8): 1862–1868.

90 Foltynowicz, Z., Bardenshtein, A., Sängerlaub, S. et al. (2017). Nanoscale, zero valent iron particles for application as oxygen scavenger in food packaging. *Food Packag. Shelf Life* 11: 74–83.

91 Yemmireddy, V.K. and Hung, Y.C. (2017). Using photocatalyst metal oxides as antimicrobial surface coatings to ensure food safety—opportunities and challenges. *Compr. Rev. Food Sci. Food Saf.* 16 (4): 617–631.

92 Hannon, J.C., Kerry, J.P., Cruz-Romero, M. et al. (2016). Human exposure assessment of silver and copper migrating from an antimicrobial nanocoated packaging material into an acidic food simulant. *Food Chem. Toxicol.* 95: 128–136. https://doi.org/10.1016/j.fct.2016.07.004.

93 Bi, F., Zhang, X., Liu, J. et al. (2020). Development of antioxidant and antimicrobial packaging films based on chitosan, D-α-tocopheryl polyethylene glycol 1000 succinate and silicon dioxide nanoparticles. *Food Packag. Shelf Life* 24 (March): 100503. https://doi.org/10.1016/j.fpsl.2020.100503.

94 Polat, S., Fenercioğlu, H., and Güçlü, M. (2018). Effects of metal nanoparticles on the physical and migration properties of low density polyethylene films. *J. Food Eng.* 229: 32–42.

95 Ana MD, Angel LD. ZnO-reinforced poly(3-hydroxybutyrate-co-3-hydroxyvalerate) bionanocomposites with antimicrobial function for food packaging. 2014;

96 Emamifar, A., Kadivar, M., Shahedi, M., and Solimanian-Zad, S. (2012). Effect of nanocomposite packaging containing Ag and ZnO on reducing pasteurization temperature of orange juice. *J. Food Process. Preserv.* 36 (2): 104–112.

97 Stará, H., Staraý, Z., and Münstedt, H. (2011). Silver nanoparticles in blends of polyethylene and a superabsorbent polymer: morphology and silver ion release. *Macromol. Mater. Eng.* 296 (5): 423–427.

98 Beak, S., Kim, H., and Song, K.B. (2017). Characterization of an olive flounder bone gelatin-zinc oxide Nanocomposite film and evaluation of its potential application in spinach packaging. *J. Food Sci.* 82 (11): 2643–2649.

99 Kuorwel, K.K., Cran, M.J., Orbell, J.D. et al. (2015). Review of mechanical properties, migration, and potential applications in active food Packaging systems containing nanoclays and nanosilver. *Compr. Rev. Food Sci. Food Saf.* 14 (4): 411–430.

100 Bhunia, K., Dhawan, S., and Sablani, S.S. (2012). Modeling the oxygen diffusion of nanocomposite-based food Packaging films. *J. Food Sci.* 77 (7): 29–38.

101 Zhao, L., Li, F., Chen, G. et al. (2012). Effect of nanocomposite-based packaging on preservation quality of green tea. *Int. J. Food Sci. Technol.* 47 (3): 572–578.

102 Aframehr, W.M., Molki, B., Heidarian, P. et al. (2017). Effect of calcium carbonate nanoparticles on barrier properties and biodegradability of polylactic acid. *Fibers Polym.* 18 (11): 2041–2048.

103 Dadbin, S., Noferesti, M., and Frounchi, M. (2008). Oxygen barrier LDPE/LLDPE/organoclay nano-composite films for food packaging. *Macromol. Symp.* 274 (1): 22–27.

104 Zhang, J., Cao, C., Zheng, S. et al. (2020). Poly(butylene adipate-co-terephthalate)/magnesium oxide/silver ternary composite biofilms for food packaging application. *Food Packag. Shelf Life* 24 (December 2019): 100487. https://doi.org/10.1016/j.fpsl.2020.100487.

105 Sungkaworn, T., Triampo, W., Nalakarn, P. et al. (2007). The effects of TiO_2 nanoparticles on tumor cell colonies: fractal dimension and morphological properties. *Int. J. Biomed. Sci.* 2 (1): 67–74.

106 Tang, Z.X. and Lv, B.F. (2014). MgO nanoparticles as antibacterial agent: preparation and activity. *Braz. J. Chem. Eng.* 31 (3): 591–601.

107 Gan, I. and Chow, W.S. (2018). Antimicrobial poly(lactic acid)/cellulose bionanocomposite for food packaging application: a review. *Food Packag. Shelf Life* 17 (June): 150–161. https://doi.org/10.1016/j.fpsl.2018.06.012.

108 Shahmohammadi Jebel, F. and Almasi, H. (2016). Morphological, physical, antimicrobial and release properties of ZnO nanoparticles-loaded bacterial cellulose films. *Carbohydr. Polym.* 149: 8–19. https://doi.org/10.1016/j.carbpol.2016.04.089.

109 Dhapte, V., Gaikwad, N., More, P.V. et al. (2015). Transparent ZnO/polycarbonate nanocomposite for food packaging application. *Nanocomposites* 1 (2): 106–112.

110 Esmailzadeh, H., Sangpour, P., Shahraz, F. et al. (2016). Effect of nanocomposite packaging containing ZnO on growth of *Bacillus subtilis* and *Enterobacter aerogenes*. *Mater. Sci. Eng., C* 58: 1058–1063. https://doi.org/10.1016/j.msec.2015.09.078.

111 Meshram, J.V., Koli, V.B., Phadatare, M.R., and Pawar, S.H. (2017). Anti-microbial surfaces: An approach for deposition of ZnO nanoparticles on PVA-gelatin composite film by screen printing technique. *Mater. Sci. Eng., C* 73: 257–266. https://doi.org/10.1016/j.msec.2016.12.043.

112 Al-Naamani, L., Dobretsov, S., and Dutta, J. (2016). Chitosan-zinc oxide nanoparticle composite coating for active food packaging applications. *Innovative Food Sci. Emerg. Technol.* 38: 231–237. https://doi.org/10.1016/j.ifset.2016.10.010.

113 Abreu, A.S., Oliveira, M., De Sá, A. et al. (2015). Antimicrobial nanostructured starch based films for packaging. *Carbohydr. Polym.* 129: 127–134. https://doi.org/10.1016/j.carbpol.2015.04.021.

114 De Moura, M.R., Mattoso, L.H.C., and Zucolotto, V. (2012). Development of cellulose-based bactericidal nanocomposites containing silver nanoparticles and their use as active food packaging. *J. Food Eng.* 109 (3): 520–524.

115 Mohamed, R.R. and Sabaa, M.W. (2014). Synthesis and characterization of antimicrobial crosslinked carboxymethyl chitosan nanoparticles loaded with silver. *Int. J. Biol. Macromol.* 69: 95–99. https://doi.org/10.1016/j.ijbiomac.2014.05.025.

116 Jokar, M., Abdul Rahman, R., Ibrahim, N.A. et al. (2012). Melt production and antimicrobial efficiency of low-density polyethylene (LDPE)-silver nanocomposite film. *Food Bioprocess Technol.* 5 (2): 719–728.

117 Zhang, X., Xiao, G., Wang, Y. et al. (2017). Preparation of chitosan-TiO_2 composite film with efficient antimicrobial activities under visible light for food packaging applications. *Carbohydr. Polym.* 169: 101–107. https://doi.org/10.1016/j.carbpol.2017.03.073.

118 Lian, Z., Zhang, Y., and Zhao, Y. (2016). Nano-TiO_2 particles and high hydrostatic pressure treatment for improving functionality of polyvinyl alcohol and chitosan composite films and nano-TiO_2 migration from film matrix in food simulants. *Innovation Food Sci. Emerging Technol.* 33: 145–153. https://doi.org/10.1016/j.ifset.2015.10.008.

119 Shankar, S., Wang, L.F., and Rhim, J.W. (2017). Preparation and properties of carbohydrate-based composite films incorporated with CuO nanoparticles. *Carbohydr. Polym.* 169: 264–271. https://doi.org/10.1016/j.carbpol.2017.04.025.

120 Peighambardoust, S.J., Peighambardoust, S.H., Mohammadzadeh Pournasir, N., and Pakdel, P. (2019). Properties of active starch-based films incorporating a combination of Ag, ZnO and CuO nanoparticles for potential use in food packaging applications. *Food Packag. Shelf Life* 22 (December 2018): 100420. https://doi.org/10.1016/j.fpsl.2019.100420.

121 Shankar, S. and Rhim, J.W. (2017). Facile approach for large-scale production of metal and metal oxide nanoparticles and preparation of antibacterial cotton pads. *Carbohydr. Polym.* 163: 137–145. https://doi.org/10.1016/j.carbpol.2017.01.059.

122 Dallas, P., Sharma, V.K., and Zboril, R. (2011). Silver polymeric nanocomposites as advanced antimicrobial agents: classification, synthetic paths, applications, and perspectives. *Adv. Colloid Interface Sci.* 166 (1–2): 119–135. https://doi.org/10.1016/j.cis.2011.05.008.

123 Morones, J.R., Elechiguerra, J.L., Camacho, A. et al. (2005). The bactericidal effect of silver nanoparticles. *Nanotechnology* 16 (10): 2346–2353.

124 Teodoro, J.S., Simões, A.M., Duarte, F.V. et al. (2011). Assessment of the toxicity of silver nanoparticles in vitro: a mitochondrial perspective. *Toxicol. in Vitro* 25 (3): 664–670.

125 Le, A.T., Huy, P.T., Tam, P.D. et al. (2010). Green synthesis of finely-dispersed highly bactericidal silver nanoparticles via modified Tollens technique. *Curr. Appl Phys.* 10 (3): 910–916.

126 Shrivastava, S., Bera, T., Singh, S.K. et al. (2009). Characterization of antiplatelet properties of silver nanoparticles. *ACS Nano* 3 (6): 1357–1364.

127 Siqueira, M.C., Coelho, G.F., de Moura, M.R. et al. (2014). Evaluation of antimicrobial activity of silver nanoparticles for carboxymethylcellulose film applications in food packaging. *J. Nanosci. Nanotechnol.* 14 (7): 5512–5517.

128 Azlin-Hasim, S., Cruz-Romero, M.C., Cummins, E. et al. (2016). The potential use of a layer-by-layer strategy to develop LDPE antimicrobial films coated with silver nanoparticles for packaging applications. *J. Colloid Interface Sci.* 461: 239–248.

129 Mohammed Fayaz, A., Balaji, K., Girilal, M. et al. (2009). Mycobased synthesis of silver nanoparticles and their incorporation into sodium alginate films for vegetable and fruit preservation. *J. Agric. Food. Chem.* 57 (14): 6246–6252.

130 Becaro, A.A., Puti, F.C., Panosso, A.R. et al. (2016). Postharvest quality of fresh-cut carrots packaged in plastic films containing silver nanoparticles. *Food Bioprocess Technol.* 9 (4): 637–649.

131 Appierot, G., Lipovsky, A., Dror, R. et al. (2009). Enhanced antibacterial activity of nanocrystalline ZnO due to increased ROS-mediated cell injury. *Adv. Funct. Mater.* 19 (6): 842–852.

132 Indumathi, M.P., Saral Sarojini, K., and Rajarajeswari, G.R. (2019). Antimicrobial and biodegradable chitosan/cellulose acetate phthalate/ZnO nano composite films with optimal oxygen permeability and hydrophobicity for extending the shelf life of black grape fruits. *Int. J. Biol. Macromol.* 132: 1112–1120. https://doi.org/10.1016/j.ijbiomac.2019.03.171.

133 He, L., Liu, Y., Mustapha, A., and Lin, M. (2011). Antifungal activity of zinc oxide nanoparticles against *Botrytis cinerea* and *Penicillium expansum*. *Microbiol. Res.* 166 (3): 207–215. https://doi.org/10.1016/j.micres.2010.03.003.

134 Morsy, M.K., Khalaf, H.H., Sharoba, A.M. et al. (2014). Incorporation of essential oils and nanoparticles in pullulan films to control foodborne pathogens on meat and poultry products. *J. Food Sci.* 79 (4): 675–684.

135 Shankar, S., Wang, L.F., and Rhim, J.W. (2018). Incorporation of zinc oxide nanoparticles improved the mechanical, water vapor barrier, UV-light barrier, and antibacterial properties of PLA-based nanocomposite films. *Mater. Sci. Eng., C* 93 (October 2017): 289–298.

136 Bogdanović, U., Lazić, V., Vodnik, V. et al. (2014). Copper nanoparticles with high antimicrobial activity. *Mater. Lett.* 128: 75–78.

137 Saravanakumar, K., Sathiyaseelan, A., Mariadoss, A.V.A. et al. (2020). Physical and bioactivities of biopolymeric films incorporated with cellulose, sodium alginate and copper oxide nanoparticles for food packaging application. *Int. J. Biol. Macromol.* 153: 207–214. https://doi.org/10.1016/j.ijbiomac.2020.02.250.

138 Yoon, K.Y., Hoon Byeon, J., Park, J.H., and Hwang, J. (2007). Susceptibility constants of *Escherichia coli* and *Bacillus subtilis* to silver and copper nanoparticles. *Sci. Total Environ.* 373 (2–3): 572–575.

139 Beigmohammadi, F., Peighambardoust, S.H., Hesari, J. et al. (2016). Antibacterial properties of LDPE nanocomposite films in packaging of UF cheese. *LWT Food Sci. Technol.* 65: 106–111. https://doi.org/10.1016/j.lwt.2015.07.059.

140 Wang, Y., Wu, Y., Yang, H. et al. (2016). Doping TiO_2 with boron or/and cerium elements: effects on photocatalytic antimicrobial activity. *Vaccum* 131: 58–64. https://doi.org/10.1016/j.vacuum.2016.06.003.

141 Priyanka, K.P., Sukirtha, T.H., Balakrishna, M., and Varghese, T. (2016). Microbicidal activity of TiO_2 nanoparticles synthesised by sol – gel method. *IET Nanobiotechnol.* 10: 81–86.

142 El-maghrabi, H.H., Barhoum, A., Nada, A.A. et al. (2018). Synthesis of mesoporous core-shell CdS@TiO_2 (0D and 1D) photocatalysts for solar-driven hydrogen fuel production. *J. Photochem. Photobiol., A* 351: 261–270. https://doi.org/10.1016/j.jphotochem.2017.10.048.

143 Pagno, C.H., Costa, T.M.H., De Menezes, E.W. et al. (2015). Development of active biofilms of quinoa (*Chenopodium quinoa* W.) starch containing gold nanoparticles and evaluation of antimicrobial activity. *Food Chem.* 173: 755–762.

144 Mohamed, M.M., Fouad, S.A., Elshoky, H.A. et al. (2017). Antibacterial effect of gold nanoparticles against *Corynebacterium pseudotuberculosis*. *Int. J. Vet. Sci. Med.* 5 (1): 23–29. https://doi.org/10.1016/j.ijvsm.2017.02.003.

145 Rai, A., Prabhune, A., and Perry, C.C. (2010). Antibiotic mediated synthesis of gold nanoparticles with potent antimicrobial activity and their application in antimicrobial coatings. *J. Mater. Chem.* 20 (32): 6789–6798.

146 Gordon, T., Perlstein, B., Houbara, O. et al. (2011). Synthesis and characterization of zinc/iron oxide composite nanoparticles and their antibacterial properties. *Colloids Surf., A* 374 (1–3): 1–8.

147 Dey, A. and Neogi, S. (2019). Oxygen scavengers for food packaging applications: a review. *Trends Food Sci. Technol.* 90 (April): 26–34.

148 Yu, J., Liu, R.Y.F., Poon, B. et al. (2004). Polymers with palladium nanoparticles as active membrane materials. *J. Appl. Polym. Sci.* 92 (2): 749–756.

149 Ahmed, S., Rasul, M.G., Brown, R., and Hashib, M.A. (2011). Influence of parameters on the heterogeneous photocatalytic degradation of pesticides and phenolic contaminants in wastewater: a short review. *J. Environ. Manage.* 92 (3): 311–330.

150 Bogdan, J., Zarzyńska, J., and Pławińska-Czarnak, J. (2015). Comparison of infectious agents susceptibility to photocatalytic effects of nanosized titanium and zinc oxides: a practical approach. *Nanoscale Res. Lett.* 10 (1): 1–15.

151 Chawengkijwanich, C. and Hayata, Y. (2008). Development of TiO_2 powder-coated food packaging film and its ability to inactivate *Escherichia coli* in vitro and in actual tests. *Int. J. Food Microbiol.* 123 (3): 288–292.

152 Yemmireddy, V.K., Farrell, G.D., and Hung, Y.C. (2015). Development of titanium dioxide (TiO_2) nanocoatings on food contact surfaces and method to evaluate their durability and Photocatalytic bactericidal property. *J. Food Sci.* 80 (8): N1903–N1911.

153 Kumar, J.S. (2017). The psychology of colour influences Consumers' buying behaviour – a diagnostic study. *Ushus - J. Bus. Manage.* 16 (4): 1–13.

154 Jiang, X., Valdeperez, D., Nazarenus, M. et al. (2015). Future perspectives towards the use of nanomaterials for smart food packaging and quality control. *Part. Part. Syst. Char.* 32 (4): 408–416.

155 Karimi, M., Sadeghi, R., and Kokini, J. (2018). Human exposure to nanoparticles through trophic transfer and the biosafety concerns that nanoparticle-contaminated foods pose to consumers. *Trends Food Sci. Technol.* 75 (February): 129–145. https://doi.org/10.1016/j.tifs.2018.03.012.

156 Bopp, A.F. (2019). The evolution of food preservation and packaging. *ACS Symp. Ser.* 1314: 211–228.

157 Magnuson, B.A., Jonaitis, T.S., and Card, J.W. (2011). A brief review of the occurrence, use, and safety of food-related nanomaterials. *J. Food Sci.* 76 (6): 126–133.

158 Song, H., Li, B., Lin, Q.B. et al. (2011). Migration of silver from nanosilver-polyethylene composite packaging into food simulants. *Food Addit. Contam., Part A* 28 (12): 1758–1762.

159 Metak, A.M. (2015). Effects of nanocomposite based nano-silver and nano-titanium dioxideon food packaging materials. *Int. J. Appl. Sci. Technol.* 5 (2). Available from: www.ijastnet.com.

160 Carrero-Sánchez, J.C., Elías, A.L., Mancilla, R. et al. (2006). Biocompatibility and toxicological studies of carbon nanotubes doped with nitrogen. *Nano Lett.* 6 (8): 1609–1616.

161 Kim, Y.S., Kim, J.S., Cho, H.S. et al. (2008). Twenty-eight-day oral toxicity, genotoxicity, and gender-related tissue distribution of silver nanoparticles in Sprague-Dawley rats. *Inhalation Toxicol.* 20 (6): 575–583.

162 Baltic, M., Boskovic, M., Ivanovic, J. et al. (2013). Nanotechnology and its potential applications in meat industry. *Tehnol. Mesa* 54 (2): 168–175.

163 FDA (2015). Inventory of effective food contact substance (FCS) notifications.

164 FDA (2015). List of indirect additives used in food contact substances. http://www.fda.gov/%0AFood/IngredientsPackagingLabeling/PackagingFCS/IndirectAdditives/ucm115333.%0Ahtm

165 PJP, E., Otoni, C.G., and NFF, S. (2016). Zinc oxide nanoparticles for food packaging applications. In: *Antimicrobial Food Packaging* (ed. J. Barros-Velázquez), 425–431. Elsevier Inc.

166 EFSA (2018). Guidance on risk assessment of the application ofnanoscience and nanotechnologies in the food and feedchain: Part 1, human and animal health. https://efsa.onlinelibrary.wiley.com/doi/epdf/10.2903/j.efsa.2018.5327

167 Commission Regulation (EU) (2016). 2016/1416 of 24 August 2016 amending and correcting Regulation (EU) No 10/2011 on plastic materials and articles intended to come into contact with food. https://eur-lex.europa.eu/legal-content/EN/TXT/PDF/?uri=CELEX:32016R1416&from=EN

168 EFSA (2012). Scientific opinion on the safety evaluation of the substance, titanium nitride, nanoparticles, for use in food contact materials. European Food Safety Authority Panel on food contact materials, enzymes, flavourings and processing aids. https://efsa.onlinelibrary.wiley.com/doi/abs/10.2903/j.efsa.2012.2641

169 EFSA (2004). Opinion of the scientific panel on food additives, flavourings, processing aids and materials in contact with food (AFC) on a request from the commission related to a 3rd list of substances for food contact materials. *EFSA J.* 2 (3): 1–13. https://doi.org/10.2903/j.efsa.2004.24a.

170 Calmon-Decriaud, A., Bellon-Maurel, V., and Silvestre, F. (1998). Standard methods for testing the aerobic biodegradation of polymeric materials. Review and perspectives. In: *Blockcopolymers-Polyelectrolytes-Biodegradation*, 207–226. Springer.

171 Ramos, M., Fortunati, E., Peltzer, M. et al. (2014). Influence of thymol and silver nanoparticles on the degradation of poly(lactic acid) based nanocomposites: thermal and morphological properties. *Polym. Degrad. Stab.* 108: 158–165.

172 Kotharangannagari, V.K. and Krishnan, K. (2016). Biodegradable hybrid nanocomposites of starch/lysine and ZnO nanoparticles with shape memory properties. *Mater. Des.* 109: 590–595.

173 Zhou, J.J., Wang, S.Y., and Gunasekaran, S. (2009). Preparation and characterization of whey protein film incorporated with TiO_2 nanoparticles. *J. Food Sci.* 74 (7): 50–56.

174 Part, F., Berge, N., Baran, P. et al. (2018). A review of the fate of engineered nanomaterials in municipal solid waste streams. *Waste Manage. (Oxford)* 75: 427–449. https://doi.org/10.1016/j.wasman.2018.02.012.

175 Sánchez, C., Hortal, M., Aliaga, C. et al. (2014). Recyclability assessment of nano-reinforced plastic packaging. *Waste Manage. (Oxford)* 34 (12): 2647–2655.

176 Bertolini, M., Bottani, E., Vignali, G., and Volpi, A. (2016). Comparative life cycle assessment of packaging systems for extended shelf life milk. *Packag. Technol. Sci.* 29 (10): 525–546.

177 Silvestre, C., Duraccio, D., and Cimmino, S. (2011). Food packaging based on polymer nanomaterials. *Prog. Polym. Sci.* 36 (12): 1766–1782.

178 McMillin, K.W. (2008). Where is MAP going? A review and future potential of modified atmosphere packaging for meat. *Meat Sci.* 80 (1): 43–65.

179 Nowack, B., Krug, H.F., and Height, M. (2011). *120 Years of Nanosilver History: Implications for Policy Makers*. ACS Publications.

180 Zhang, B.H., Hortal, M., Dobon, A. et al. (2017). Selection of nanomaterial-based active agents for packaging application: using life cycle assessment (LCA) as a tool. *Packag. Technol. Sci.* 30: 575–586.

8

Fabrication of Intelligent Packaging Systems Using Nano-Indicators and Sensors

Xiuting Hu[1,2] and Ming Miao[2]

[1]Nanchang University, State Key Laboratory of Food Science and Technology, 235 Nanjing East Road, Nanchang 330047, P.R. China
[2]Jiangnan University, State Key Laboratory of Food Science and Technology, 1800 Lihu Road, Wuxi 214122, P.R. China

8.1 Introduction of Intelligent Packaging

Packaging of food is an essential part in the modern food industry, aiming to assure integrity, safety, wholesomeness, and quality of the food product [1], which is also one of the most important means of advertising. Thus, packaging is a coordinated system designed for storage, distribution, transport, retailing, and subsequent end use of goods. Conventional food packaging, the container of food, is used to keep the packaged food free of the external environmental hazards and prolong the shelf life of the food product. Nevertheless, it is not able to provide users with information regarding quality deterioration induced by fluctuation of temperature, change of the gas concentration, and growth of microorganism inside the package environment. Take fresh fruit and vegetables for instance, they still respire after harvesting. Accordingly, their quality gradually decreases. Thus, an intelligent system is required to continuously detect the gas composition or concentration within the package environment and inform users of the quality characteristics of the fruit and vegetables on time. In addition, microorganisms may proliferate fast in the packaged protein-based food products. As a result, a new branch of packaging emerges, namely, intelligent packaging, which provides references for assessing the quality status of the packaged food. In addition to being a container of food, intelligent packaging is able to transmit colorimetric, electric, or other signals in real-time in response to the changes of the initial packaging conditions or food quality [2].

An academic definition of intelligent packaging is a packaging system that is able to perform intelligent functions, including tracing, detecting, sensing, recording, and communicating [3]. Thus, intelligent packaging provides the basis for decisions on how to deal with the food product and alerts on potential problems. Intelligent packaging and active packaging are two terms that are easy to confuse. Active packaging positively alters the conditions inside the package to retard decline in quality

Nanotechnology-Enhanced Food Packaging, First Edition. Edited by Jyotishkumar Parameswaranpillai, Radhakrishnan Edayileveettil Krishnankutty, Aswathy Jayakumar, Sanjay Mavinkere Rangappa, and Suchart Siengchin.

of the food product and prolong the shelf life [4], while intelligent packaging is able to track the product, sense the environmental factors of the package interior and exterior, and communicate with the users. Thus, an intelligent packaging can monitor the quality status of a food product and early inform the users, while an active packaging takes some measures, such as using antioxidant or an antimicrobial agent, to protect the food product and extend the shelf life. The basic functions of packaging are divided into containment, convenience, protection, and communication [3]. Obviously, the distinct function of intelligent packaging is communication, while the distinct function of active packaging is protection. However, intelligent packaging and active packaging are not mutually exclusive as some packaging systems contain both [3]. Actually, they can work synergistically. For instance, intelligent and active films were developed in which bacterial cellulose was modified using polypyrrole–zinc oxide nanocomposite [5]. It was found that this nanocomposite film increased the shelf life and stabilized rheological properties of chicken thigh through antioxidant and antimicrobial activity, which suggested that this film worked as the active packaging. Moreover, the good correlation between the storage time/temperature and change of film electrical resistance provided the basis to estimate the storage time/temperature of chicken thigh, which suggested that this film also acted as the intelligent packaging.

Generally, an intelligent packaging system is constructed through internally or externally incorporating an external component in the final package. The external component can be classified into three types, including data carriers, indicators, and sensors [6]. Data carriers mainly refer to barcodes and radiofrequency identification tags (RFIDs), which are specifically intended for tracing packaged food and recording some basic information. Indicators are usually colorimetric systems that are sensitive to temperature, humidity, time, damage, pH, or freshness. As a result, they provide information about changes that happen in the packaged product or in the environment in which the product is stored. The indicators usually have a distinct characteristic, namely, qualitatively or semiquantitatively monitoring the environmental factors that must be controlled or the specificity of a particular product. Sensors are rigorously defined as the devices that can detect one form of signal and convert this signal to other form of signal through a transducer [1]. Thus, sensors can rapidly and definitely quantify the target analytes. The original signal refers to the product information, quality changes, or the environmental conditions. The final signal is usually optical or electric. For the optical colorimetric sensors, the final signal is change of the color that the naked eye can easily distinguish. Thus, the colorimetric indicators and optical colorimetric sensors work similarly, and many references confuse these two items.

With development of nanotechnology, nanomaterials have been applied in packaging, including intelligent packaging. For common packaging, the application of nanotechnology is usually used to enhance the mechanical properties, gas barrier characteristics, and thermal stability of package. For instance, different organic nanoparticles (NPs), such as starch NPs, and inorganic NPs, such as ZnO NPs, have widely been added into the polymer matrix to enhance mechanical and barrier properties of films [7, 8]. On the other hand, increasing nanotechnology-based indicators/sensors are developed and incorporated into the intelligent packaging of food, where they interact with food components, internal, or external environmental factors and perform intelligent functions [9]. On the other hand, NPs are also used in intelligent packaging as the nanofiller instead of the indicator or sensor. For instance, an intelligent packaging with pH sensitivity was developed through incorporating anthocyanins from black carrot and bentonite into the chitosan/PVA films [10]. Herein, anthocyanins from black carrot worked as the pH indicator, and bentonite, the nanofiller, enhanced the barrier properties of the films. Similarly, a green pH-sensitive nanocomposite film was prepared in which both anthocyanins from black rice bran and oxidized chitin nanocrystals were incorporated in the gelatin matrix [11]. The color of the nanocomposite film remarkably changed at various pHs due to the pH sensitivity of anthocyanins from black rice bran, and oxidized chitin nanocrystals improved mechanical and barrier properties and thermal stability of the film. Traditional indicators, data carriers, and sensors have been summarized by a few reviews [1–4, 12–18]. Thus, this chapter focuses on nanotechnology-based indicators and sensors that have been already used in intelligent packaging or have great potential to be used in intelligent packaging.

8.2 Nanoparticle-Based Temperature Indicators

Temperature has the most profound influence on the shelf life of food after packaging of food, particularly frozen food, and thus it is of great importance to monitor and control it [19]. Two types of temperature indicators are developed, including simple temperature indicators and time–temperature integrators (TTIs) [12]. The former are able to demonstrate whether products have stayed above or below a critical temperature and thus warn consumers about the possibility of food deterioration. Since both temperature and time influence the kinetics of food deterioration, TTIs have attracted more attention [6]. TTIs can record the thermal history and indicate the remaining shelf life of perishable food products, since the cumulative effects of temperature and time usually result in the irreversible color change of TTIs. Obviously, when the packaging of food contains the TTI, users can easily judge whether the food has deteriorated according to the color change by eyes, which helps to ensure the safety and quality of packaged food. Nowadays, thermochromic nanomaterials have been applied as the temperature indicators.

8.2.1 Silver Nanoparticle-Based TTI

Triangular Ag nanoplates were developed and used as the TTI based on localized surface plasmon resonance (LSPR) of Ag nanoplates [20]. Specifically, the Ag nanoplate displayed an in-plane dipole resonance mode in the visible region, whose peak position was strongly sensitive to the corner sharpness. As the corners of triangular Ag nanoplates became increasingly round, the dipole resonance peak position on the crystal surface gradually shifted to lower wavelengths. External thermal energy influenced the Gibbs free energy of the surface atoms of Ag nanoplates, thus changing the surface morphology of Ag NPs and resulting in the shift of the maximal wavelength and the change of color. With increasing of the ambient temperature, the atoms on the edges of the Ag nanoplates migrated due to the highest free energy, and the migration rate was positively correlated to the temperature. Accordingly, when the temperature increased, the formation of the round Ag NPs became faster. Thus, Ag NPs could be used as the TTI.

Obviously, the mechanism of the TTI was that high temperature changed triangular Ag nanoplates into round particles, thus changing the color. Therefore, effects of poly(styrene-4-sulfonate) (PSS) and potassium persulfate (KPS) on the degradation behavior of the triangular Ag nanoplates were investigated [21]. Herein, a one-pot water-based synthetic method was used to prepare triangular Ag nanoplates, in which two ligands, polyacrylamide and acetonitrile, were introduced to form good coordination complexes with Ag^+. These two ligands worked synergistically to reduce the reduction rate of Ag^+, thus providing much greater control in the kinetic of the reaction and increasing the yield of Ag nanoplates. In the presence of PSS, the Ag nanoplates were irreversibly degraded, and the triangular Ag nanoplates changed into oblate spheroidal Ag nanoplates at room temperature. However, in the presence of KPS, the etching degree of the triangular Ag nanoplates depended on the ambient temperature, and thus the color of triangular Ag nanoplate solutions was different corresponding to different ambient temperatures. Therefore, it was believed that triangular Ag nanoplate solutions supplemented with KPS could be applied for colorimetrical sensing of the thermal history.

Later, the application of triangular Ag nanoplates as the TTI was also confirmed by another study [22]. Triangular Ag nanoplates with the diameter between 40 and 160 nm were produced. Accordingly, the triangular Ag nanoplate suspension had a characteristic absorption peak at 420 nm. The emission wavelength of this suspension showed bigger shift after exposure to 22 °C compared with after exposure to 4 °C for the same time. The NP size decreased from 164 to 44 nm after storage at 22 °C for five hours, which induced shift in the emission wavelength at 20 nm.

Moreover, Zhang et al. found that triangular Ag nanoplates could be face-to-face assembled after addition of citrate (Figure 8.1) [23]. The face-to-face assembled triangular Ag nanoplates quickly disassembled after treatment at a wide temperature range from 30 to 80 °C. Accordingly, a red shift occurred to the LSPR peak of the triangular Ag nanoplates, and the shift of the peak position was linearly related to the heating temperature. Accordingly, the color changed. Moreover, the spectrum peak of triangular Ag nanoplates did not significantly shift in eight hours after placing triangular Ag nanoplates from high temperature to room temperature, which confirmed that the disassembled triangular Ag nanoplate suspension was very stable. Thus, a plasmonic temperature indicator could be established using the assembled triangular Ag nanoplates based on the mentioned phenomenon, whose operating temperature ranged from 30 to 80 °C. However, the naked eye could not well distinguish the color change of triangular Ag nanoplates at low temperature, and only the color change induced by ≥50 °C could be easily discerned by the naked eye.

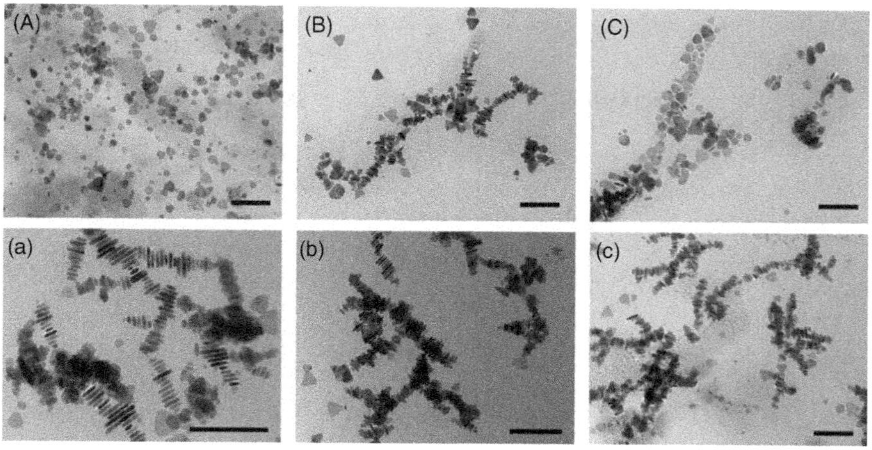

Figure 8.1 TEM images of assembled and disassembled Ag triangular nanoplates: (A) original Ag triangular nanoplates; (B) face-to-face assembled Ag triangular nanoplates in the presence of 15.0 mM citrate; (a–c), the same sample observed at different areas and magnification; (C) disassembled Ag triangular nanoplates after treatment at 70 °C for five minutes. Source: Zhang et al. [23]/with permission from Elsevier.

8.2.2 Gold Nanoparticle-Based TTIs

Likewise, Au NPs were also used as the TTI [24–27]. Au NPs produced different colorimetric signals, which were dependent on their surface morphology, shape, and size. When the size of Au NPs grew irreversibly with time, the Au NPs also generated and grew with the increasing of temperature, accompanied by the color change. Importantly, the color change was dependent on both the storage temperature and time [25]. The increase in size of Au NPs was mainly ascribed to higher storage temperature, whereas longer storage time mainly resulted in the shape change. As a result, higher temperature treatments induced a more significant color change of Au NPs due to the LSPR effect. In addition, the capping agent significantly affected synthesis and stabilization of Au NPs, thus affecting the sensitivity of Au NPs as the TTI. It was found that gelatin was able to act as both the reductant and the stabilizer to synthesize Au NPs via mixing gelatin and the precursor of gold, $HAuCl_4$ [24, 25]. The gelatin concentration and pH of the gelatin solution and the heating temperature and time had significant effects on the size and morphology of gelatin-templated Au NPs, thus affecting their color signal. The color intensity of the Au NPs reached the peak value with addition of 2% gelatin. The Au NPs synthesized at pH 3 were bigger and less regular in shape than those synthesized at pH 5 or 9. Thus, the color was developed at the largest rate at pH 3. When the gelatin-templated Au NPs (2%, pH 5) in a simulated frozen storage were exposed to 30 °C, the red color signal developed, which was irreversible. In addition, the intensity of the red color was dependent on the duration of exposure. Thus, the change of color was easily observed by the naked eye. However, the color of samples kept at frozen storage did not change. Thus, the temperature abuse could be easily observed by the color change. Au NPs was also synthesized *in situ* in alginate as the TTI [26]. The color intensity of the TTI was positively correlated with the storage temperature and/or time. Specifically, the gray TTI became red with time of exposure at 40 °C, which suggested that the alginate-templated Au NPs had higher color change sensitivity than the gelatin-templated Au NPs. Moreover, the temperature range that alginate-templated Au NPs could monitor was much wider. Particularly, the alginate-templated Au NPs could be made into a solid hydrogel by addition of Ca^{2+}, and the resultant solid hydrogel stilled remained the sensitivity to storage temperature and time. Nanocomposite of chitosan and Au NPs was used as the TTI [27]. Similar to the previously mentioned alginate-templated Au NPs, the pink color of the chitosan-templated Au NPs became irreversibly more intense with the duration when subjected to room temperature or higher temperature. Particularly, the pink chitosan-templated Au NPs became dark gray upon freezing for one day, which suggested that chitosan-templated Au NPs could be also applied as a frozen detector to indicate the quality of frozen products. Gelation, alginate, and chitosan had good film-forming ability and good compatibility with other polymers. Therefore, it is inferred that these biopolymer-templated Au NPs can be directly incorporated within the polymer matrix to construct intelligent packaging.

Moreover, Zhang et al. designed and prepared a plasmonic TTI using the chronochromic evolution from Au nanorods to Au/Ag nanorods, which gave a sharp contrast in red-to-green color change (Figure 8.2) [28]. The Au nanorods dispersed in cetyltrimethylammonium chloride solution were red and exhibited two extinction bands at approximately 510 and 880 nm resulting from the transverse and longitudinal plasmon resonances, respectively (Figure 8.2e). When the precursor of Ag, $AgNO_3$, reacted with ascorbic acid, the reducing agent, the produced Ag atoms were epitaxially deposited on the Au nanorods, yielding core/shell-structured Au/Ag nanorods (Figure 8.2a–c). As the Ag shell thickened, there was a blue shift in the extinction band of the longitudinal plasmon resonance. The initial red colloidal solution became orange, yellow, greenish-yellow, and further green, as the maximum wavelength of the longitudinal plasmon peak shifted to ~690 nm (Figure 8.2d–f). Higher temperature accelerated the color change of this TTI (Figure 8.2g,f). Moreover, the kinetic rate and apparent activation energy of the evolution from Au nanorods to Au/Ag nanorods were widely tunable by simply changing the concentrations of the relevant reagents, which guaranteed a large flexibility to develop the TTI.

8.2.3 Polydiacetylene/Silica Nanocomposite-Based TTI

The polydiacetylene/silica (PDA/SiO_2) nanocomposites were also developed as the TTI, since the blue PDA/silica nanocomposites became red after storage at different temperatures [29]. In addition, a diffusion-type TTI was developed using the reaction between the PDA/SiO_2 nanocomposite and Tween 20 [30]. This novel diffusion-type TTI consisted of a reservoir and a track strip. The former contained

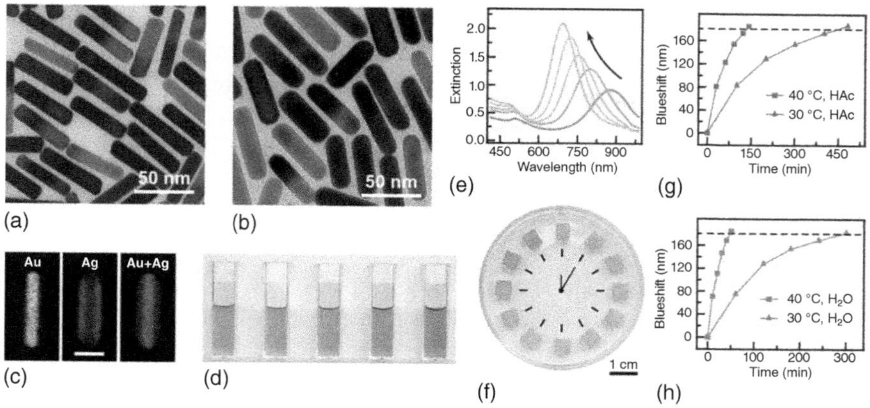

Figure 8.2 Chronochromic evolution from Au nanorods to Au/Ag nanorods: (a) TEM image of the Au nanorods; (b) TEM image of the Au/Ag nanorods; (c) energy-dispersive X-ray elemental mapping images of a typical Au/Ag nanorod (The scale bar represents 20 nm); (d) the color change of the self-evolving colloidal solution during reaction (from left to right: red, orange, yellow, greenish-yellow, and green); (e) extinction spectra of the samples in (d); (f) photograph of the hydrogel cubes (~5 × 5 × 3 mm^3) arranged in a manner resembling a clock in a Petri dish; (g, h) plots of the blueshifts of the longitudinal plasmon band as a function of time for the acidified sample and the blank sample at 40 and 30 °C [28]. Source: Zhang et al. [28]/with permission from American Chemical Society.

Tween 20 that started moving along the adjacent strip after TTI was activated by removing a film gate. The strip had a top microporous sheet as a diffusion path and a bottom self-adhesive base layer. The PDA/SiO_2 nanocomposite solution was injected into a predetermined end of the diffusion path. When Tween 20 arrived at the blue test line containing the PDA/SiO_2 nanocomposite, the line became red, which indicated the TTI end point. Therefore, TTIs for indicating different temperatures could be designed by determining the diffusion rates of Tween 20 under different isothermal conditions.

8.2.4 Nanofiber-Based TTIs

Interestingly, Choi et al. designed a highly sensitive TTI using the electrospun self-healing nanofiber nonwoven material (polyurethane) [31]. Two different versions of the opaque TTIs, including a frozen food sensor (FFS) and a chilled food sensor (CFS), were developed, which were refrigerated at −20 and 2 °C, respectively. They gradually become transparent at room temperature through interfibrillar fusion induced by self-healing, which was irreversible. Thus, a warning sign "!" appeared. The timescale and the standby temperature were tunable by altering the T_g of the polymer composition and the nonwoven thickness. For application in intelligent packaging of food, the electrospun self-healing polyurethane was fully coagulated and separated from the collector after accumulation by a water coagulation bath. The average fiber diameter of FFS and CFS nanofibers was 1920 and 960 nm, respectively. The opaque FFS gradually became transparent after exposure to 20 °C within 24 hours, and the hidden warning sign appeared. Accordingly, the light transmittance of FFS and CFS gradually increased over the duration time. However, the light transmittance of CFS did not increase when CFS was subjected to −20 and 2 °C for four weeks, which indicated that it was exclusively applicable as the CFS rather than as the FFS.

8.3 Nanomaterial-Based Humidity Sensors

High moisture is able to accelerate deterioration of food products, particularly those that have low or intermediate moisture contents [32]. Therefore, monitoring the humidity can provide information about the integrity of the package and actual status regarding the quality of the packaged food items. Recently, a variety of nanomaterials-based humidity sensors have been fabricated due to superior characteristics of nanomaterials, including low cost and the high surface ratio.

8.3.1 ZnO Nanoparticle-Based Humidity Sensors

Among these nanomaterials, ZnO nanomaterials are one of the most widely used materials for development of the sensor, because they can grow in different morphologies. For instance, Zhang et al. prepared ZnO nanorod and nanowire films on the Si substrates with comb type Pt electrodes and found that the resistance of the films was negatively correlated with the relative humidity (RH), which confirmed their humidity sensitivity [33]. The ZnO nanowire and nanorod films had more than 4 and 2 orders of magnitude of resistance change, respectively, when exposed to a moisture pulse of 97% RH at room temperature. In addition, these nanomaterial films showed good stability and short response time. Wang et al. fabricated a humidity sensor in which a quartz crystal worked as the transducer and ZnO nanotetrapods acted as the sensing component [34]. The mechanism of the quartz crystal microbalance (QCM) for detection of humidity was that the shift of the crystal resonance frequency was linearly related to the mass of water molecules adsorbed on a rigid surface attached to the QCM. Herein, the maximum frequency sensitivity of the ZnO nanotetrapod-based QCM sensors was close to 100 Hz/% RH in the humidity range of 40–80% RH. The sensors also had good frequency stability and reproducibility. Similarly, Zhou et al. developed a wireless humidity sensor prototype based on QCM coated by nanostructured flowerlike ZnO with legs as the sensing component [35]. Besides the probe, the wireless QCM sensor system required a receiver. The probe sensed the RH, displayed the frequency signals, and transmitted the signals to the receiver. The receiver connected to the computer received and recorded the signals of the sensor. The interaction distance between the probe and the receiver could reach 300 m. Moreover, they fabricated quartz tuning forks (QTFs) coated with nanocrystalline ZnO films as a humidity sensor [36]. The QTFs were extremely sensitive to change of mass, and the frequency shift was negatively correlated with the mass loading on the QTF beams, which make the QTFs a sensitive humidity sensor. In this work, it was the nanocrystalline ZnO film instead of the QTFs that sensed response to the RH. Thus, the thickness of the nanocrystalline ZnO film coated on the QTF significantly influenced the sensitivity of the sensors. Erol et al. studied the humidity adsorption and desorption kinetics of ZnO NPs via QCM measurements at room temperature, which exhibited about 40 Hz/% sensitivity for environments of 45–88% RH [37]. Later, they developed ZnO nanowires films as the humidity sensor [38]. Under different RH conditions, ZnO nanowire films coated on the quartz crystal had a positive QCM frequency shift, which was due to viscoelastic change in the mechanical stiffness of the ZnO nanowire films on the QCM electrodes during adsorption. The sensitivity of the ZnO nanowire films reached 3 Hz/% RH in the range of 48–88% RH. On the other hand, Hong et al. synthesized ZnO nanorods and coated them on nanocrystalline ZnO/AlN/Si-layered structures, thus obtaining ZnO nanorod-based surface acoustic wave (SAW) humidity sensors [39]. The sensitivity response of this structure reached the shift of the center frequency (around 750 kHz) over an RH range of 10–90%, which was much higher than that of nanocrystalline ZnO films (110 kHz). In conclusion, ZnO NPs could be used to fabricate the humidity sensors, and the morphology of ZnO NPs significantly affected the sensitivity to humidity of the sensors.

8.3.2 Other Metallic Nanoparticle-Based Humidity Sensors

Other metallic nanosized materials were also developed for detection of humidity. For example, Kuang et al. developed an SnO_2 nanowire-based humidity sensor using a field effect transistor [40]. A single SnO_2 nanowire with the diameter at 250 nm was placed between two Au electrodes of 100 nm thickness, and Pt was deposited as the top electrode to enhance contact. To detect the current signals through the SnO_2 nanowire, two Au electrodes were connected with a support chip by an Au wire binding technique. The SnO_2 nanowire-based humidity sensor presented a linear response to RH in air using response sensitivity as a function of RH in air, which indicated that this sensor was effective in humidity detection. The mechanism was that the pre-adsorbed O_2 on the surface of the SnO_2 nanowire was replaced by adsorbed water molecules, thus increasing the conductivity of the SnO_2 nanowire.

Moreover, a novel extremely sensitive Ag-nanocomposite-based QCM sensor for detection of humidity was successfully developed [41]. Herein, carboxymethyl cellulose (CMC) acted as the reducing and stabilizing agent to synthesize Ag NPs. As a result, the synthesized spherical Ag NPs whose size ranged 13–28 nm were well-dispersed, and a long-term stable colloidal suspension was obtained. The immobilized metallic nanostructures greatly decreased changes of viscoelastic properties and increased surface area and surface local charge density of the CMC. As a result, performances of the sensor were greatly improved, including better stability even at higher RH, good reproducibility, and linear response toward humidity (11–98% RH), low hysteresis, short response time (14 seconds), and short recovery time (6 seconds). In summary, this Ag-nanocomposite-coated QCM could be used as sensitive layer for detection of humidity.

Later, an impedance-type nonoxide humidity sensor based on SiC nanosheets was developed through carbothermal reduction between silicon powder and graphene oxide [42]. The final product with a high surface area of 90.1 m^2/g contained SiC with a thin oxide layer on the surface, and some remained as reduced graphene oxide. The prepared SiC nanosheets exhibited excellent sensing properties, such as a high sensitivity to humidity (16 991.1 at 95% RH), fast response (three seconds), fast recovery (three seconds), and good linearity in a wide range of RH (11–95%), which might be ascribed to the advantage of the nanosheet structure, the fast electron transfer rate in the SiC semiconductor, and the donor effect of electrons.

In addition, it was found that porous metal organic framework MIL-101(Cr) NPs had great potential as the humidity-sensing materials [43]. The main charge carrier of the sensor was H_3O^+ at low humidity. With the increasing of RH, more H^+ was yielded via Grotthuss chain reaction, and H_3O^+ and H^+ worked cooperatively as main charge carriers. Thus, the impedance of MIL-101(Cr) humidity sensor decreased gradually with the increasing of RH due to the increasing number of carriers. As a result, the sensor was very sensitive to humidity and showed fast response and recovery. Specifically, the impedance changed more than 3 orders of magnitude in the range of 33–95% RH at 100 Hz. The response and recovery time was 17 and 90 seconds, respectively.

8.3.3 Polymeric Nanocomposite-Based Humidity Sensors

Nanoscale polymeric composite humidity sensors were also developed. For instance, the *N,N*-dimethylformamide solution containing polyaniline (PANi), poly(vinyl butyral) (PVB), and poly(ethylene oxide) (PEO) was electrospun to prepare PANi nanofiber-based humidity sensors [44]. Formation of beads in the nanofibers had a great effect on their adhesion to the substrates and the nanostructures. Through changing the composition of the electrospinning solution, PANi nanofibers containing beads not only exhibited good adhesion to the electrode and thus good electrical contact but also retained relatively high specific surface area. Moreover, the hydrophilicity of the nanofibers could be modulated by the ratio of PVB to PEO of the electrospinning solution, which influenced their sensing properties. Impedance of the nanofiber sensors altered 3 orders of magnitude when the RH increased from 20% to 90%, which revealed that they were much more sensitive than their film counterparts. In addition, they had the advantages of good sensing linearity, fast response ($t_{90\%}$ of ~8 seconds and 6 seconds for humidification and desiccation adsorption and desorption, respectively), good repeatability, and small hysteresis.

Similarly, Lin et al. deposited PANi composite nanofibers on the SAW resonator with a central frequency of 433 MHz to prepare the humidity sensor [45]. They found that the mixture of PVB and PANi was a suitable matrix to form nanofibers through electrospinning. The resultant PANi/PVB nanofibers displayed a core–sheath structure and possessed very high sensitivity of ~75 kHz/%RH from 20% to 90% RH, ultrafast response (one and two seconds for adsorption and desorption, respectively), and good sensing linearity. Particularly, this sensor was able to detect humidity as low as 0.5%RH, indicating its potential for detection of low humidity.

The nanoporous LiCl–poly(divinylbenzene) (PDVB) composites were prepared as the humidity sensor [46]. Addition of LiCl greatly enhanced the humidity sensitivity of the composites compared with pure PDVB and the optimum addition of LiCl was 5 wt%. The impedance of the optimum sensor changed 3 orders of magnitude over the whole humidity range. The response time and the recovery time of 5 wt% LiCl/PDVB sensor were 6 and 30 seconds, respectively, when this sensor was subjected to the change from 95% to 11% RH. Moreover, the nanocomposites had a good linearity and little hysteresis. The mechanism by the complex impedance plots suggested that the protons were the domination carrier at low RH, while ions, including Li^+ and Cl^-, were the domination carriers at high RH. Likewise, LiCl–Pebax 2533 composite nanofibers containing different amounts of $LiCl{\cdot}H_2O$ were successfully constructed by electrospinning [47]. It was found that LiCl-added material exhibited higher humidity sensitivity and the optimum addition of LiCl was 44.4 wt%. The impedance of the sensor changed about 4 orders of magnitude when the RH increased from 11% to 95%. Especially, the hysteresis could be neglected. However, the response and recovery time of the sensor were 30 and 80 seconds, respectively.

8.4 Nanomaterial-Based pH Indicators and Sensors

Within packaging of food, certain microorganisms can grow during storage of the food products. Their main metabolites include organic acids, such as acetic acid and lactic acid, which can decrease the pH of the food samples. Moreover, CO_2 may be yielded from microbial respiration and plant respiration, which can be dissolved in food samples and produce carbonic acid. The dissociation of carbonic acid also decreases the pH. On the other hand, alkaline substances resulting from food spoilage, such as biogenic amines, increase the pH. Therefore, intelligent packaging sensitive to pH change can potentially monitor the quality and freshness of packaged food items.

The pH-sensitive pigments that are extracted from natural sources or synthetic sources have been widely applied as the pH indicator in intelligent packaging [2]. Nanomaterial-based pH indicators are also fabricated. For instance, Tang et al. produced cellulose nanocrystal@polyrhodanine core–shell NPs as the pH indicator [48]. The polyrhodanine shell was synthesized by *in situ* polymerization of rhodanine monomer with cellulose nanocrystals as the core (Figure 8.3). In addition, Fe^{3+} acted as the initiator, the oxidant, and counterion to dope the synthesized polyrhodanine. Since Fe^{3+} was deposited on the surface of cellulose nanocrystals via the electrostatic interaction, the polymerization preferentially occurred around cellulose nanocrystals, forming the shell on the cellulose nanocrystal core. The core–shell NPs altered from the pale red color to the blue violet color, when the pH changed from 2.04 to 12.04, which confirmed their potential as the optical pH indicator. Particularly, this pH sensitivity was maintained when the NPs were configured into flat films or incorporated into chitosan hydrogel beads. Thus, it is inferred that the cellulose nanocrystal@polyrhodanine core–shell NPs have great potential to construct intelligent packaging.

Different from the mentioned study, Patra et al. entrapped the pigment, quercetin, in calcium phosphate NP and thus obtained calcium phosphate–quercetin

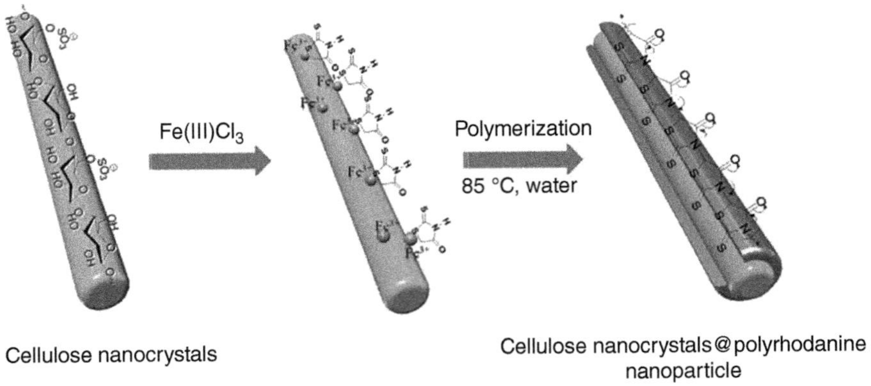

Figure 8.3 Synthesis of cellulose nanocrystal@polyrhodanine nanoparticles through *in situ* polymerization [48].

nanocomposite [49]. The nanocomposite suspension varied from the light yellow color to the dark brown color when the pH varied from 4.5 to 10.5. Thus, the NP could be used as a pH indicator. Moreover, when the NP-coated blotting strips were in contact with drops of different pH buffers from 4.5 to 10.5, the colors of strips were different and changed from pale yellow to deep brown. However, the yellow quercetin-coated strips had no marked change in the pH range 4.5–10.5, but only the color intensity altered slightly. These results demonstrated that the calcium phosphate–quercetin nanocomposite was more sensitive to pH than quercetin. Similarly, bacterial cellulose nanofibers doped with anthocyanins, and silica NPs doped with anthocyanin were designed for monitoring pH [50, 51]. Likewise, Kuntzler et al. developed a colorimetric pH indicator through encapsulating the microalga *Spirulina* sp. LEB 18 in nanofibers of poly(lactic acid) (PLA) and polyethylene oxide (PEO), since *Spirulina* sp. LEB 18 consisted of several pH-sensitive green–blue pigments [52]. Specifically, the high surface area of the polymer nanofibers increased the reactivity of *Spirulina* sp. LEB 18 and accelerated the adsorption of H^+ or OH^-, which improved the sensitivity of *Spirulina* sp. LEB 18 to pH. Moreover, the nanofibers showed hydrophilicity, which allowed ions to interact with the biomass. Thus, the color of this material showed a fast response to pH. When exposed to different pHs, indicators with ≥1% (w/v) *Spirulina* sp. LEB 18 had ΔE values >12, which indicated that difference in color was significant. Therefore, it is inferred that this innovative material has the great potential to be a pH-sensitive indicator in intelligent packaging.

However, some papers considered polymer nanofibers encapsulating the pH-sensitive pigments as the pH sensors instead of pH indictors. For instance, Schueren et al. reported a nanofibrous pH sensor by electrospinning polyamide 6.6 polymer with incorporated pH-indicator dyes [53]. It was observed that incorporation of soluble dyes had no evident influence on the process parameters of electrospinning or the nanofiber size. However, the nanofibrous structures had some droplets in the presence of poorly soluble dyes. The halochromic behavior of the bromocresol purple and brilliant yellow nonwovens suggested that the dyes were still pH sensitive after incorporated into the nano nonwoven. However, this halochromic behavior was different from the pH-indicator behavior of the dye in an aqueous solution because of certain dye–fiber interactions. Similarly, nanofibers from poly(acrylonitrile-*co*-vinyl acetate)/phenolphthalein (PAN/phenolphthalein) were prepared as the pH sensor [54]. Addition of phenolphthalein into the PAN solution decreased the viscosity and led to nanofibers with smaller size under the same electrospinning conditions. Halochromic nanofiber webs showed color change at different pHs and were pH sensitive in the pH range of 8–13. The nanofibers with smaller size showed more significant change in color because of the higher surface contact area, while the nanofibers with bigger size had lower dye release. Moreover, these nanofibers could be repetitively used more than 200 times as the pH sensor.

Actually, halochromic nanosensors had been applied in monitoring fish freshness. For instance, Aghaei et al. described a real-time halochromic sensor containing cellulose acetate nanofibers and alizarin as an indicator of fish spoilage, since the color change of this on-packaging sensor was visual as the amount of total volatile basic nitrogen (TVB-N) and pH of fish increased [55]. Herein, the halochromic behavior of the nanosensor was studied at 4 °C with rainbow trout fillets as the packaged object, and TVB-N, pH, and total viable count (TVC) of rainbow trout fillets and the color of the sensor were determined. The TVC in rainbow trout increased with the storage time, thus increasing the amount of enzymes. With the activity of proteolytic enzymes, the TVB-N generated in the fish resulted in increasing of pH. According to the TVB-N, TVC, and pH values, the shelf life of fresh rainbow trout stored at 4 °C was approximately six days. Accordingly, the color of this nanosensor did not change within 48 hours but became light brick on the 4th day. This color became darker on the sixth day and tended toward violet on the 12th day, which suggested that rainbow trout became deteriorated on the sixth day. These results suggested that the colorimetric results were well correlated with pH changes as well as microbial and chemical changes in the fish. Therefore, this halochromic nanosensor could be applied in the headspace of food packaging for monitoring food freshness by the color change. Later, they also designed a protein-based halochromic nanosensor to monitor the freshness of rainbow trout fillets [56]. Herein, zein nanofibers containing alizarin were electrospun. Thus, this fabricated halochromic nanosensor could also assess fish freshness in real-time according to change of color. In conclusion, nanomaterial-based pH indicators and sensors are usually designed through loading pH-sensitive pigments onto the nanomaterials. As a result, the resultant nanomaterial-based pH indicators usually have the advantages of higher sensitivity and stability than the pigments as the pH indicators.

8.5 Nanoparticle-Based O_2 Indicators

Oxygen is essential for certain biological activities, which may lead to the decline of food quality and even food spoilage. Several examples of these activities include respiration, fat oxidation, browning reactions, pigment oxidation, growth of aerobic microbial, etc. Thus, food packaging systems with limited O_2 concentration are preferred, which belongs to modified atmosphere packaging (MAP). The headspace O_2 of MAP is either deliberately removed to an optimal level or totally removed according to the requirement of food products. However, damage may occur anytime during the transportation and storage of MAP, usually leading to increasing of the oxygen concentration. Therefore, oxygen indication is essential for nondestructive testing of MAP.

The UV-based colorimetric O_2 indicator is the most widely used NPs-based O_2 indicator. In 2005, Lee et al. developed an O_2 indictor that used TiO_2 NPs to photosensitize the reduction of methylene blue by triethanolamine in a polymer encapsulation medium [57]. When the indicator was exposed to UV radiation, the electrons in valence band were promoted to the conduction band. As a result, electron–hole pairs were produced on the surface of the TiO_2 semiconductor. The sacrificial electron donor, such as glycerol, donated electrons to the photogenerated holes in the valence band, thus inhibiting the recombination of the excited electrons with the holes. This led to accumulation of the excited electrons in the conduction band, making them available to reduce the methylene blue dye. As a result, the dye was reduced to a leuko state. Thus, the indicator bleached and remained colorless in the dark. In this case, the O_2 indicator was activated. Upon subsequent exposure to O_2, the leuko dye was oxidized and recovered its original blue color. Moreover, the intensity of color recovery was proportional to the O_2 concentration. The indicator could be activated in package after sealing, as long as UV radiation could penetrate into the package. Thus, this type of UV-activated indicator is useful to prevent premature activation of O_2 measurement. Particularly, the indicator could be reusable in the presence of the sacrificial electron donor.

Similarly, a colorimetric O_2 indicator was developed, which consisted of nanocrystalline SnO_2 as photosensitizer, methylene blue as a redox dye, glycerol as the electron donor, and hydroxyethyl cellulose as the encapsulating polymer [58]. In the absence of O_2, the film stayed bleached but recovered its original color upon exposure to O_2. Interestingly, different from the above TiO_2-based predecessor, this indicator was activated by UVB light rather than by UVA light, which promoted the control in activating the indicator. In another study, TiO_2 NPs, glycerol, and methylene blue were dispersed in PEO solution with aqueous ethanol as the solvent to synthesize a UV-activated oxygen indicator [59]. Particularly, the solution was electrospun to produce O_2-sensitive nanofibrous membrane, and the electrospun indicators were four to five times more sensitive to UV irradiation than continuous film indicators obtained by traditional casting. This difference was mainly ascribed to their morphological differences.

Later, intelligent packaging with a suitable transparent multilayer structure was prepared by inserting a TiO_2 NPs-based O_2 indicator into a structural polymer matrix, ethylene-vinyl alcohol copolymer [60]. In the final intelligent packaging, the indicators were entrapped by the outer layer and the inner layer. The outer layer protected the indicators from the external atmosphere and allowed visualization of the color. The inner layer was opaque for reading of the indictor color from the exterior and thick enough to prevent direct contact between the indictors and food. In addition, the inner layer was oxygen permeable to expose the indictors to the internal package atmosphere.

Recently, a new visible colorimetric O_2 indicator was reported [61]. The indicator consisted of an Ag-loaded TiO_2 nanotube, methylene blue, glycerol, and hydroxyethyl cellulose. Compared with the mentioned TiO_2 NPs, the Ag-loaded TiO_2 nanotube had efficient photocatalytic performance, enhanced specific surface area, and an increased band gap, which evidently improved the performance of Ag-loaded TiO_2 nanotube as the O_2 indicator. In addition, it had two other outstanding advantages of inhibiting response of the indicator to natural light and accelerating the bleaching process during activation.

8.6 Nanomaterial-Based CO_2 Sensors

CO_2 inhibits bacterial and fungal growth, since it is able to produce an anaerobic environment. Besides, high CO_2 concentration is beneficial for storage of fruit and vegetables. Thus, the MAP is typically composed of a high CO_2 concentration and a low O_2 concentration to extend the shelf life of packaged food products. Food quality may be reflected by any changes to the gas concentration/composition. On the other hand, CO_2 is also the spoilage metabolite. Therefore, the CO_2 concentration is an important indicator for leaking of MAP and freshness of some food. As stated earlier, CO_2 can decrease the pH. The pH indicators and detectors have been discussed previously. Thus, this part focuses on the nanomaterial-based sensors that can detect gaseous CO_2 rather than pH indicators or detectors. On the other hand, because CO_2 is the main greenhouse gas, many CO_2 sensors based on metal oxide NPs have been developed for application in the field of environmental science [58]. However, these sensors should work at very high temperature (>100 °C), which cannot be used in intelligent packaging of food. Therefore, those sensors are not introduced here.

Nanomaterial-based CO_2 sensor used in intelligent packaging is rarely reported. Shang et al. fabricated a sensor through attaching CO_2-sensitive spirogyras groups onto versatile photo-cross-linked poly(glycidyl methacrylate) precursor nanofibers using a nucleophilic ring-opening reaction (Figure 8.4) [62]. The spiropyran functional nanofibers could detect and sense gaseous and dissolved CO_2 and display a corresponding color change. Specifically, the yellow indicator became orange and red to dark red when exposed to increased CO_2, which suggested that the color intensity of spiropyran nanofibers was positively correlated with the CO_2 concentration. Particularly, this color change of spiropyran functional nanofibers was completely reversible, and different color states could be obtained by altering the CO_2 concentration. Moreover, these nanofibers responded very quickly (<2 seconds) to dissolved CO_2. Hence, it is believed that the CO_2 ultrasensitive nanofibers can serve as sensors in intelligent packaging to visually monitor gaseous and dissolved CO_2.

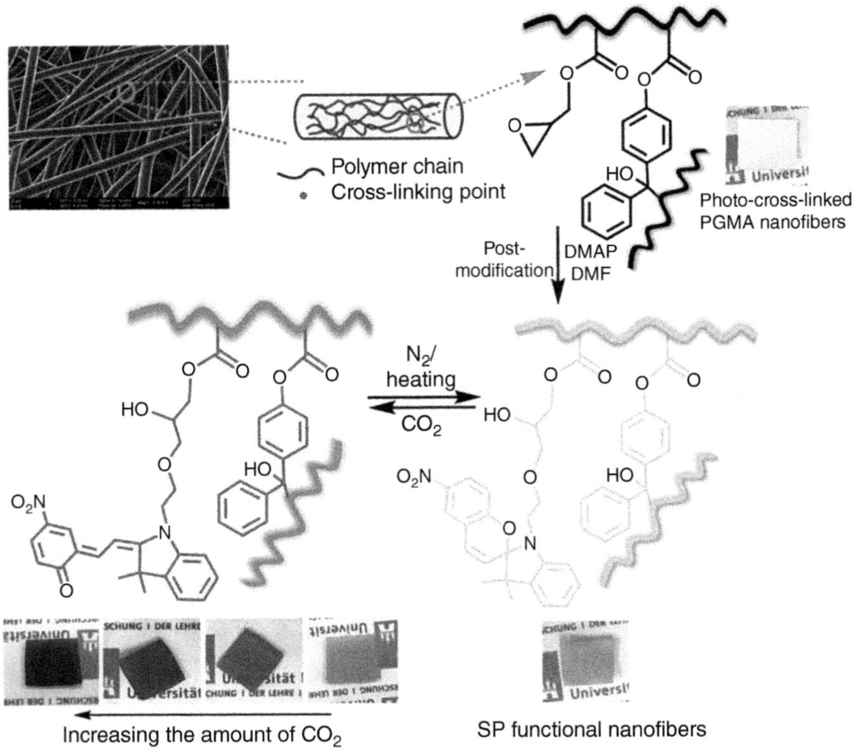

Figure 8.4 Synthesis of a spirogyras functional nanofibrous mat for visually monitoring CO_2. Source: Shang et al. [62]/with permission from Royal Society of Chemistry.

8.7 Nanomaterial-Based Freshness Sensors

Freshness sensors are devices that are used to provide direct information about quality and freshness of the packaged foods through physical or chemical changes, or microbial growth [63]. Therefore, various freshness sensors have been designed based on detection of the concentration changes in metabolites, including biogenic amines, biogenic sulfides, and degradation products of adenosine triphosphate (ATP) and bacterial counts. These changes can be observed by indicators, and translated into a signal, usually change in color, or other signal that can simply be detected and correlated with freshness of food products. As stated previously, some metabolites can change the pH. Thus, monitoring the pH change can also reflect the freshness and quality of food products. The nanomaterial-based pH indicators and sensors are introduced earlier and thus not discussed here. Moreover, although some Au NPs-based sensors had been developed to detect harmful microorganisms, most of them depended on the antibody or polymerase chain reaction [64], which were too costly to be applied in food packaging. Therefore, these sensors are not introduced.

8.7.1 Freshness Sensors Based on Detection of Biogenic Amines

Biogenic amines are organic bases that can be found in some food, including aliphatic amines (cadaverine, putrescine, spermidine, and spermine), aromatic amines (tyramine and phenylethylamine), and heterocyclic amines (tryptamine and histamine) [65]. The biogenic amine concentration of the food is greatly improved due to thermal or bacterial enzymatic decarboxylation of free amino acids during storage. Therefore, biogenic amines are one of the most important volatile compounds generated by food spoilage and are usually employed as the indicators of protein-based food deteriorations. In terms of biogenic amine nanosensor, Lin et al. explored an Au nanorod-based nanosensor to visually monitor biogenic amines based on the LSPR adsorption of Au nanorods [66]. Besides Au nanorods, the sensor system consisted of $AgNO_3$ and resorcinol monoacetate in agarose hydrogel. Biogenic amines hydrolyzed resorcinol monoacetate and produced the reducing product, resorcinol, which reduced Ag^+ to Ag and deposited Ag nanoshells on the surface of Au nanorods (Figure 8.5). As a result, the surface composition and aspect ratio of Au nanorods changed. Accordingly, there was a blue shift in the longitudinal LSPR peak and a distinctive multiple color change from peach to brick red. Moreover, the blue shift of longitudinal LSPR peak and color change of Au nanorods was positively correlated to the biogenic amine concentration, namely, the spoilage degree. To confirm the practical application of this sensor, the sensory hydrogels were used to monitor spoilage of fresh beef meat or salmon fish under different storage conditions. It was found that the color change of sensory hydrogels was dependent on the storage temperature and time. The sensory hydrogels had different color from the references, when beef and salmon meat were stored at 25 °C for one day and at 4 °C for three days, respectively. Thus, users could easily discern

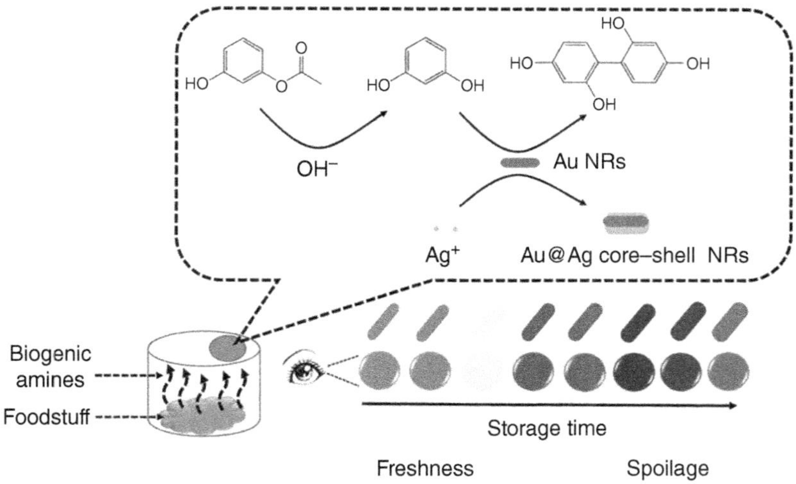

Figure 8.5 The color responses of Au@Ag core–shell nanorods to different levels of biogenic amines released from food spoilage (from left to right: peach, orange, yellow, green, bluish-green, blue, violet, brick red, in print edition shown by different shades of gray) [66]. Source: Reprinted with permission from the reference Lin et al. [66]. Copyright (2016) American Chemical Society.

the freshness of food during transportation and storage via the naked eye. Moreover, the reference sensory hydrogel placed in the exterior of the package remained peach under the previously mentioned storage conditions, which confirmed that the color change of sensory hydrogel placed in the interior of the package was induced by volatile biogenic amines produced from spoilage of the packaged food and could provide the end users with the reliable visual information on food freshness.

Similarly, Tseng et al. constructed paper-based plasmonic refractometric sensors of volatile biogenic amines through embedding metal NPs onto flexible papers (Figure 8.6) [67]. Commercial inkjet papers had high reflectance and smooth surfaces, which provided obvious optical signals for reflection-mode plasmonic refractometric sensing and high particle transfer efficiency, respectively. Thus, the inkjet papers served as substrates of sensor herein. Solid Ag NPs, solid Au NPs and hollow Au–Ag alloyed NPs (HGNs) were immobilized on a solid mold and then embedded directly onto the surfaces of softened inkjet paper. During detection, the gaseous biogenic amines attached spontaneously onto the surface of the NPs through their amino groups and formed a dense layer surrounding the NPs, which was stabilized via intermolecular interactions. Thus, the adsorption of biogenic amines altered the local refractive index surrounding the NPs, leading to a redshift of the LSPR wavelength and change of color. Moreover, the paper embedded by HGNs had a more significant wavelength dip shift after spontaneously adsorbing biogenic amines than the paper embedded by Ag NPs or Au NPs, which indicated that HGNs was a better probe to fabricate a refractometric paper sensor than Ag NPs or Au NPs. Higher refractometric sensitivity of the HGNs might be due to their

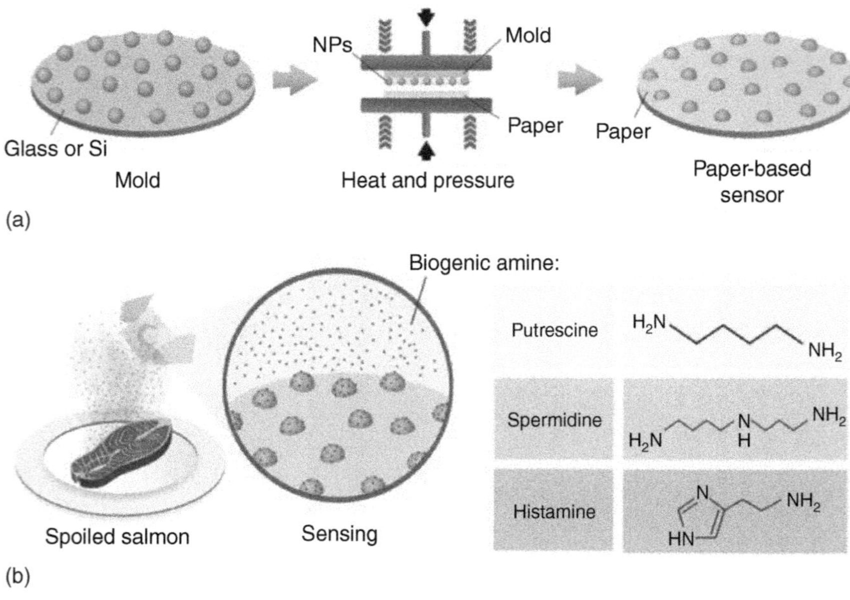

Figure 8.6 (a) Preparation of refractometric sensors through embedding metal NPs onto an inkjet paper and (b) NPs-embedded paper working as a sensor of biogenic amines generated in spoiled food [67]. Source: Reprinted with permission from the reference Tseng et al. [67]. Copyright (2017) American Chemical Society.

porous, hollow structures. The hollow structure allowed biogenic amines to diffuse freely into the interior of the HGNs and form a dense layer on both the interior and exterior of the shells [68, 69]. Moreover, the paper had no visible spectroscopic responses to other gases, such as N_2, CO_2, air, and water vapor, which indicated high selectivity of the sensors. When the plasmonic sensing paper with the HGNs was used to detect the freshness of salmon stored at 25 °C, its center-of-mass wavelength shift ($\Delta\lambda c$) was positively correlated with the accumulation of biogenic amines in salmon during storage. The previously stated results indicated that the plasmonic papers had the ability to assess the freshness of fish through detecting volatile biogenic amines and it was facile to apply the papers in intelligent packaging.

8.7.2 Freshness Sensors Based on Detection of Biogenic Sulfides

Biogenic sulfides are a typical class of thiol-containing compounds with low molecular weight, including hydrogen sulfide (H_2S), methanethiol (CH_3SH), dimethyl disulfide (CH_3SSCH_3), dimethyl trisulfide (CH_3SSSCH_3), and dimethyl sulfide (CH_3SCH_3) [70]. They are also the main volatile compounds released from spoilage of meat and seafood products [71–73]. Thus, biogenic sulfides are also used as a freshness indicator of these products. Ryspayeva et al. manufactured an optical sensor using polyetherimide (PEI) and Ag NPs for packaging and detecting freshness of food [74]. PEI/Ag films were prepared and exposed to 0.01–0.005 M 2-mercaptoethanol to test their practicability as optical sensors. Thiol groups of 2-mercaptoethanol reacted on the surface of the sensors and reduced Ag^+ to Ag NPs, which led to the change of color. The real-time detection of freshness was carried on salmon, chicken, turkey, and beef by measuring the changes in color and UV–vis absorbance over duration of 80 hours. The results showed that the PEI/Ag sensor showed a good response to the gases released from meat. Therefore, it is inferred that this nanosensor can be employed to detect food spoilage in the intelligent packaging.

Similarly, a colorimetric sensor of H_2S was fabricated using gellan gum-templated Ag NPs to real-time detect meat spoilage [75]. The mechanism of this colorimetric sensor was based on the fact that Ag had the ultrastrong ability to bind with H_2S and form Ag_2S. Thus, when exposed to H_2S, the yellow gellan gum-templated Ag NPs gradually became colorless with a limited detection at 0.81 μM H_2S. Moreover, it had a strong anti-interference ability in the presence of other volatile compounds released from deteriorated chicken breast and silver carp. As the color change was easily discerned by the naked eye, the gum-templated Ag NPs could be used in intelligent packaging as a colorimetric H_2S sensor to detect meat spoilage.

Moreover, a biogenic amines- and sulfides-responsive colorimetric sensor based on Au NPs was prepared to detect the spoilage of raw meat, fish, crustaceans, and preserved meat [76]. During detection, the non-aggregated form of Au NPs aggregated when exposed to biogenic amines or sulfides, resulting in the shift of the LSPR peak and the change of the color. The detection limit for dimethyl sulfide, the representative of biogenic sulfides, and histamine, the representative of biogenic amines, was 0.5 and 0.035 μg/ml, respectively. In addition, the response of the sensor to biogenic amines was not affected by other volatile substances commonly produced by spoiled meat and seafood. Thus, this sensor could serve as a nondestructive detector to real-time monitor spoilage of meat and seafood.

8.7.3 Freshness Sensors Based on Detection of ATP Degradation Products

During fish and meat spoilage, the existing ATP in muscle degrades in the following order: adenosine diphosphate, adenosine phosphate, inosine monophosphate (IMP), inosine, hypoxanthine (Hx), xanthine (X), and uric acid [77, 78]. Thus, determination of the Hx or X concentration can monitor freshness of fish and meat. The biosensors are one of the commonly employed analytical methods, since biosensors have the advantages of fast response and simplicity of operation and construction. Xanthine oxidase (XOD) can catalyze the oxidation of Hx to X and the oxidation of X to uric acid, accompanied by the reduction of O_2 to H_2O. Thus, XOD is widely used in development of Hx or X biosensors. The basic principle and electrochemical reactions of amperometric biosensors containing XOD are shown in the following, in which H_2O_2 is generated from oxidation of xanthine catalyzed by XOD and then degraded into $2H^+$, $2e^-$, and O_2 at potential to produce flow of electrons, namely, the current [79]:

$$\text{Hypoxanthine} \rightarrow \text{Xanthine}$$

$$\text{Xanthine} + O_2 + H_2O \rightarrow \text{Uric acid} + H_2O_2$$

$$H_2O_2 \rightarrow 2H^+ + 2e^- + O_2$$

$$2e^- \rightarrow \text{electrode}$$

Recently, NPs have been used in fabrication of electrochemical biosensors to enhance their sensing performance, since NPs have large surface area, high surface activity, and catalytic efficiency. Furthermore, NPs are able to accelerate the electron transfer between electrode and the active site of enzymes [80]. Thus, several biosensors have been developed based on immobilization of XOD onto nanomaterials. For instance, Liu et al. immobilized XOD onto the phosphonic acid-functionalized silica (SiO_2-P) film coated on the surface of glassy carbon electrode to design a novel amperometric Hx biosensor [81]. In this study, the entrapped XOD preserved its activity and created a high formal potential of −0.37 V in neutral phosphate buffer solution due to the advantages of SiO_2-P NPs. It was observed that the biosensor had a wide linearity with Hx in the range of 1.00–261 μM and the detection limit of 0.233 μM at a signal-to-noise ratio (S/N) of 3. In addition, this sensor also had a strong sensitivity to Hx against uric acid and ascorbic acid. When the sensor was applied in detecting Hx in fish samples, the assay results of the sensor were consistent with the reference values. Later, Dervisevic et al. designed a novel nanocomposite host matrix for immobilization of XOD, in which multi-walled carbon nanotube (MWCNT) was incorporated in the poly(GMA-*co*-VFc) copolymer film. The resultant biosensor had a good response to X with three linear ranges (2–28, 28–46, and 46–86 μM), very low detection limit at 0.12 μM, and sensitivity of

16 mA/M. Moreover, the biosensor exhibited very good stability and sensitivity to X against other components, including glucose, ascorbic acid, uric acid, and sodium benzoate. Similarly, an Hx nanosensor was fabricated to evaluate meat freshness in which a hydrothermally synthesized graphene–titanium dioxide nanocomposite was used as platform to immobilize XOD [82]. This sensor demonstrated excellent linearity with the Hx concentration in the range of 20–512 μM, detection limit at 9.5 μM, and sensitivity of 4.1 nA/μM. Similarly, the biosensor also exhibited strong sensitivity to Hx against other components, including glucose, uric acid, and ascorbic acid. However, the presence of X had a minimal interference at ~7%. In addition, the biosensor showed good repeatability (4.3% RSD) and good reproducibility (3.8% RSD). When the biosensor was used to detect Hx in pork tender loins stored at room temperature for seven days, the biosensor response was linearly related ($r = 0.9795$) with the results measured using a standard enzymatic colorimetric method.

Different from the mentioned studies, XOD was covalently immobilized onto boronic acid functionalized gold-coated iron NPs coated on pencil graphite electrode and chitosan bound gold-coated iron NPs coated on the surface of pencil graphite electrode, respectively, thus obtaining amperometric biosensors for determination of Hx or X [83, 84]. The former biosensor responded to Hx at the fastest rate (<3 seconds) at pH 7.2 and 30 °C and exhibited electro catalytic activity to Hx with linear range of 0.05–150 μM and detection limit of 0.05 μM [83]. In addition, when used to detect Hx of fish, chicken, pork, and beef meat, 50% initial activity was kept in the biosensor after its 200 uses over 100 days (stored at 4 °C). The later one also exhibited optimal current response to X within three seconds, but the working range (0.1–300 μM) was different. Its sensitivity was 0.001 169 mA/μM/cm^2 with detection limit of 0.1 μM (S/N = 3). This biosensor kept 75% initial activity after it was used for 100 times over 100 days (stored at 4 °C). It was obvious that electrochemical biosensors resulting from covalently immobilization of XOD were more efficient than those resulting from non-covalent immobilization of XOD. Moreover, the nanomaterial used also affected the performance of electrochemical biosensors.

Moreover, due to production of H_2O_2 during oxidation of Hx by XOD, an economical visual multicolor biosensor was developed based on the etching of Au nanorods to monitor freshness of fish by the naked eye [85]. Au nanorods were quickly etched by H_2O_2 when there was Fe^{2+}. As a result, the LSPR of Au nanorods was regulated, and the color of the system distinctly changed, which were easily distinguished by the naked eye (Figure 8.7). Thus, the Hx content in the samples and food freshness could be easily detected according to the working curves and standard color card. Take *Miichthys miiuy* for example, the sample was brown in the first hour, which indicated that the Hx concentration ranged 0.05–0.15 mM through the standard color card of Hx. Subsequently, the Hx concentration was positively correlated with the storage time when *M. miiuy* was stored at 25 °C. When the fish was stored for 10 hours, the solution became blue and the Hx concentration ranged 0.375–0.45 mM, which suggested that the fish began to deteriorate. After the fish was stored for 20 hours, the Hx concentration kept increasing and the solution became purple,

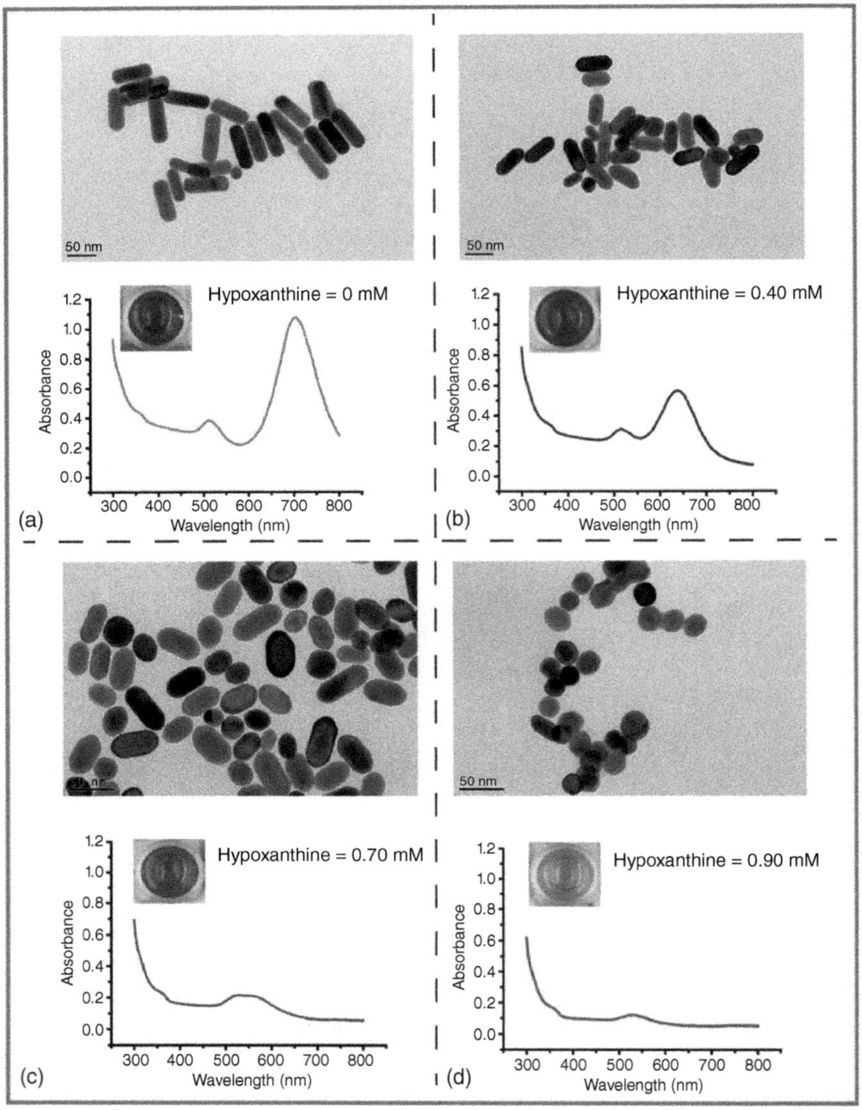

Figure 8.7 TEM images and the UV–vis absorption spectra of Au nanorods at different Hx concentrations. (a) 0 mM; (b) 0.40 mM; (c) 0.7 mM; and (d) 0.9 mM. Source: Chen et al. [85]/ with permission from Elsevier.

which suggested that the fish was fully deteriorated. Therefore, the fish freshness could be semiquantitatively assessed by the naked eye.

In another study, a fluorescence biosensor was proposed for rapid detection of Hx [86]. This biosensor consisted of XOD, *o*-phenylenediamine, and Pt NPs, which were used to mimic the peroxidase. Firstly, Hx reacted with O_2 under the catalysis of XOD, thus producing uric acid and H_2O_2. Catalyzed by Pt NPs, *o*-phenylenediamine was oxidized by H_2O_2, and 2,3-phenylenediamine was produced, which exhibited bright yellow fluorescence [87]. Therefore, the amount of 2,3-phenylenediamine was positively correlated with the Hx concentration. Accordingly, the fluorescence intensity of this sensor had a linear response to Hx ranging 8–2500 μM, and the detection limit was 2.88 μM (S/N = 3). Moreover, due to its excellent selectivity, the biosensor could quickly detect the Hx concentration in fish, shrimp, and squid with good recovery rate (103.94–109.00%). Also, the Pt NPs used in the biosensor was reusable, whose recovery rate remained 91% after three cycles.

8.8 Conclusions and Perspectives

Nanomaterials have been deeply used for the development of different indicator and sensors, and the utilization of nanomaterial exhibits better performance in the communication function, compared with the traditional indicators and sensors. Furthermore, some nanomaterial-based indicators and sensors have been already applied in intelligent packaging of food. Obviously, optical colorimetric sensors have greater potential to be applied in the market than other types of sensors due to the convenience and low cost. Among the nanomaterials, metal NPs, particularly Ag NPs and Au NPs, are attractive due to their LSPR properties. Their LSPR peaks usually have visible spectroscopic responses to changes of their compositions, morphologies, and the surrounding medium. Accordingly, their color is sensitive to those variations, which allow them to be developed as different optical colorimetric sensors. It should be notable that construction of metal NPs-based sensors in a suitable morphology can greatly enhance their sensing performance. Moreover, it seems that hybrid NPs or biopolymer-templated metal NPs also show better sensing performance than single pure NPs. Despite the unique advantages, there are still challenges to use NPs as sensors. For instance, the safety and the cost should be concerned.

On the other hand, when nanoscale polymers are used in indicators or sensors, electrospinning is a simple, facile, and efficient method to construct nanoscale fibers. The properties of the electrospinning solution and processing parameters usually influence the morphology and size of electrospun nanofibers, thus affecting the performance of the obtained sensors. Nanofibrous structures have advantages of high surface area, small pore size, high porosity, and a high absorbance capacity. Therefore, their application can improve the performance of intelligent packaging. Further efforts are required to explore new sensing materials, control the size of the electrospun nanofibers, and improve the sensitivity to different analytes.

In summary, most of the previously mentioned advances still do not achieve commercial application. Therefore, it is essential to propose various new intelligent packaging systems with the economic and technical feasibility, so that these systems can be applied in the market. Construction of intelligent packaging involves professional knowledge in different fields, including food science, material science and physics, etc. Thus, it is suggested that scientists in these fields work synergistically to develop new intelligent packaging that can be applied in the market. Moreover, combination of intelligent packaging and active packaging may be a better choice for distribution and storage of food.

References

1 Kalpana, S., Priyadarshini, S.R., Maria Leena, M. et al. (2019). Intelligent packaging: trends and applications in food systems. *Trends Food Sci. Technol.* 93: 145–157.

2 Balbinot-Alfaro, E., Craveiro, D.V., Lima, K.O. et al. (2019). Intelligent packaging with pH indicator potential. *Food Eng. Rev.* 11 (4): 235–244.

3 Yan, K.L., Takhistov, P.T., and Miltz, J. (2005). Intelligent packaging: concepts and applications. *J. Food Sci.* 70: R1–R10.

4 Janjarasskul, T. and Suppakul, P. (2018). Active and intelligent packaging: the indication of quality and safety. *Crit. Rev. Food Sci. Nutr.* 58 (5): 808–831.

5 Pirsa, S. and Shamusi, T. (2019). Intelligent and active packaging of chicken thigh meat by conducting nano structure cellulose-polypyrrole-ZnO film. *Mater. Sci. Eng., C Mater. Biol. Appl.* 102: 798–809.

6 Ghaani, M., Cozzolino, C.A., Castelli, G., and Farris, S. (2016). An overview of the intelligent packaging technologies in the food sector. *Trends Food Sci. Technol.* 51: 1–11.

7 Ali, A., Xie, F., Yu, L. et al. (2018). Preparation and characterization of starch-based composite films reinforced by polysaccharide-based crystals. *Composites Part B* 133: 122–128.

8 Hu, X., Jia, X., Zhi, C. et al. (2019). Improving the properties of starch-based antimicrobial composite films using ZnO-chitosan nanoparticles. *Carbohydr. Polym.* 210: 204–209.

9 Sharma, C., Dhiman, R., Rokana, N., and Panwar, H. (2017). Nanotechnology: an untapped resource for food packaging. *Front. Microbiol.* 8: 1735.

10 Koosha, M. and Hamedi, S. (2019). Intelligent chitosan/PVA nanocomposite films containing black carrot anthocyanin and bentonite nanoclays with improved mechanical, thermal and antibacterial properties. *Prog. Org. Coat.* 127: 338–347.

11 Ge, Y., Li, Y., Bai, Y. et al. (2020). Intelligent gelatin/oxidized chitin nanocrystals nanocomposite films containing black rice bran anthocyanins for fish freshness monitorings. *Int. J. Biol. Macromol.* 155: 1296–1306.

12 Ahvenainen, R. and Hurme, E. (1997). Active and smart packaging for meeting consumer demands for quality and safety. *Food Addit. Contam.* 14 (6–7): 753–763.

13 Vanderroost, M., Ragaert, P., Devlieghere, F., and De Meulenaer, B. (2014). Intelligent food packaging: the next generation. *Trends Food Sci. Technol.* 39 (1): 47–62.

14 Fuertes, G., Soto, I., Carrasco, R. et al. (2016). Intelligent packaging systems: sensors and nanosensors to monitor food quality and safety. *J. Sens.* 2016: 1–8.

15 Fang, Z., Zhao, Y., Warner, R.D., and Johnson, S.K. (2017). Active and intelligent packaging in meat industry. *Trends Food Sci. Technol.* 61: 60–71.

16 Kuswandi, B. and Jumina (2020). Active and intelligent packaging, safety, and quality controls. In: *Fresh-Cut Fruits and Vegetables: Technologies and Mechanisms for Safety Control* (ed. M.W. Siddiqui), 243–294. Academic Press.

17 Gao, T., Tian, Y., Zhu, Z., and Sun, D.-W. (2020). Modelling, responses and applications of time-temperature indicators (TTIs) in monitoring fresh food quality. *Trends Food Sci. Technol.* 99: 311–322.

18 Wu, D., Zhang, M., Chen, H., and Bhandari, B. (2020). Freshness monitoring technology of fish products in intelligent packaging. *Crit. Rev. Food Sci. Nutr.*: 1–14.

19 Wang, S., Liu, X., Yang, M. et al. (2015). Review of time temperature indicators as quality monitors in food packaging. *Packag. Technol. Sci.* 28 (10): 839–867.

20 Zeng, J., Roberts, S., and Xia, Y. (2010). Nanocrystal-based time-temperature indicators. *Chemistry (Easton)* 16 (42): 12559–12563.

21 Wijaya, Y.N., Kim, J., Choi, W.M. et al. (2017). A systematic study of triangular silver nanoplates: one-pot green synthesis, chemical stability, and sensing application. *Nanoscale* 9 (32): 11705–11712.

22 Lanza, G.A., Perez-Taborda, J.A., and Avila, A. (2019). Time temperature indicators (TTIs) based on silver nanoparticles for monitoring of perishables products. *J. Phys. Conf. Ser.* 1247: 012055.

23 Zhang, X.Q., Ling, J., Liu, C.J. et al. (2018). An irreversible temperature indicator fabricated by citrate induced face-to-face assembly of silver triangular nanoplates. *Mater. Sci. Eng., C Mater. Biol. Appl.* 92: 657–662.

24 Lim, S., Gunasekaran, S., and Imm, J.Y. (2012). Gelatin-templated gold nanoparticles as novel time-temperature indicator. *J. Food Sci.* 77 (9): N45–N49.

25 Wang, Y.-C., Lu, L., and Gunasekaran, S. (2015). Gold nanoparticle-based thermal history indicator for monitoring low-temperature storage. *Microchim. Acta* 182 (7–8): 1305–1311.

26 Wang, Y.C., Lu, L., and Gunasekaran, S. (2017). Biopolymer/gold nanoparticles composite plasmonic thermal history indicator to monitor quality and safety of perishable bioproducts. *Biosens. Bioelectron.* 92: 109–116.

27 Wang, Y.-C., Mohan, C.O., Guan, J. et al. (2018). Chitosan and gold nanoparticles-based thermal history indicators and frozen indicators for perishable and temperature-sensitive products. *Food Control* 85: 186–193.

28 Zhang, C., Yin, A.-X., Jiang, R. et al. (2013). Time–temperature indicator for perishable products based on kinetically programmable Ag overgrowth on Au nanorods. *ACS Nano* 7 (5): 4561–4568.

29 Nopwinyuwong, A., Boonsupthip, W., Pechyen, C., and Suppakul, P. (2013). Formation of polydiacetylene/silica nanocomposite as a colorimetric indicator: effect of time and temperature. *Adv. Polym. Tech.* 32 (S1): E724–E731.

30 Suppakul, P., Kim, D.Y., Yang, J.H. et al. (2018). Practical design of a diffusion-type time-temperature indicator with intrinsic low temperature dependency. *J. Food Eng.* 223: 22–31.

31 Choi, S., Eom, Y., Kim, S.M. et al. (2020). A self-healing nanofiber-based self-responsive time-temperature indicator for securing a cold-supply chain. *Adv. Mater.* 32 (11): e1907064.

32 Mihindukulasuriya, S.D.F. and Lim, L.T. (2014). Nanotechnology development in food packaging: a review. *Trends Food Sci. Technol.* 40 (2): 149–167.

33 Zhang, Y., Yu, K., Jiang, D. et al. (2005). Zinc oxide nanorod and nanowire for humidity sensor. *Appl. Surf. Sci.* 242 (1–2): 212–217.

34 Wang, X.-H., Ding, Y.-F., Zhang, J. et al. (2006). Humidity sensitive properties of ZnO nanotetrapods investigated by a quartz crystal microbalance. *Sens. Actuators, B* 115 (1): 421–427.

35 Zhou, X., Zhang, J., Jiang, T. et al. (2007). Humidity detection by nanostructured ZnO: a wireless quartz crystal microbalance investigation. *Sens. Actuators, A* 135 (1): 209–214.

36 Zhou, X., Jiang, T., Zhang, J. et al. (2007). Humidity sensor based on quartz tuning fork coated with sol–gel-derived nanocrystalline zinc oxide thin film. *Sens. Actuators, B* 123 (1): 299–305.

37 Erol, A., Okur, S., Comba, B. et al. (2010). Humidity sensing properties of ZnO nanoparticles synthesized by sol–gel process. *Sens. Actuators, B* 145 (1): 174–180.

38 Erol, A., Okur, S., Yağmurcukardeş, N., and Arıkan, M.Ç. (2011). Humidity-sensing properties of a ZnO nanowire film as measured with a QCM. *Sens. Actuators, B* 152 (1): 115–120.

39 Hong, H.-S., Phan, D.-T., and Chung, G.-S. (2012). High-sensitivity humidity sensors with ZnO nanorods based two-port surface acoustic wave delay line. *Sens. Actuators, B* 171-172: 1283–1287.

40 Kuang, Q., Lao, C., Wang, Z.L. et al. (2007). High-sensitivity humidity sensor based on a single SnO_2 nanowire. *J. Am. Chem. Soc.* 129: 6070–6071.

41 Marzouk, W., Sakly, N., Roudesli, S. et al. (2016). Ag-nanocomposite based on carboxymethylcellulose for humidity detection: green synthesis and sensing performances. *J. Appl. Polym. Sci.* 133: 43686.

42 Sun, L., Wang, B., and Wang, Y. (2018). A novel silicon carbide nanosheet for high-performance humidity sensor. *Adv. Mater. Interfaces* 5 (6): 1701300.

43 Zhang, J., Sun, L., Chen, C. et al. (2017). High performance humidity sensor based on metal organic framework MIL-101(Cr) nanoparticles. *J. Alloys Compd.* 695: 520–525.

44 Lin, Q., Li, Y., and Yang, M. (2012). Polyaniline nanofiber humidity sensor prepared by electrospinning. *Sens. Actuators, B* 161 (1): 967–972.

45 Lin, Q., Li, Y., and Yang, M. (2012). Highly sensitive and ultrafast response surface acoustic wave humidity sensor based on electrospun polyaniline/poly(vinyl butyral) nanofibers. *Anal. Chim. Acta* 748: 73–80.

46 Fei, T., Jiang, K., Liu, S., and Zhang, T. (2014). Humidity sensors based on Li-loaded nanoporous polymers. *Sens. Actuators, B* 190: 523–528.

47 Liang, S., He, X., Wang, F. et al. (2015). Highly sensitive humidity sensors based on LiCl–Pebax 2533 composite nanofibers via electrospinning. *Sens. Actuators, B* 208: 363–368.

48 Tang, J., Song, Y., Berry, R.M., and Tam, K.C. (2014). Polyrhodanine coated cellulose nanocrystals as optical pH indicators. *RSC Adv.* 4 (104): 60249–60252.

49 Patra, M., Mukherjee, R., Banik, M. et al. (2017). Calcium phosphate-quercetin nanocomposite (CPQN): a multi-functional nanoparticle having pH indicating, highly fluorescent and anti-oxidant properties. *Colloids Surf., B* 154: 63–73.

50 Pourjavaher, S., Almasi, H., Meshkini, S. et al. (2017). Development of a colorimetric pH indicator based on bacterial cellulose nanofibers and red cabbage (*Brassica oleracea*) extract. *Carbohydr. Polym.* 156: 193–201.

51 Ha, C.T., Lien, N.T.H., Anh, N.D., and Lam, N.L. (2017). Development of natural anthocyanin dye-doped silica nanoparticles for pH and borate-sensing applications. *J. Electron. Mater.* 46 (12): 6843–6847.

52 Kuntzler, S.G., Costa, J.A.V., Brizio, A., and Morais, M.G. (2020). Development of a colorimetric pH indicator using nanofibers containing *Spirulina* sp. LEB 18. *Food Chem.* 328: 126768.

53 Van der Schueren, L., Mollet, T., Ceylan, Ö., and De Clerck, K. (2010). The development of polyamide 6.6 nanofibres with a pH-sensitive function by electrospinning. *Eur. Polym. J.* 46 (12): 2229–2239.

54 Sharifabad, A.N. and Bahrami, S.H. (2016). Halochromic chemosensor from poly(acrylonitrile)/phenolphthalein nanofibers as pH sensor. *IEEE Sens. J.* 16: 873–880.

55 Aghaei, Z., Emadzadeh, B., Ghorani, B., and Kadkhodaee, R. (2018). Cellulose acetate nanofibres containing alizarin as a halochromic sensor for the qualitative assessment of rainbow trout fish spoilage. *Food Bioprocess Technol.* 11 (5): 1087–1095.

56 Aghaei, Z., Ghorani, B., Emadzadeh, B. et al. (2020). Protein-based halochromic electrospun nanosensor for monitoring trout fish freshness. *Food Control* 111: 107065.

57 Lee, S.-K., Sheridan, M., and Mills, A. (2005). Novel UV-activated colorimetric oxygen indicator. *Chem. Mater.* 17: 2744–2751.

58 Mills, A. and Hazafy, D. (2009). Nanocrystalline SnO_2-based, UVB-activated, colourimetric oxygen indicator. *Sens. Actuators, B* 136 (2): 344–349.

59 Mihindukulasuriya, S.D.F. and Lim, L.-T. (2013). Oxygen detection using UV-activated electrospun poly(ethylene oxide) fibers encapsulated with TiO_2 nanoparticles. *J. Mater. Sci.* 48 (16): 5489–5498.

60 Lopez-Carballo, G., Muriel-Galet, V., Hernandez-Munoz, P., and Gavara, R. (2019). Chromatic sensor to determine oxygen presence for applications in intelligent packaging. *Sensors (Basel)* 19 (21): 4684.

61 Wen, J., Huang, S., Jia, L. et al. (2019). Visible colorimetric oxygen indicator based on Ag-loaded TiO_2 nanotubes for quick response and real-time monitoring of the integrity of modified atmosphere packaging. *Adv. Mater. Technol.* 4 (9): 1900121.

62 Shang, J., Lin, S., and Theato, P. (2017). Fabrication of color changeable CO_2 sensitive nanofibers. *Polym. Chem.* 8 (48): 7446–7451.

63 Ahmed, I., Lin, H., Zou, L. et al. (2018). An overview of smart packaging technologies for monitoring safety and quality of meat and meat products. *Packag. Technol. Sci.* 31 (7): 449–471.

64 Bumbudsanpharoke, N. and Ko, S. (2018). Nanomaterial-based optical indicators: promise, opportunities, and challenges in the development of colorimetric systems for intelligent packaging. *Nano Res.* 12 (3): 489–500.

65 Santos, M.H.S. (1996). Biogenic amines: their importance in foods. *Int. J. Food Microbiol.* 29: 213–231.

66 Lin, T., Wu, Y., Li, Z. et al. (2016). Visual monitoring of food spoilage based on hydrolysis-induced silver metallization of Au Nanorods. *Anal. Chem.* 88 (22): 11022–11027.

67 Tseng, S.Y., Li, S.Y., Yi, S.Y. et al. (2017). Food quality monitor: paper-based plasmonic sensors prepared through reversal nanoimprinting for rapid detection of biogenic amine odorants. *ACS Appl. Mater. Interfaces* 9 (20): 17306–17316.

68 Sun, Y.G. and Xia, Y.N. (2002). Increased sensitivity of surface plasmon resonance of gold nanoshells compared to that of gold solid colloids in response to environmental changes. *Anal. Chem.* 74: 5297–5305.

69 Cheng, C.S., Chen, Y.Q., and Lu, C.J. (2007). Organic vapour sensing using localized surface plasmon resonance spectrum of metallic nanoparticles self assemble monolayer. *Talanta* 73 (2): 358–365.

70 Chow, C.F., Ho, P.Y., Sun, D. et al. (2017). Development of sensitive and selective food sensors using new Re(I)-Pt(II) bimetallic complexes to detect volatile biogenic sulfides formed by meat spoilage. *Food Chem.* 216: 382–389.

71 Vasta, V., Ratel, J., and Engel, E. (2007). Mass spectrometry analysis of volatile compounds in raw meat for the authentication of the feeding background of farm animals. *J. Agric. Food. Chem.* 55: 4630–4639.

72 Lovestead, T.M. and Bruno, T.J. (2010). Detection of poultry spoilage markers from headspace analysis with cryoadsorption on a short alumina PLOT column. *Food Chem.* 121 (4): 1274–1282.

73 Varlet, V. and Fernandez, X. (2010). Review. Sulfur-containing volatile compounds in seafood: occurrence, odorant properties and mechanisms of formation. *Food Sci. Technol. Int.* 16 (6): 463–503.

74 Ryspayeva, A. Jones, T.D.A., Hughes, P.A. et al. (2018). PEI/Ag as an optical gas nano-sensor for intelligent food packaging. *18th IEEE International Conference on Nanotechnology (IEEE-NANO).* Tyndall Natl Inst, Cork: IEEE.

75 Zhai, X., Li, Z., Shi, J. et al. (2019). A colorimetric hydrogen sulfide sensor based on gellan gum-silver nanoparticles bionanocomposite for monitoring of meat spoilage in intelligent packaging. *Food Chem.* 290: 135–143.

76 Chow, C.F. (2020). Biogenic amines- and sulfides-responsive gold nanoparticles for real-time visual detection of raw meat, fish, crustaceans, and preserved meat. *Food Chem.* 311: 125908.

77 Lawal, A.T. and Adeloju, S.B. (2012). Progress and recent advances in fabrication and utilization of hypoxanthine biosensors for meat and fish quality assessment: a review. *Talanta* 100: 217–228.

78 Dervisevic, M., Custiuc, E., Cevik, E., and Senel, M. (2015). Construction of novel xanthine biosensor by using polymeric mediator/MWCNT nanocomposite layer for fish freshness detection. *Food Chem.* 181: 277–283.

79 Pundir, C.S., Devi, R., Narang, J. et al. (2012). Fabrication of an amperometric xanthine biosensor based on polyvinylchloride membrane. *J. Food Biochem.* 36 (1): 21–27.

80 Pandey, P., Datta, M., and Malhotra, B.D. (2008). Prospects of nanomaterials in biosensors. *Anal. Lett.* 41 (2): 159–209.

81 Liu, M., Chen, S., Zhao, X. et al. (2013). Biocompatible phosphonic acid-functionalized silica nanoparticles for sensitive detection of hypoxanthine in real samples. *Talanta* 117: 536–542.

82 Albelda, J.A.V., Uzunoglu, A., Santos, G.N.C., and Stanciu, L.A. (2017). Graphene-titanium dioxide nanocomposite based hypoxanthine sensor for assessment of meat freshness. *Biosens. Bioelectron.* 89 (Pt 1): 518–524.

83 Devi, R., Yadav, S., Nehra, R., and Pundir, C.S. (2013). An amperometric hypoxanthine biosensor based on Au@FeNPs for determination of hypoxanthine in meat samples. *Int. J. Biol. Macromol.* 62: 629–635.

84 Devi, R., Yadav, S., Nehra, R. et al. (2013). Electrochemical biosensor based on gold coated iron nanoparticles/chitosan composite bound xanthine oxidase for detection of xanthine in fish meat. *J. Food Eng.* 115 (2): 207–214.

85 Chen, Z., Lin, Y., Ma, X. et al. (2017). Multicolor biosensor for fish freshness assessment with the naked eye. *Sens. Actuators, B* 252: 201–208.

86 Chen, J., Lu, Y., Yan, F. et al. (2020). A fluorescent biosensor based on catalytic activity of platinum nanoparticles for freshness evaluation of aquatic products. *Food Chem.* 310: 125922.

87 Sun, J., Wang, B., Zhao, X. et al. (2016). Fluorescent and colorimetric dual-readout assay for inorganic pyrophosphatase with $Cu(^{2+})$-triggered oxidation of o-phenylenediamine. *Anal. Chem.* 88 (2): 1355–1361.

9

Nanostructure-Based Edible Coatings as a Function of Food Preservation

Anna Rafaela C. Braga[1], *Josemar G. Oliveira Filho*[2], *Ailton C. Lemes*[3], *and Mariana B. Egea*[4]

[1] *Universidade Federal de São Paulo (UNIFESP), Department of Chemical Engineering, Rua São Nicolau 201, Diadema, São Paulo 09972-270, Brazil*
[2] *São Paulo State University (UNESP), School of Pharmaceutical Sciences, Rodovia Araraquara - Jaú Km 1, Araraquara, São Paulo 14800-903, Brazil*
[3] *Federal University of Rio de Janeiro (UFRJ), School of Chemistry, Department of Biochemical Engineering, Av. Athos da Silveira Ramos, 149 21941-909, Rio de Janeiro, Brazil*
[4] *Goiano Federal Institute of Education, Science and Technology, Campus Rio Verde, Rodovia Sul Goiana, Km 01, Rural Area 75901-970, Rio Verde, Goiás, Brazil*

9.1 Nanotechnology in Food Packaging: Principles and Applications

In recent decades, technological advances in materials and processes have established the necessity of the expansion of packaging fabrication with more specific actions and functions. Another factor that has contributed to the advancement of the packaging sector is the need for less polluting materials [1, 2].

Food packaging production applies numerous types of materials alone as well as the combination of one or more materials. Thus, the polymeric materials have improved to obtain "perfect" packaging that combines technological, economic, environmental, and social properties [3, 4].

In this sense, nanotechnology has the possibility for application in food packaging, especially concerning the development of new and more specific packages from the use of different nanocomposites, such as nanoparticles, nanodispersions, nanolaminates, and nanotubes. Nanocomposites can provide differentiated functions with better application properties when designing the composition of the materials to meet the specific application and to optimize the food properties. Additionally, the use of nanomaterials can improve fabrication processes to deliver better product properties and new feed functionality [5–8]. Figure 9.1 shows the principles of packaging and the expected functions for these materials.

The request for biocompatible materials provided by renewable resources is increasing, and a feasible option is the production of constituents for food packaging applications. Nanotechnology allows selective manipulation of material at the nanoscale to modify the characteristics in an advantageous way [9].

Nanotechnology-Enhanced Food Packaging, First Edition. Edited by Jyotishkumar Parameswaranpillai, Radhakrishnan Edayileveettil Krishnankutty, Aswathy Jayakumar, Sanjay Mavinkere Rangappa, and Suchart Siengchin.

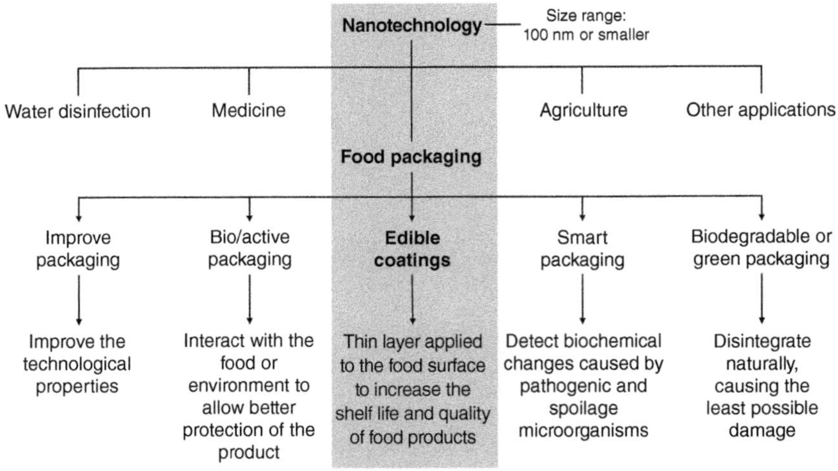

Figure 9.1 Uses of nanotechnology in different food packaging.

Nanomaterials can be tools for the development of food processes in areas where stability and shelf life are sought alongside customer sensory acceptance, showing possibilities to unravel the most persistent problems of the food industry. The nanotechnology in food packaging finds its primary purpose to be the protection of products against biological action, defense against metallic components, and improvement of technological properties. Moreover, nano-production can deliver functional food constituents, thereby increasing the bioaccessibility, bioavailability, nutritional quality, and technological features, making these compounds more viable for industrial applications [10].

The primary purpose linked to the compounds of nanomaterial for food packaging applications is the fact that particle size is reduced to the nanoscale, which increases surface/volume ratio, causing an increase in the reactivity and altering mechanical, electrical, and optical properties [11, 12].

The inclusion of food ingredients into nanostructures can also optimize the use of packaging with the addition of compounds presenting different bioactive properties such as antimicrobials and antioxidants. Bioactive nanoparticles and nanocomposites are capable of keeping bioactive compounds intact until their delivery to the food product, which improves flexibility, barrier properties, gases, and humidity as well as the UV radiation absorption of the materials and the temperature stability [13].

In practice, nanotechnology possesses potential to be useful in all stages of the food processing chain. It can be used to monitor the quality and deliver its functionality to the consumer's table, including packaging products that interact with foods to modify their properties or prevent spoilage [14].

9.2 Edible Coatings

Edible coatings are described as a thin layer of edible material, commonly not surpassing 0.3 mm, applied to the food surface, in addition to or as a replacement for natural protection coverings. Edible coatings are frequently applied to increase the shelf life and quality of food products throughout the storage time [15, 16]. This edible coating system intended for food must be able to improve the quality of the products that it coats, acting as a carrier for ingredients, such as food coloring, flavoring, antimicrobials, and antioxidants, among others [17, 18].

Another positive characteristic attributed to edible coatings is the reduction of synthetic residues, mainly of fossil origin, since the material applied in its elaboration must be produced from nontoxic material and be generally considered as safe (GRAS). In addition, their application is aimed at reducing microbial (deteriorating or pathogenic) growth on the surface of food [19].

The ideal edible coating is one that is able to create suitable conditions for the group of foods for which it was developed and improved, not only creating favorable conditions for the preservation of the characteristics expected for the products that it coats but also creating conditions capable of controlling chemical and enzymatic reactions. In addition, ideally, coatings need to present the characteristics required in terms of barriers against gases, humidity, and microorganisms, among others, as well as the general requirements of another packaging used for food [20].

9.2.1 Chemical Characteristics of Edible Coatings

In order to perform functions appropriate to the products to which they are applied, edible coatings are developed with characteristics that allow interaction with food and, also, favor the preservation of its original characteristics throughout its shelf life. Thus, edible coatings are commonly classified according to the molecular structure of the film-forming composition. The materials applied in edible coatings can be classified into two categories [21–23]:

- *Hydrophilic*: Materials with structures in which there is a predominance of amino or hydroxyl and carboxyl groups (OH, COO–, NH_3) characterized with the presence of polar covalent bonds. Due to the characteristics of these groups, the carbon chain has partially positively charged sites and others that are negatively charged. This characteristic of the chemical structure favors the accumulation and rearrangement of polar molecules and, especially, water, around these sites. Hydrophilic materials usually present good solubility in aqueous media, favoring a better dispersion of the solute and a more homogeneous formation of the film. Depending on the chemical structure, gels may form or even require chemical changes for complete solubilization. Hydrophilic materials used in food packaging include cellulose, chitin, xanthan gum, guar gum, pectin, starch, and polyelectrolytic polysaccharides, such as carboxymethylcellulose, chitosan, and alginate.

- *Hydrophobic*: Materials characterized by molecules in which structures predominate whose bonds tend to be electrically neutral, that is, they do not form defined polar regions. They include alkyl (CH_3, CH_2 $-CH_3$, etc.) and aromatic groups. In the presence of water, these materials tend to agglomerate and exclude polar molecules from their surroundings. For this class of materials, suitable solvents must be more nonpolar, such as alcohol, acetone, and hexane. The hydrophobic material used in food packaging includes hydrophobic proteins (predominance of nonpolar substituents), waxes, kinds of paraffin, oils and fatty acids, and aliphatic chain substituents in these oils and fatty acids.

Hydrophilic or hydrophobic materials will influence the characteristics of edible coatings when applied to food. Good adhesion of the coating to the food surface is essential for it to perform its function [24]. Wettability is used to quantify the interfacial interaction that occurs between the food surface and the coating. This variable must be taken into account when assessing the performance of the coating solution on the food surface [25]. Wettability can be defined as a dispersion coefficient (W_s), affected by adhesion (W_a) and cohesion (W_c), which interrelate in equations [24, 26]. The dispersion coefficient measures the tendency of the liquid phase to diffuse over a second liquid or solid phase. Enhanced adhesion of the coating on the surface is measured (W_s) as closer to zero, denoting that the surface is well coated. If the W_s presents positive values, the phase is considered to be spread. Nevertheless, if it is negative, the wetting is not complete. The reduction of cohesive forces leads to an upsurge in wettability [25].

9.2.2 Methods to Apply Edible Coatings

The immersion (Figure 9.2a) and spraying (Figure 9.2b) techniques are the most common methods to apply edible coats on foods. Dripping and foaming techniques are also available, but these methods are rarely practiced in commercial and laboratory scales. The adherence of the edible coating on the food surface depends on the viscosity and density of the solution, and surface tension of the food, which can be applied as a single or multilayer coating [25, 27].

Immersion is the procedure in which the entire surface of the food comes into contact with the film-forming solution, allowing complete wetting on the surface. After that, the coating solution is drained to eliminate the overload coating from the food surface. Lastly, the food is dried to produce a remarkably intact coating with the surface. The immersion technique can be used to apply filmogenic solutions that have a wide viscosity range [28, 29]. By promoting the direct immersion of the food in the coating solution, the immersion method can confer gradual contamination during the process, giving a high microbial load to products coated with this solution. Thus, proper sanitation of the product to be covered must be considered, as well as the replacement of the solution frequently as a preventive measure.

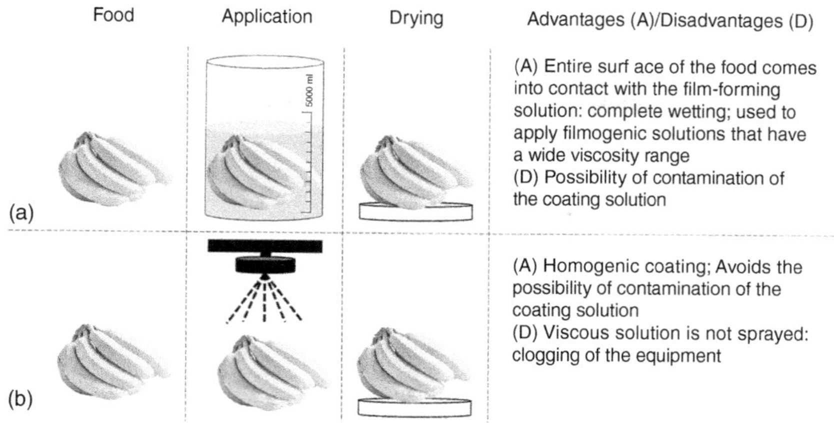

Figure 9.2 Immersion (a) and spraying (b) techniques to apply edible coats on foods.

Spraying technique provides a homogenic coating and additionally avoids the possibility of contamination of the coating solution [30]. In the spraying procedure, the food product is placed on a plate rotating at a coordinated speed under a manual spray gun. This procedure is repeated until the preferred coating thickness is reached [27]. A critical point of this system is that viscous solution cannot be sprayed and produces clogging of the equipment [31].

The method of application of the edible coating influences the dispersion mechanism of the solution on the food to be coated, as well as the formation and characteristics given to the coating at the end of the application [32]. Thus, during the establishment of the process and the choice of the covering method, one must take into account which mode has the best efficiency in relation to the food that will be subjected to the coating, as well as the cost and the availability of equipment for operationalizing the process, among other factors.

Additionally, in the scale-up case, the adoption of additional steps to the process, such as the drying step [33], should be taken into account, in order to make the coating more feasible and with greater standardization and operability.

9.2.3 Materials Used in the Edible Coatings

The compounds used in the preparation of the coating solution must be safe for the consumer and the environment, presenting characteristics such as low cost, easy obtainability, and absence of toxicity. The coating solution also needs to confer the appropriate properties and functionality required for application in food.

The raw materials utilized in the development of edible coatings can be of animal origin, vegetable origin, or a combination of both. The four main structural materials usually applied for coating elaboration include polysaccharides, lipids, proteins, and compounds of more than one (composite). It is important to highlight that most of these materials can be obtained from agro-industrial residues, which, after appropriate treatments, can be used as raw material to produce the coating solution. Additionally, raw materials from agro-industrial residues have the

Table 9.1 Summary of diverse structural materials frequently used for edible coating.

Raw material	Main matrices	Positive points	Negative points	References
Polysaccharide	Starch, chitosan, alginate, cellulose and its derivatives, and pectin	Good gas and mechanical barrier properties	Low moisture barrier due to hydrophilic nature	[34, 35]
Lipid	Animal and vegetable waxes, vegetable oil, and fatty acids	Good moisture barriers with a shiny appearance	Undesirable mechanical and gas barrier properties	[27, 36]
Protein	Gelatin, casein, whey protein, zein, soy protein, myofibrillar protein, and quinoa protein	Desirable gas barrier properties without aerobic conditions	Brittle and susceptible to cracking	[37]
Composite	Combination of polysaccharide and/or protein with lipids	Good moisture and gas barrier	Formation of homogeneous emulsion	[35, 38, 39]

advantage of contributing to the reduction of environmental impacts, are found in large quantities, have low cost, and can provide greater added value. The summary of diverse structural materials frequently used for edible coating as well as their main properties are shown in Table 9.1.

Polysaccharides (starches, pectins, chitosan, gums, and cellulose derivatives) are widely reported in the literature for edible coating applications [27]. Polysaccharides are extremely available, allergen-free, and typically water-soluble, producing suitable coating materials. These compounds possess outstanding mechanical properties and gas barrier characteristics as a result of a well-ordered and well-compacted hydrogen-bonded network structure [34, 35]. Lipids (fatty acids, vegetable oil, or waxes) are hydrophobic and produce coatings with excellent moisture barrier properties. However, these coatings do not have good mechanical and gas barrier properties [27, 36]. On the other hand, proteins (wheat gluten, gelatin, corn zein, whey, and soy) provide good mechanical properties and gas barrier properties. The disadvantage of protein-based coatings is that they form brittle and crack-susceptible materials [35, 37]. Composite (or emulsion) coatings are a combination of hydrocolloid components (proteins or polysaccharides) and lipids. The combination of the advantages of hydrocolloid and lipid components in composite coatings can mask or reduce the limitations of each isolated component [40].

9.2.4 Incorporation of Nanomaterials in Edible Coatings

The physical and mechanical changes in the edible coatings depend not only on the material incorporated but also on the polymer matrix used in the production of the coating, as well as their interaction. Among these materials, protein-based (e.g. corn zein, wheat gluten, and soy protein isolate) or polysaccharide-based (e.g. starch, cellulose, carrageenan, and pectin) films are the most used as a polymeric matrix. In this sense, to improve the mechanical and barrier properties of the coatings, and to confer different bioactivities or functionality, such as antimicrobial activities, antioxidants, among others, the incorporation of nanosized filling materials in the biopolymer has been carried out to create a bionanocomposite polymer [41]. The added nanocomposites have the ability to improve the functional properties, morphology, and stability of the polymeric matrix used as a coating [42], being able to produce a variety of materials with improved characteristics and required for application in the food industry, especially in the edible coating of certain food products.

Nanostructured coatings perform an essential role in food because they contain biocomposites that act in the protection or control of the alteration in food. In addition, nanotechnology improves this role since packaging also has the function of physically protecting food against mechanical impact during transport, storage, and marketing [43].

Several nanocomposites have already been presented as potential fillers in edible coatings, such as nano-titanium dioxide (nano-TiO_2), nano-silver (AgNP), and silica dioxide (nano-silica, nano-SiO_2), as well as chitosan and cellulose nanofibers. Nano-TiO_2 is a chemically inert material extensively used in food, medical, and biological products. TiO_2 produces hydroxyl radicals and reactive oxygen species (ROS) under UV light and then reacts with organic molecules to disseminate antimicrobial and ethylene photodegradation activity [44]. AgNPs exhibit a unique biological property: high antimicrobial activity [45]. Ahmed et al. [46] reported that AgNPs show resistance to several microorganisms, including bacteria, fungi, and viruses. However, AgNPs cause toxicity when they easily migrate from the packaging matrix to the food, therefore limiting their food applications [47, 48]. Nano-SiO_2 can increase the water and gas barrier properties of the coating, maintaining quality and prolonging food life [49, 50].

Chitosan is a mucopolysaccharide narrowly related to cellulose and acquired by deacetylation of the compound chitin (found in the exoskeletons of crustaceans) [51]. Chitosan has been utilized for the elaboration and formulation of nanoparticles for ionotropic gelation due to its beneficial physicochemical and biological properties. In this same sense, cellulose also seems to be a good alternative for the production of nanocomposites. Nanocellulose is manufactured via acid hydrolysis and integrated with base matrices through wet processing. The tensile strength test, food protection, transparency in visible and UV, and water contact angle are also performed on the nanocomposite films [52].

In this sense, films developed and filled with nanomaterials serve as a theoretical and practical basis for the development of edible coatings, through the determination of their properties, interactions, functionalities, and challenges to be overcome. Table 9.2 presents different polymeric matrices produced from the incorporation of nanomaterials in edible films and coatings, their characteristics, and their application.

Table 9.2 Polymeric matrix with the nanomaterial-filled and edible coatings, their characteristics, and their application.

Filled	Polymeric matrix	Food application	Characteristics obtained of nanocomposites	References
Films				
Nano-SiO_2 (NS)	Whey protein concentrate	–	↑ Tensile strength ↓ Elongation at break and water vapor permeability Uniform distribution of NS into a polymer matrix	[53]
Chitosan nanofiber and nano-formulated cinnamon oil (CiEO)	Whey protein isolate (WPI)	–	↓ Water solubility, water vapor permeability of WPI-based film, tensile strength Good complexation between WPI and chitosan nanofiber with fiber networks well distributed throughout the films	[35]
Edible coatings				
Chitosan nanoparticles (CN)	Chitosan	Banana fruits	↑ Shelf life (when compared to uncoated fruit) ↓ Fruit ripening, weight loss, and deterioration process CN did not affect the sensory quality of treated fruits	[54]
ZnO	Chitosan	Okra	↓ Fungi and bacteria growth after 12 d of storage Not affecting the quality attributes of okra, such as pH, total soluble solids, moisture content, and weight loss	[55]
Nano-chitosan	Chitosan	Peach fruits Cv. desert Red	↓ Fruit decay percentage and total soluble solids/acid ratio compared with other treatments, and fruit weight losses Nano-chitosan maintained fruit pulp firmness	[56]
Chitosan/silver nano	Chitosan	Apples and tomatoes	↑ Shelf life ↓ Ripening of the fruits Promote moisture content control between fruit and external environment, and acting in the prevention of contamination by fungi	[57]

Silver–montmorillonite (combined with modified atmosphere packaging)	Montmorillonite	Fior di latte cheese	↑ Shelf life ↓ Proliferation of *Pseudomonas* spp. The coating needs to be removed before eating	[58]
Chitosan-based nanoparticles	Chitosan	Grapes	↓ Ripening process of the grapes, weight loss, soluble solids, and sugar content Preserved the sensory properties	[59]
Chitosan-based nanoparticles	Chitosan	Freshly cut apples	↑ Antimicrobial activity (mesophilic and psychrotrophic bacteria, molds, and yeasts) ↓ Darkness in coated apples	[60]
Nano-ZnO	Carrageenan	Fresh mangoes	↑ Tensile strength, antimicrobial properties against *Escherichia coli* and qualities and shelf life of fresh mangoes ↓ Rate of water vapor transmission of the film	[61]
Ag/chitosan	Chitosan	Freshly processed papaya	↑ Sensory quality of coated melons ↓ Respiration rate and ethylene production of freshly processed papaya samples, mesophyll count during the last stages of melon storage, and translucency and texture/firmness degradation after 13 d at 5 °C	[62]
Cellulose nanocrystal	Chitosan/oleic acid	Pears	↑ Post-harvest preservation for pears ↓ Respiration rate and peak ethylene production of coated pears during ripening under environmental conditions over 2 mo Maintain the titratable acidity and total soluble solids of coated pears after 7 mo	[63]
Cellulose nanocrystal	Starch	Pears	↑ Barrier agent in coating ↓ Activities of peroxidase and polyphenol oxidase of the pears during the storage Not change the color of the film; maintain the color, texture, permeability of the cell membrane, the total content of total phenolics, soluble solids, and titratable acids in the pears	[64]

(Continued)

Table 9.2 (Continued)

Filled	Polymeric matrix	Food application	Characteristics obtained of nanocomposites	References
Nano-TiO_2	Chitosan	Mangoes	↑ Time before the appearance of the respiratory peak (climacteric peak) of mangoes stored at 13 ° C, firmness ↑ The activity of the peroxidase (POD) and polyphenol oxidase (PPO) enzymes, the total phenolic and flavonoid contents (when compared to uncoated fruit) ↓ Soluble solids and the malonaldehyde content of mangoes	[44]
Nano-SiO_2 (NS)	Chitosan	Fresh longan fruits (*Dimocarpus longan* Lour. cv Shijia)	↑ Shelf life ↓ Browning index by decreasing the activity of peroxidases and polyphenol oxidases; mass, soluble solids, and vitamin C content Inhibited the increase in the amount of malondialdehyde in fresh longan fruits (*Dimocarpus longan* Lour. cv Shijia)	[49]
AgNPs	Carboxymethyl-cellulose/guar gum	Kinnow fruit (*Citrus reticulata*)	↑ Ascorbic acid, total phenolics, and antioxidant activity when the fruits were stored at 4 °C, as well as rot in fruits stored at 10 °C ↓ Soluble solids, total sugars, mass of the fruits of *Citrus reticulata* cv. Blanco Total aerobic psychrotrophic bacteria and yeasts and molds were observed in all treatments during storage, and growth was not affected in fruits coated at 4 °C	[65]
Nano-SiO_2 (NS)	Konjac glucomannan (KGM)/ carrageenan (KC)	White mushrooms (*Agaricus bisporus*)	↑ Absorbance of the UV light (300 nm) by the KGM/KC threefold ↓ Gas permeability of the coatings and effect of the UV light on the food quality Effective application for extending the storage time of white mushrooms (*Agaricus bisporus*)	[66]
Nano-TiO_2	Chitosan	*Ginkgo biloba* seeds	↑ Mechanical properties of the composite films ↓ Water vapor and gas permeability, decay rate of *Ginkgo biloba* seeds, shrinkage rate, respiration rate, ethylene production rate, electrolyte leakage rate, and superoxide anion (O_2^-) production rate and the accumulation of malondialdehyde (MDA) content	[67]

In general, the incorporation of nanostructured materials in polymeric matrices (Table 9.2) is able to improve characteristics considered important for the production of edible food packaging and coatings, such as increasing the material's resistance; changing permeability; protecting against microbial growth, including deteriorating and pathogenic; and preserving the original characteristics of the tested food products (color, texture, and chemical composition, for example), which lead to prolonged shelf life with reduction of the respiratory rate and ethylene production in vegetables.

In this sense, since nanostructure is a new technology not yet implemented in the food industry, development care, assessment of potential risks, and regulation must be taken into account, as the safety of consumers and protection of the environment must be guaranteed.

9.3 Safety of Nanocomposite for Application of Edible Coatings

Nanocomposites have several properties, in addition to their reduced dimensions, that allow them to perform enhanced functions when appropriate [68], allowing the design and incorporation of these materials in packaging and edible coatings to foods [69, 70]. The use of the nanostructured materials requires, like any other new material, more specific tests to identify possible impacts on safety or efficiency of the product. In addition, it is necessary to establish procedures to identify the forms of exposure, distribution, metabolism, excretion, and toxicological effects [71].

The procedures are necessary since the bioavailability of nanomaterials can be higher when compared with conventional materials, due to their smaller size and consequently larger surface area, which directly influences their resistance, reactivity, and other properties that can also indicate higher absorption capacity for cells, tissues, and organs [72, 73]. When misused and without proper study, the nanomaterials can present a potential for penetration into the human body in a disordered way, avoiding the control and protection mechanisms and crossing membranes and causing potential damage. On the other hand, when produced in an appropriate, controlled manner and by processes of optimization of size, shape, and/or surface chemistry, they have adequate effects on uptake, cell targeting, and improved functionalities [74].

The use of nanomaterials in food systems has possibilities that range from direct incorporation into food to incorporation into packaging, both during food processing. The application of nanomaterials in packaging systems, when in contact with food, must follow the parameters and requirements already adopted for other packaging, regardless of the type of material. The requirements are as follows: be nontoxic and compatible with the product; provide sanitary protection; protect against the passage of moisture, gases, and light and resist physical impacts; have a good appearance and good impression of the necessary information; present ease of opening and subsequent closing; make it possible to limit weight, shape, and size; be transparent in some cases and opaque in others; be quickly eliminated to avoid environmental problems; have economic viability; and have unique characteristics [75–77].

Nanocomposites can also be successfully applied in the development of edible coatings to improve properties and also incorporated with substances with distinct properties (hydrophilic and lipophilic) and can be released during the storage period, increasing shelf life of products, acting directly in the protection of oxidative processes, microbial, and enzymatic control, among others, without modifying the original sensory characteristics of the products. In the same way, for the development of edible nanostructured packaging, it is also necessary to adopt nontoxic components, derived from natural, economic, and readily available sources. These are the same requirements used for conventional packaging, concerning quality, safety, and food stability [78, 79].

Despite the myriad of properties of nanomaterials in packaging systems, these same characteristics require the adoption of care and complete determinations so that they can be used safely, which requires adequate studies of their properties, interactions, and migration. Different conditions can influence the migration of nanostructured components present in food packaging such as temperature, the mechanical stress on the packaging, the characteristics of the packaging related to the polymer structure, viscosity, and pH. Thus, the nanoparticle migration can also be influence by the food composition and the contact time, as well as the concentration, particle size, and molecular properties of nanoparticles, and the weight, solubility, and diffusivity of the polymer [80, 81].

One of the problems in determining the biological/toxicological effects is the diversity of physical and chemical properties that nanoparticles make possible through their use. Thus, it is necessary to establish predictive toxicity methods in order to characterize potential risks of the rapid and growing number of new nanoparticles available for application [82]. The expression of the toxicological activity of the nanoparticles in relation to reference nanoparticles that have already been subjected to rigorous tests and were considered to be of low or very high toxicity has already been addressed as one of the alternatives to solve the problem of lack of information and data on toxicity. In this way, this procedure would classify nanoparticles of unknown toxicity and, thus, provide necessary information for the identification of hazards. Other possibilities for determining the safety of these new nanomaterials, with potential applications in packaging systems, include the use of *in vitro* tests validated by *in vivo* tests [82] and also the adoption of a hazard scale based on greater reactivity (or effect) per unit surface area [83].

Few data regarding the toxicity of nanomaterials used in food packaging or edible coatings are available in the literature. However, there is a consensus that consumers should be aware and informed about the issues of functionality and possible dangers of this technology, as they have a significant impact on the development and implementation of emerging technologies, especially those employed in the food sector. In addition, regardless of the application, it is known that the use of nanoparticles must present a high level of protection of public health and consumer safety, as well as protection of the environment.

9.4 Nanotechnology Regulation

The application of components in food and packaging is governed by control measures to guarantee the safety and protection of consumers, workers, and the environment, also serving to control the actions of public or private companies and, furthermore, to standardize the productive process through consistent criteria and safer routines.

The adoption of regulations, especially regarding the implementation of new technologies, is critical, taking into account that every new material contains numerous uncertainties and properties that need to be understood before its effective implementation. Thus, considering the advances in the area of nanomaterials for food packaging and edible coatings, and even so that society can benefit from these new applications, there is a great challenge in order to overcome barriers related to the development of regulatory frameworks and certain requirements of security, being indisputable that these are applied and developed together.

In this sense, despite the positive evidence presented about the enhancement of mechanical, technological, and functional properties of packaging and edible coatings made with nanocomposites and also the application in products, the survey of risks and functionality of these materials is still neglected by consumers, researchers, and producers, as well legal frameworks. According to FAO [82], the acceptance of products with nanocomposites by consumers depends on the degree of transparency and confidence in regulatory systems as well as the ability to protect them against possible risks. Thus, it is important to highlight that the regulatory frameworks should be created and constantly updated to fill the gaps created by this new technology.

In general, there is no consensus or global regulation of the sector, with few protocols to assess the environmental impact, as well as established exposure limits, as it exists for other environmental components, such as heavy metals and gases, among others. In Brazil no approved legislation directed to the sector of nanostructured products is verified, and the products that present this type of material in their composition need to be evaluated on a case-by-case basis, according to the responsible body, which creates a barrier and subjectivity to the process. From 2005 to 2015, several projects were proposed for consideration in the Chamber of Deputies but were filed, not being included as a bill for the regulation of the sector. Among the proposed projects were as follows: (i) about research and the use of nanotechnology in the country and creation of the National Technical Commission on Nanosecurity–CTNano with instituted the Nanotechnology Development Fund (PL 5076/2005), (ii) about the regulation of the labeling of nanotechnology products (PL 5133/2013), and (iii) about the national nanotechnology policy, research, production, the disposal of tailings, and the use of nanotechnology in the country (PL 6741/2013) [84].

More recently, in 2019, the ordinance n° 3.459 was created by the Ministry of Science, Technology, Innovations, and Communications of the Brazilian government, which institutes the Brazilian Nanotechnology Initiative as a strategic program to encourage nanotechnology inside the country. The ordinance aims to structure governance and (i) coordinate the State's efforts on the subject of nanotechnology; (ii) promote scientific and technological advancement, strengthening, and national innovation related to the properties of matter on a nanometer scale; (iii) encourage the development of new technologies and the transfer of knowledge and technologies, associated with nanotechnology, from academia to the public and private sectors, with a view to generating wealth, employment, and national growth; (iv) mobilize, articulate, and foster national public and private actors to act in a coordinated manner in the development of processes, products, instrumentation, and innovations in the nanotechnology area; (v) guarantee universal access to advanced infrastructure in the nanotechnology area for production, characterization, scaling, and technological development for the scientific community and for the public and private sectors; and (vi) promote the training, qualification and fixing of human resources, education in nanotechnology, and its dissemination. In this ordinance, the following themes were prioritized: nanomaterials and nanocomposites, nanosensors and nanodevices, bio-based nanomaterials, nanopharmaceuticals and nanomedicine, nanosecurity, health, environment, agribusiness and food, energy, defense and national security, and urban mobility and infrastructure.

FDA-mandated regulation [85], in the United States, focuses on understanding the interactions of nanomaterials with biological systems and the appropriateness of testing approaches to assess the safety, efficacy, and quality of products containing nanomaterials. In addition, it considers that these nanomaterials have the potential to exhibit new or changed characteristics, allowing the development of new products and, therefore, they need tests that determine the effects on safety, efficacy, or other attributes of the product. The assessment is carried out in accordance with legal standards and may vary between the various product categories that the FDA regulates. For technical evaluation, the agency considers the attributes related to the technical characteristics of the products, considering the specific effects on each product and its intended use, as well as the evaluation of variations in legal standards for different classes of products, considering safety or efficacy standards. In addition, some products are subject to pre-market review, requiring applicants to submit additional data to answer questions related to product safety, effectiveness, or regulatory status. For products that are not categorize as a pre-market review, the FDA recommends consultation to reduce the risk of unintended damage to human or animal health. Post-market monitoring will be continuous for products that contain nanomaterials, and measures are taken when necessary to protect consumers. Despite post-market control by the FDA, industries remain responsible for meeting all legal requests, comprising safety standards, regardless of the emerging nature of a technology involved in the manufacture of a product.

The European Union (EU), in turn, has regulated the use of nanomaterials by updating its documents (EU 2015/2283) and facilitating the production of new foods and the use of innovative ingredients on the EU market that have food security for being consumed. Products resulting from state-of-the-art production processes and practices, including nanomaterials, were included in the category of new foods (products that were not produced or used before 1997). Moreover, EU [86] requires food products that incorporate nanomaterials to have the word "nano" on the label for each nanoscale ingredient (EU 1169/2011).

Despite the data presented previously, there are few regulations for products that have nanomaterials in their composition, even in countries that conduct most of the development and application of these materials. However, several challenges need to be overcome to ensure the effective regulation of food products developed using nanotechnologies and to promote a standardized approach among countries, such as the assessment of risk and safety of nanoparticles, terminologies, definitions, traceability of nanomaterials, and regulations for labeling.

9.5 Final Considerations and Outlook

Consumers are increasingly looking for foods with improved physicochemical and sensory properties, as well as products with greater safety and, consequently, longer shelf life. To this end, several alternative methods, such as the application of edible coatings on food, can be used to achieve these requirements, replacing methods that are more expensive and difficult to access. The use of edible coatings with properties enhanced by the use of nanocomposites also gains special attention due to the possibility of developing coatings with different properties, which can provide better functionality due to the addition of components with specific activities, such as antioxidants and antimicrobials, or with specific functional characteristics.

In addition, the use of agro-industrial waste as a source of raw material for the construction of nano-polymer matrixes for the edible covering of certain food groups stands out, which can contribute to reduce environmental impacts, add value to a raw material that would be discarded, and also reduce operational costs for the implementation of the technique.

Finally, for the effective implementation of the edible coating technique for foods, including the incorporation of nanocomposites, the global trend will be the adoption of clearer criteria to assess the toxicity or not of the nanomaterials used, leaving the researchers to establish the parameters and criteria of security and, also, the governments and regulatory agencies to carry out the standardization and inspection of these products.

References

1 Aguirre-Joya, J.A., De Leon-Zapata, M.A., Alvarez-Perez, O.B. et al. (2018). Basic and applied concepts of edible packaging for foods, Chapter 1. In: *Food Packaging and Preservation* (eds. A.M. Grumezescu and A.M. Holban), 1–61. Academic Press.

2 Idumah, C.I., Hassan, A., and Ihuoma, D.E. (2019). Recently emerging trends in polymer nanocomposites packaging materials. *Polym. Plast. Technol. Mater.* 58 (10): 1054–1109.

3 García-Moreno, P.J., Mendes, A.C., Jacobsen, C., and Chronakis, I.S. (2018). Biopolymers for the nano-microencapsulation of bioactive ingredients by electrohydrodynamic processing. In: *Polymers for Food Applications* (ed. T.J. Gutiérrez), 447–479. Cham: Springer International Publishing.

4 Rostami, M., Yousefi, M., Khezerlou, A. et al. (2019). Application of different biopolymers for nanoencapsulation of antioxidants via electrohydrodynamic processes. *Food Hydrocolloids* 97: 105170.

5 Almeida, A.C.S., Franco, E.A.N., Peixoto, F.M. et al. (2015). Aplicação de nanotecnologia em embalagens de alimentos. *Polímeros* 25: 89–97.

6 Sánchez-Safont, E.L., Aldureid, A., Lagarón, J.M. et al. (2018). Biocomposites of different lignocellulosic wastes for sustainable food packaging applications. *Compos. Part B* 145: 215–225.

7 Wróblewska-Krepsztul, J., Rydzkowski, T., Borowski, G. et al. (2018). Recent progress in biodegradable polymers and nanocomposite-based packaging materials for sustainable environment. *Int. J. Polym. Anal. Charact.* 23 (4): 383–395.

8 Zubair, M. and Ullah, A. (2020). Recent advances in protein derived bionanocomposites for food packaging applications. *Crit. Rev. Food Sci. Nutr.* 60 (3): 406–434.

9 Kalita, D. and Baruah, S. (2019). The impact of nanotechnology on food, Chapter 11. In: *Nanomaterials Applications for Environmental Matrices* (eds. N. RFd, O.P. Ferreira, A.J. De Paula and S.N. VdO), 369–379. Elsevier.

10 Pund, S., Joshi, A., and Banerjee, R. (2020). Engineered nanomaterials in functional Foods. In: *Reference Module in Food Science II*, 1–19. Elsevier.

11 Neethirajan, S. and Jayas, D.S. (2011). Nanotechnology for the food and bioprocessing industries. *Food Bioprocess Technol.* 4 (1): 39–47.

12 Ragavan, K.V. and Neethirajan, S. (2019). Nanoparticles as biosensors for food quality and safety assessment, Chapter 7. In: *Nanomaterials for Food Applications* (eds. A. López Rubio, M.J. Fabra Rovira, M. Martínez Sanz and L.G. Gómez-Mascaraque), 147–202. Elsevier.

13 Bao, C., Jiang, P., Chai, J. et al. (2019). The delivery of sensitive food bioactive ingredients: absorption mechanisms, influencing factors, encapsulation techniques and evaluation models. *Food Res. Int.* 120: 130–140.

14 Aditya, N.P., Espinosa, Y.G., and Norton, I.T. (2017). Encapsulation systems for the delivery of hydrophilic nutraceuticals: food application. *Biotechnol. Adv.* 35 (4): 450–457.

15 Embuscado, M. and Huber, K. (2009). *Edible Films and Coatings for Food Applications*, 403 p. New York: Springer-Verlag.

16 Maringgal, B., Hashim, N., Mohamed Amin Tawakkal, I.S., and Muda Mohamed, M.T. (2020). Recent advance in edible coating and its effect on fresh/fresh-cut fruits quality. *Trends Food Sci. Technol.* 96: 253–267.

17 Eça, K.S., Sartori, T., and Menegalli, F.C. (2014). Films and edible coatings containing antioxidants – a review. *Braz. J. Food Technol.* 17: 98–112.

18 Gniewosz, M., Alicja, S., and Marta, D. (2009). Application of edible coatings with antimicrobial activity for food preservation. *Biotechnologia*: 40–53.

19 Dhall, R.K. (2013). Advances in edible coatings for fresh fruits and vegetables: a review. *Crit. Rev. Food Sci. Nutr.* 53 (5): 435–450.

20 Barman, K., Sharma, S., and Siddiqui, M.W. (eds.) (2018). *Emerging Postharvest Treatment of Fruits and Vegetables*, 1e. Apple Academic Press https://doi.org/10.1201/9781351046312.

21 Kurek, M., Ščetar, M., and Galić, K. (2017). Edible coatings minimize fat uptake in deep fat fried products: a review. *Food Hydrocolloids* 71: 225–235.

22 Zaritzky, N. (ed.) (2011). *Edible Coatings to Improve Food Quality and Safety*. New York, NY: Springer New York.

23 Assis, O.B.G. and Dd, B. (2014). Revisão: coberturas comestíveis protetoras em frutas: fundamentos e aplicações. *Braz. J. Food Technol.* 17: 87–97.

24 Sapper, M. and Chiralt, A. (2018). Starch-based coatings for preservation of fruits and vegetables. *Coatings* 8: 152.

25 Nor, S. and Ding, P. (2020). Trends and advances in edible biopolymer coating for tropical fruit: a review. *Food Res. Int.* 134: 109208.

26 Parreidt, T.S., Schott, M., Schmid, M., and Müller, K. (2018). Effect of presence and concentration of plasticizers, vegetable oils, and surfactants on the properties of sodium-alginate-based edible coatings. *Int. J. Mol. Sci.* 19 (3): 742.

27 Eddin, A.S., Ibrahim, S.A., and Tahergorabi, R. (2019). Egg quality and safety with an overview of edible coating application for egg preservation. *Food Chem.* 296: 29–39.

28 Cisneros-Zevallos, L. and Krochta, J.M. (2003). Dependence of coating thickness on viscosity of coating solution applied to fruits and vegetables by dipping method. *J. Food Sci.* 68 (2): 503–510.

29 Menezes, J. and Athmaselvi, K. (2016). Polysaccharide based edible coating on sapota fruit. *Int. Agrophys.* 30.

30 Zhong, Y., Cavender, G., and Zhao, Y. (2014). Investigation of different coating application methods on the performance of edible coatings on mozzarella cheese. *LWT Food Sci. Technol.* 56 (1): 1–8.

31 Nouri, J.M., Mackenzie, S., Gaskell, C., and Dhunput, A. (2012). Effect of viscosity, temperature and nozzle length-to-diameter ratio on internal flow and cavitation in a multi-hole injector. In: *Fuel Systems for IC Engines* (ed. E. Imech), 265–278. Woodhead Publishing.

32 Atieno, L., Owino, W., Ateka, E.M., and Ambuko, J. (2019). Influence of coating application methods on the postharvest quality of cassava. *Int. J. Food Sci. Technol.* 2019: 1–16.

33 Lin, D. and Zhao, Y. (2007). Innovations in the development and application of edible coatings for fresh and minimally processed fruits and vegetables. *Compr. Rev. Food Sci. Food Saf.* 6 (3): 60–75.

34 Thakur, R., Pristijono, P., Scarlett, C. et al. (2019). Starch-based films: major factors affecting their properties. *Int. J. Biol. Macromol.* 132.

35 Mohamed, S.A.A., El-Sakhawy, M., and El-Sakhawy, M.A.-M. (2020). Polysaccharides, protein and lipid-based natural edible films in food packaging: A Review. *Carbohydr. Polym.* 238: 116178.

36 Dehghani, S., Hosseini, S.V., and Regenstein, J.M. (2018). Edible films and coatings in seafood preservation: a review. *Food Chem.* 240: 505–513.

37 Arnon-Rips, H. and Poverenov, E. (2018). Improving food products' quality and storability by using layer by layer edible coatings. *Trends Food Sci. Technol.* 75: 81–92.

38 Dhumal, C.V. and Sarkar, P. (2018). Composite edible films and coatings from food-grade biopolymers. *J. Food Sci. Technol.* 55 (11): 4369–4383.

39 Galus, S. and Kadzińska, J. (2015). Food applications of emulsion-based edible films and coatings. *Trends Food Sci. Technol.* 45 (2): 273–283.

40 Yousuf, B., Qadri, O.S., and Srivastava, A.K. (2018). Recent developments in shelf-life extension of fresh-cut fruits and vegetables by application of different edible coatings: a review. *LWT* 89: 198–209.

41 Kumar, S., Mukherjee, A., and Dutta, J. (2020). Chitosan based nanocomposite films and coatings: emerging antimicrobial food packaging alternatives. *Trends Food Sci. Technol.* 97: 196–209.

42 Slavutsky, A.M. and Bertuzzi, M.A. (2014). Water barrier properties of starch films reinforced with cellulose nanocrystals obtained from sugarcane bagasse. *Carbohydr. Polym.* 110: 53–61.

43 Preethi, R., Maria Leena, M., Moses, J.A., and Anandharamakrishnan, C. (2020). Biopolymer nanocomposites and its application in food processing. In: *Green Nanomaterials: Processing, Properties, and Applications* (eds. S. Ahmed and W. Ali), 283–317. Singapore: Springer Singapore.

44 Xing, Y., Yang, H., Guo, X. et al. (2020). Effect of chitosan/nano-TiO_2 composite coatings on the postharvest quality and physicochemical characteristics of mango fruits. *Sci. Hortic.* 263: 109135.

45 Shankar, S. and Rhim, J.-W. (2015). Amino acid mediated synthesis of silver nanoparticles and preparation of antimicrobial agar/silver nanoparticles composite films. *Carbohydr. Polym.* 130: 353–363.

46 Ahmed, S., Ahmad, M., Swami, B.L., and Ikram, S. (2016). A review on plants extract mediated synthesis of silver nanoparticles for antimicrobial applications: a green expertise. *J. Adv. Res.* 7 (1): 17–28.

47 Mackevica, A., Olsson, M.E., and Hansen, S.F. (2016). Silver nanoparticle release from commercially available plastic food containers into food simulants. *J. Nanopart. Res.* 18 (1): 5.

48 Wang, Z., Xia, T., and Liu, S. (2015). Mechanisms of nanosilver-induced toxicological effects: more attention should be paid to its sublethal effects. *Nanoscale* 7 (17): 7470–7481.

49 Shi, S., Wang, W., Liu, L. et al. (2013). Effect of chitosan/nano-silica coating on the physicochemical characteristics of longan fruit under ambient temperature. *J. Food Eng.* 118 (1): 125–131.

50 Song, H., Yuan, W., Jin, P. et al. (2016). Effects of chitosan/nano-silica on postharvest quality and antioxidant capacity of loquat fruit during cold storage. *Postharvest Biol. Technol.* 119: 41–48.

51 Nagpal, K., Singh, S.K., and Mishra, D.N. (2010). Chitosan nanoparticles: a promising system in novel drug delivery. *Chem. Pharm. Bull. (Tokyo).* 58 (11): 1423–1430.

52 Noorbakhsh-Soltani, S.M., Zerafat, M.M., and Sabbaghi, S. (2018). A comparative study of gelatin and starch-based nano-composite films modified by nanocellulose and chitosan for food packaging applications. *Carbohydr. Polym.* 189: 48–55.

53 Hassannia-Kolaee, M., Khodaiyan, F., Pourahmad, R., and Shahabi-Ghahfarrokhi, I. (2016). Development of ecofriendly bionanocomposite: Whey protein isolate/pullulan films with nano-SiO_2. *Int. J. Biol. Macromol.* 86: 139–144.

54 Lustriane, C., Dwivany, F.M., Suendo, V., and Reza, M. (2018). Effect of chitosan and chitosan-nanoparticles on post harvest quality of banana fruits. *J. Plant Biotechnol.* 45 (1): 36–44.

55 Al-Naamani, L., Dutta, J., and Dobretsov, S. (2018). Nanocomposite zinc oxide-chitosan coatings on polyethylene films for extending storage life of okra (*Abelmoschus esculentus*). *Nanomaterials (Basel)* 8 (7).

56 Gad, M., Zagzog, O., and Mohamed Hemeda, O. (2016). Development of nano-chitosan edible coating for peach fruits cv. Desert Red. *Int. J. Environ.* 05: 43–55.

57 Nisha, V., Monisha, C., Ragunathan, R., and Jesteena, J. (2018). Use of chitosan as edible coating on fruits and in micro biological activity – an ecofriendly approach. *Int. J. Pharm. Sci. Invent.* 5 (8): 7–14.

58 Gammariello, D., Conte, A., Buonocore, G.G., and Del Nobile, M.A. (2011). Bio-based nanocomposite coating to preserve quality of Fior di latte cheese. *J. Dairy Sci.* 94 (11): 5298–5304.

59 Melo, N.F.C.B., BLdM, S., Diniz, K.M. et al. (2018). Effects of fungal chitosan nanoparticles as eco-friendly edible coatings on the quality of postharvest table grapes. *Postharvest Biol. Technol.* 139: 56–66.

60 Pilon, L., Spricigo, P., Miranda, M. et al. (2014). Chitosan nanoparticle coatings reduce microbial growth on fresh-cut apples while not affecting quality attributes. *Int. J. Food Sci. Technol.* 50.

61 Meindrawan, B., Suyatma, N.E., Wardana, A.A., and Pamela, V.Y. (2018). Nanocomposite coating based on carrageenan and ZnO nanoparticles to maintain the storage quality of mango. *Food Packag. Shelf Life* 18: 140–146.

62 Ortiz-Duarte, G., Pérez-Cabrera, L., Artés-Hernández, F., and Martínez-Hernández, G. (2018). Ag-chitosan nanocomposites in edible coatings affect the quality of fresh-cut melon. *Postharvest Biol. Technol.* 147: 174–184.

63 Rosenbloom, R., Wang, W., and Zhao, Y. (2020). Delaying ripening of 'Bartlett' pears (*Pyrus communis*) during long-term simulated industrial cold storage: mechanisms and validation of chitosan coatings with cellulose nanocrystals Pickering emulsions. *LWT* 122: 109053.

64 Dai, L., Zhang, J., and Cheng, F. (2020). Cross-linked starch-based edible coating reinforced by starch nanocrystals and its preservation effect on graded Huangguan pears. *Food Chem.* 311: 125891.

65 Shah, S.W., Jahangir, M., Qaisar, M. et al. (2015). Storage stability of Kinnow fruit (*Citrus reticulata*) as affected by CMC and guar gum-based silver nanoparticle coatings. *Molecules (Basel, Switzerland)* 20 (12): 22645–22661.

66 Zhang, R., Wang, X., Li, L. et al. (2019). Optimization of konjac glucomannan/carrageenan/nano-SiO(2) coatings for extending the shelf-life of Agaricus bisporus. *Int. J. Biol. Macromol.* 122: 857–865.

67 Tian, F., Chen, W., Wu, C. et al. (2019). Preservation of Ginkgo biloba seeds by coating with chitosan/nano-TiO(2) and chitosan/nano-SiO(2) films. *Int. J. Biol. Macromol.* 126: 917–925.

68 Müller, K., Bugnicourt, E., Latorre, M. et al. (2017). Review on the processing and properties of polymer nanocomposites and nanocoatings and their applications in the packaging, automotive and solar energy fields. *Nanomaterials (Basel).* 7 (4): 74.

69 Kuswandi, B. Chapter 6 - Nanotechnology in food packaging. In: *Nanoscience in Food and Agriculture*, 1e (eds. S. Ranjan, N. Dasgupta and E. Lichtfouse), 151–183. Springers.

70 Lemes, A.C., Sala, L., Ores Jda, C. et al. (2016). A review of the latest advances in encrypted bioactive peptides from protein-rich waste. *Int. J. Mol. Sci.* 17 (6): 1–24.

71 Forbe, T., García, M., and Gonzalez, E. (2011). Potential risks of nanoparticles. *Food Sci. Technol.* 31: 835–842.

72 Martioli, A.C.S. and Pinheiro, F. (2015). Nanotecnologia: uma breve discussão sobre os impactos da nanotecnologia à saúde humana e ao meio ambiente. *Rev. Acad. Oswal. Cruz.* 8: 1–13.

73 Khan, I., Saeed, K., and Khan, I. (2019). Nanoparticles: properties, applications and toxicities. *Arabian J. Chem.* 12 (7): 908–931.

74 Hoshyar, N., Gray, S., Han, H., and Bao, G. (2016). The effect of nanoparticle size on in vivo pharmacokinetics and cellular interaction. *Nanomedicine (Lond).* 11 (6): 673–692.

75 Mihindukulasuriya, S.D.F. and Lim, L.T. (2014). Nanotechnology development in food packaging: a review. *Trends Food Sci. Technol.* 40 (2): 149–167.

76 Trinetta, V. Definition and function of food packaging. In: *Reference Module in Food Sciences*, 1–2. https://doi.org/10.1016/B978-0-08-100596-5.03319-9.

77 Marsh, K. and Bugusu, B. (2007). Food packaging—roles, materials, and environmental issues. *J. Food Sci.* 72 (3): R39–R55.

78 Zambrano-Zaragoza, M.L., González-Reza, R., Mendoza-Muñoz, N. et al. (2018). Nanosystems in edible coatings: a novel strategy for food preservation. *Int. J. Mol. Sci.* 19 (3): 705.

79 Sharma, C., Dhiman, R., Rokana, N., and Panwar, H. (2017). Nanotechnology: an untapped resource for food packaging. *Front. Microbiol.* 8: 1735.

80 Honarvar, Z., Hadian, Z., and Mashayekh, M. (2016). Nanocomposites in food packaging applications and their risk assessment for health. *Electr. Phys.* 8 (6): 2531–2538.

81 Huang, C., Qian, X., and Yang, R. (2018). Thermal conductivity of polymers and polymer nanocomposites. *Mater. Sci. Eng. R: Rep.* 132: 1–22.

82 FAO (2010). editor Report of Technical Round Table Sessions. International Conference on Food and Agriculture Applications of Nanotechnologies. Sao Pedro, SP, Brazil: Food and Agriculture Organization of the United Nations (FAO).

83 Rushton, E.K., Jiang, J., Leonard, S.S. et al. (2010). Concept of assessing nanoparticle hazards considering nanoparticle dosemetric and chemical/biological response metrics. *J. Toxicol. Environ. Health Part A* 73 (5): 445–461.

84 Brasil. (2020). Projetos de Lei e Outras Proposições (PL) - PL 5076/2005 / PL 5133/2013 / PL 6741/2013. Chamber of Deputies. In Union Official Journal of Federative Republic of Brazil. Available: www.camara.leg.br.

85 Information for access: FDA. (2018). FDA's Approach to Regulation of Nanotechnology Products. Accessed: 2018. Available: https://www.fda.gov/science-research/nanotechnology-programs-fda/fdas-approach-regulation-nanotechnology-products.

86 European Union. (2011). Regulation (EU) nº 1169/2011 of the European Parliament and of the Council of 25 October 2011. Document 32011R1169. Accessed: 2019. Available: https://eur-lex.europa.eu/eli/reg/2011/1169/oj.

10

An Overview of Higher Barrier Packaging Using Nanoadditives

Johnsy George[1], Basheer Aaliya[2], Kappat V. Sunooj[2], and Ranganathan Kumar[1]

[1]*Defence Food Research Laboratory, Food Engineering and Packaging Division, Mysore, 570011, India*
[2]*Pondicherry University, Department of Food Science and Technology, Puducherry, 605014, India*

10.1 Introduction

High barrier packaging films are mainly used for decreasing the transmission rate of moisture and gases especially oxygen into food, which in turn helps to enhance the shelf life of food. Innate permeability to gas and moisture compared to glass and metal is one of the major drawbacks for plastic-based food packaging materials. The barrier properties of a plastic packaging material will be influenced by several factors such as nature and concentration gradient of the permeant molecules, thickness, area of contact, temperature, relative humidity, etc. The primary mechanism by which permeation occurs in a polymeric film is described by a diffusion model using Henry's and Fick's law. The permeation of small molecules through a packaging film is governed by adsorption, diffusion, and desorption processes wherein the permeant molecule first sorbed into the film and diffuses through it and escapes from the other end. The term high barrier package is usually used when the packaging material exhibit low or very low permeability to permeant molecules. High barrier packages are known for their strong resistance against the permeants and thereby restrict the permeant transfer. High barrier nature is undoubtedly a desirable characteristic of polymeric packaging materials that can be used for different kinds of food products.

The barrier properties of polymeric packaging materials can be widely improved by the incorporation of nanomaterials or nanoadditives, which creates a tortuous path for the permeating gas/vapor molecules by increasing the path length of diffusing molecules and thereby partially or completely blocking them from further diffusion through the polymer matrix. Nanomaterials in the form of thin, layered sheet structure with high aspect ratio are generally used to improve the barrier property of polymers. These nanomaterials can be dispersed uniformly within the polymer matrix and creates physical obstacle against permeant molecules such as

Nanotechnology-Enhanced Food Packaging, First Edition. Edited by Jyotishkumar Parameswaranpillai, Radhakrishnan Edayileveettil Krishnankutty, Aswathy Jayakumar, Sanjay Mavinkere Rangappa, and Suchart Siengchin.

oxygen (O_2), carbon dioxide (CO_2), and water (H_2O) and accomplishes effective high barrier.

Maintaining the quality of the food is the prime function of a food packaging material. For instance, a food package with good O_2 barrier property can avert the lipid oxidation in a fat-rich food product. Similarly, a high moisture barrier package deprives the water vapor transmission between environment and food material. Till date, a wide variety of polymer nanocomposites with high barrier property have been prepared by incorporating nanomaterial in the polymer matrix. Clay-incorporated polymer nanocomposites are among the first high barrier packaging developed in food packaging industry. Other nanomaterials that were successively used for improving barrier properties were metal and metal oxide nanoparticles and carbon nanotubes (CNTs). The use of nanosized materials to develop polymer nanocomposites is a promising strategy to extend the shelf life of food by increasing the barrier performance of food packaging.

The plastic materials used for packaging purposes are mostly from nonbiodegradable plastics, rendering critical global environmental issues. So edible and biodegradable packaging materials were developed from bio-based materials to reduce packaging waste. The natural or organic source of bio-based polymeric materials include those that are cost friendly and abundant for the development of food packaging materials such as cellulose, starch, chitosan, zein, gelatin, whey, and soy proteins. However, the use of biopolymers apart from its environmental-friendly characteristics showed some major problems related to processing and functionalities like poor mechanical strength and barrier properties. For example, starch is one of the widely available biopolymer that has gained substantial recognition as a biodegradable thermoplastic polymer. Yet, its limited mechanical characteristics like high brittleness and high water sensitivity due to the rapid amylose crystals formation with time when exposed to moisture adversely affect the use of such biopolymers in food packaging purposes. To combat these issues by the application of nanotechnology, novel bio-based nanocomposites or bionanocomposites are made by addition of nanomaterials having the ability to improve barrier and other functional properties of a food packaging material.

10.2 Gas and Moisture Permeability Through Polymer Packaging Materials

Permeability is the measure of transmission of permeants (gas and/or moisture) through a resisting material. If a permeate have different concentration on either side of the polymer, it then passes from a higher concentration side to a lower concentration side by a complex process involving four steps: sorption of permeate to the polymer, dissolution, diffusion, and finally desorption from the polymer (see Figure 10.1). The product of the sorption coefficient or solubility, S (in mol/m^3 Pa), and the diffusion coefficient, D (in m^2/s), is the permeation coefficient or permeability, P (in mol/Pa m s):

$$P = D \times S$$

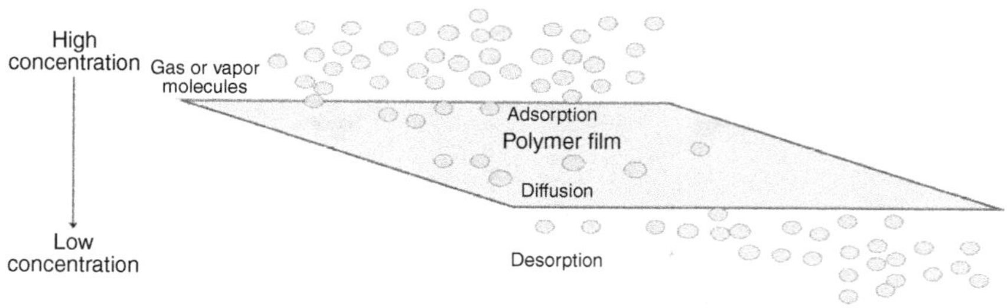

Figure 10.1 The process of permeability through a polymer film.

The coefficients "D" and "S" are independent of permeate concentration. The molecular diffusion occurring through plastic packaging materials are generally governed by Fickian behavior. However, this hypothesis is not valid in many cases such as in hydrophilic polymer film, where moisture can interact with the film and tend to show a non-Fickian characteristic. The quantification of permeability can be directly performed in a permeation cell or indirectly by any sorption/desorption experiments [1]. The gas or water vapor transmission rate is generally expressed as the volume of permeant gas/water vapor passing through the material at steady state per unit area in unit time and the unit is cm^3/m^2/day at 23 °C.

10.2.1 Permeability of Oxygen and Carbon Dioxide

Oxygen (O_2) and carbon dioxide (CO_2) are the two gases that are extremely important in deciding the shelf life of food products. O_2 is the gas that is responsible for direct and indirect food spoilage [2]. The respiration or oxidation of fresh food and development of microorganisms depends on the amount of O_2 in the environment encompassing the food material. Research have found that the interaction of higher amount of atmospheric O_2 deleteriously affects the sensory, nutritive, and microbial characteristics of food [3]. Oxidation results in the generation of peroxides, hydrogen peroxides, and epoxides. So it is a necessity to safeguard O_2-sensitive food materials in a package from O_2 ingress after sealing. The nature, respiration, and permeability of packaging material greatly affect the gas composition of packaged foods. The two food packaging methods that can control the concentration and amount of gas surrounding foods are vacuum packaging and modified atmospheric packaging (MAP). Vacuum packaging removes O_2 and other gases from the package before sealing, and MAP is a method where the food package is flushed with specific amount of gas like nitrogen (N_2) or CO_2 prior to sealing. The aim of these two methods of packaging is thus to remove or alter the air to suit the food material for longer storage and retard microbial growth. However for maintaining the fixed concentration of gases in the headspace, the packaging materials should possess higher barrier to O_2 and CO_2. The choice of materials for food packaging has to be selected based on certain requirements like method of packaging or preservation, the package form, and subsequent reuse. An appropriate packaging technology along with a suitable packaging material possessing higher barrier property effectively makes up a suitable atmosphere for varied food. The incorporation of O_2-scavenging nanomaterials into food package is another alternative method to maintain very low O_2 levels [2].

10.2.2 Permeability of Moisture

Moisture plays an important role in determining the shelf life of food. The effect of moisture content and water activity (a_w) in food affects the growth of microbes, physical and chemical properties of food, food quality, and food spoilage rates. The rates of chemical reactions like food browning, protein denaturation, lipid oxidation, etc. that affect the food quality directly depend on a_w. Each food material needs to maintain different moisture level to show its best quality. For specific foods, moisture needs to be removed to minimize many moisture-driven deterioration reactions and to prevent the development and multiplication of microbes that cause food spoilage. In some foods, the moisture needs to be maintained during production, processing, and preservation. For example, in grains, tea, nuts, etc., it is crucial to evade permeation of moisture after packaging. Yet, for fresh food, it is important to avoid dehydration.

10.3 Nanoadditives for Improving Barrier Properties

Packaging films with an oxygen transmission rate (OTR) below 5 mL/m^2 day (O_2 permeability $<14.55 \times 10^{-15}$ mL m/m^2 s Pa) can be regarded as high barrier food packaging materials [4]. Not all monolayer packaging material can meet the desired barrier requirements. So multilayer films that meet this criteria are commonly used. The materials like aluminum foil and polymers like poly(vinylidene chloride) (PVDC), ethylene vinyl alcohol (EVOH), poly(*m*-xylene adipamide) (MXD6), aliphatic PA6, etc. with high barrier properties are employed as the core layer in multilayer films. However multilayer films are having a major drawback when it comes to recycling. Many of the food packaging multilayer films especially the laminates that contain aluminum foil cause a major environmental pollution as they are nonrecyclable and nondegradable. On the other hand most of the biodegradable and environmentally degradable plastic materials are not having sufficient barrier properties to ensure the desired shelf life of food materials. Nanoadditives are best suited to fill this gap by providing sufficient barrier properties to monolayer polymeric materials. Nanoadditives are the materials that comes under the category of nanomaterials in which at least one of its dimension is below 100 nm. They are used to prepare multiphased polymeric materials called polymer nanocomposites. These type of multiphased material helps to lower gas and water transmission rate and also helps to incorporate certain additional properties such as O_2 and ethylene scavenging and antimicrobial activity that are very essential for food packaging applications [5]. Various nanoadditives such as metal, metal oxides, inorganic layered silicates, etc. have the potential for manufacturing food packaging applications.

10.4 Methods to Prepare High Barrier Packaging Materials

The major ways to manufacture high barrier packaging materials are by using nanoadditives to manufacture nanocomposites, coatings, and layer-by-layer (LbL) assemblies based on synthetic and bio-based polymer nanocomposites.

10.4.1 Polymer Nanocomposites

Even though having an inherent permeability to gases and vapors, the polymers such as poly(styrene) (PS), polyvinyl chloride (PVC), polyethylene (PE), polypropylene (PP), and poly(ethylene terephthalate) (PET) are commonly used in food packaging. For solving the aforesaid problems, inert nanoscale inorganic or organic filler is dispersed throughout a polymeric matrix to form a material called nanocomposites. Polymer nanocomposite is a mixture of nanomaterials (or fillers or nanoadditives) in discontinuous phase in a continuous phase polymer matrix. The addition of nano-sized fillers improves the barrier properties of an analogous polymer film in two distinct ways. The first way is by building a tortuous path for gas and vapor diffusion. By the addition of nanomaterials, the straight-line path that lies perpendicular to the film surface for gas and moisture diffusion is disrupted. This makes the gas and moisture to have a longer mean path to pass through the polymer matrix (see Figure 10.2).

The second way on how nanomaterials affect the barrier properties is by altering the polymeric matrix itself at the interfacial regions. When the polymer–nanomaterial interaction becomes compatible, the polymer strand locates closely to each nanoparticle, and it attenuates the hopping rates between free volume holes or alter the size and/or density of holes. This results in difficulty of the gas and water molecules to pass across the interfacial zones. Nanoclay is a widely used inorganic nanomaterial to develop high barrier packages. It possesses about 800 m^2/g specific surface area, large lateral dimensions ranging from 100 to 1000 nm, and high swelling capacity. Montmorillonite (MMT) is an exfoliated clay made of layered silicate. Exfoliated nanoclays are potent at enhancing the

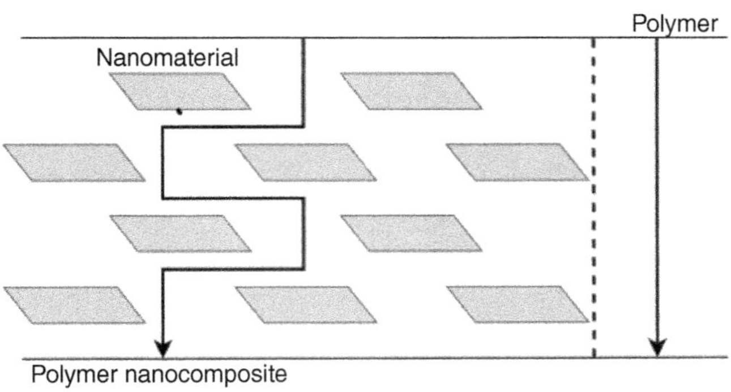

Figure 10.2 Tortuous pathway in a polymer nanocomposite.

gas barrier properties of polymers. MMT when dispersed into a polymeric matrix creates a maze structure that represents a tortuous pathway to moving gas and water vapor, effectively decelerating their permeation rate.

Attaining proper dispersion of organically modified MMT in biopolymers or water-soluble polymers are difficult. As an alternate, nanoclays functionalized with aminopropyl groups prepared by co-condensation of 3-aminopropyltriethoxysilane (APTES) and magnesium chloride is being used [6]. Polyvinyl alcohol (PVA)–aminoclay hybrid films enable higher loadings (up to 20 wt%) and show excellent retention of ductile behavior of PVA even at higher loadings [7]. The oxygen barrier properties of these types of nanocomposites were significantly improved. The ability of aminoclay to stabilize other nanomaterials such as Ag nanoparticle provides another interesting property modification. The use of aminoclay-stabilized Ag nanoparticle significantly improved the mechanical properties of the nanocomposite films [7].

10.4.2 Coating

Coating with organic and/or inorganic nanomaterial as a nanolayer over polymeric matrix has been proved to enhance the barrier properties. Methods like sol–gel coating, atomic layer deposition, evaporation, and plasma-enhanced chemical vapor deposition (PECVD) are used for coating with inorganic nanomaterials like metal and metal oxides. A very high gas barrier property can be achieved by coating with materials that possess a dense chemical structure [4].

Silicon-based thin films have been gaining attention as diffusion barriers on food packaging material. Barrier films comprising silicon oxide (SiO_2)/silicon nitride (Si_3N_4) stacks were sequentially incorporated in a polyimide (PI) matrix, and the results showed that the SiO_2 and Si_3N_4 films exhibited good moisture barrier properties [8]. The κ-aluminum oxide (κ-Al_2O_3)/resin [9] and titanium oxide (TiO_2)/polyurethane (PU) [10] composite coating on PET and high-density polyethylene (HDPE), respectively, were developed to improve the adhesion linking substrate and the protective barrier layer and to increase the gas and moisture barrier. The polymeric films can also be coated with organic nanomaterials on the substrate to increase the barrier properties. Maikap et al. [11] coated the biaxial-oriented PET with PVDC, and the water vapor transmission rate (WVTR) and OTR values reduced to 3–4 g/m^2/day and 6–7 cc/m^2/day, respectively, when compared with control PET film. The cellulose nanocrystals (CNCs) films formed by hydrolysis with hydrochloric acid and deposited by spin coating on glass substrate showed reduced homogeneity, higher roughness, improved hydrophobicity, decreased gas permeability to O_2 and N_2, and also gas selectivity to certain gas combinations [12].

10.4.3 Layer-by-Layer Assembly

Presently LbL assembly is gaining wide recognition as a cost-effective method to fabricate transparent thin film with high barrier properties. This technology can be employed to incorporate a varied range of nanomaterials to the polymeric matrix. LbL assembly is a process by which a multilayer film/coating of nanometer thickness can be prepared by successive adsorption of oppositely charged polyelectrolytes on a solid polymeric substrate [2].

LbL assembly is a cyclic deposition process to manufacture multilayer films with desired morphology and thickness. In a representative film preparation with LbL assembly, a negatively charged substrate is usually immersed into a mix of positively charged materials for specific time duration. The substrate is then removed from the mix, rinsed with deionized water (DI), dried in a stream of filtered air (or N_2), and then again immersed into a mix of negatively charged material. Finally, a bilayer or double polyelectrolyte layer is applied over the matrix after rising and drying [4]. The thickness of multilayered films is governed by factors such as temperature, buffer and pH, ionic strength, counterion, and molecular weight of the deposition mixture. For the multilayered structure preparation, hydrogen bonding, covalent bonding, hydrophobic force, and electrostatic attraction are utilized as LbL self-assembly forces.

Spraying has been impotent to replace dipping as the most common method for LbL deposition even though having clear advantages over dipping because of the discord over the spray-coated films quality. One of the major concerns is that spray-coated assemblies are thinner than their dip-coated counterparts. For preparing all-polymer thin films with a high O_2 barrier, the LbL thin film assemblies of polyethylenimine (PEI) and polyacrylic acid (PAA) were studied by Hagen et al. [13], and the OTR of 8 bilayers with 350 nm PEI/PAA film was less than 0.005 $cm^3/(m^2$ day) [14]. An OTR of 0.08 $cm^3/(m^2$ day atm) and thickness of 538 nm was achieved in a 7-bilayer PEI/PAA film produced with spraying parameters (50 psi, 0.4 g/s, 5 seconds). The same film with dipping method developed a thinner film with one order of magnitude higher OTR [15]. Lee et al. [16] conducted a study with positively charged poly(alylmine hydrochloride) and negatively charged graphene oxide (GO) monolayers and prepared an extremely thin and transparent LbL barrier films having significantly less oxygen permeability. With the barrier effect of the amino-phenyl-functionalized GO over the surface of polyimide (PI) with an electrostatic LbL assembly process, the OTR reduced from 220 to 78 $cm^3/(m^2$ day atm) [17].

The main factor for the utilization of nanoclays in barrier layers is their parallel alignment to the matrices. With strong absorption capabilities and hydrogen bonding, a self-aligned clay multilayer can be made by the absorption of PVA over MMT for improving gas barrier [18, 19]. The gas barrier characteristics and optical transparency were enhanced by the LbL deposition of oppositely charged thin films having self-assembled MMT and PVA [20, 21]. The gold(III) oxide (Au_2O_3) was formed by the drop-cast of gold(III) chloride ($AuCl_3$) over an montmorillonite(polyvinyl alcohol)-poly(diallyldimethylammonium chloride) (MMT(PVA)-PDDA) multilayer film, which help in the barrier for water vapor across the gas barrier film [22]. A quad layer was prepared with LbL assembly, which exhibited a deposition order of PEI/PAA/PEI/MMT, where anionic MMT and PAA and cationic PEI are used to develop a slight gas barrier to polymer matrices [23, 24].

Tzeng et al. [25] conducted a study and proved that 122 nm five-layer coating over a 51 μm PS exhibited better performance than metalized film and thick-laminated EVOH film and showed decreased gas permeability. Xiang et al. [26] developed a quad layer assembly comprising three repeat units of branched PEI/PAA/PEI/MMT having 24 nm thickness by subjecting one minute exposure to each composite. By decreasing the polyelectrolytes exposure time to five seconds, the OTR decreases to 0.05 $cm^3/(m^2$ day atm) and thickness of film increases to 55 nm. The OTR is 2 orders of magnitude lesser than the same film prepared by one minute exposures. Jang et al. [27] invented a potentially microwaveable thin film for replacement of aluminum foil in food packaging, by adding intercalating layers of cationic polyacrylamide and anionic sodium MMT onto a PET substrate, where the layers were held together by the electrostatic bonds. With a 30-bilayer film deposition, the OTR decreased until a negligible value less than 0.005 $cm^3/(m^2$ day atm) was achieved. The authors also suggested that to sustain a negligible OTR inside the system, it should be combined with a high moisture barrier.

10.5 Barrier Improvement by Reinforcement of Polymer Nanocomposites with Inorganic Nanoadditives

The reinforcement of nanoadditives into food packaging materials enhances the barrier properties by lowering gas and H_2O permeability. Along with high barrier properties, reinforcement of nanoadditives in the functionality of packaging materials such as mechanical strength, anti-microbial activity, and UV scavenging and ethylene scavenging ability is also improved. In food packaging, nanoadditives are classified as inorganic and organic nanoadditives. The high specific surface area and smaller size of nanosized inorganic particles render its use as excellent additives in a polymer matrix [28]. The interfacial interaction between the nanoadditive and the polymer matrix is increased by the large surface area of these particles, thereby altering the properties of the nanocomposite film. The hydrogen bonds are formed with inorganic network by the effect of nanoadditive to enhance the mixing into polymer matrix with carbonyl and hydroxyl functional groups [28].

Metal nanoparticles, metal oxide nanoparticles, nanoclays, and CNTs are the widely used inorganic nanoadditives to develop high barrier polymer nanocomposites. Presently, varied ranges of metal are engineered into nanomaterials, which are proven to show high barrier packaging ability. The metals that have been in use till date in food packaging as nanoadditives includes zinc (Zn), silver (Ag), magnesium (Mg), copper (Cu), titanium (Ti), silicon (Si), and aluminum (Al) and their oxides. Polymer reinforcements with nanoclay-based nanocomposites (MMT, halloysite, hectorite, kaolinite, etc.) have also been used since decades. Recently more research is focused on carbon-based nanocomposites (graphene nanostructures, CNTs, etc.) due to their exceptional properties and wider application in the field of paper and petroleum-based industrial products, along with high barrier properties.

10.5.1 Metal and Metal Oxides Nanomaterials

Currently, a variety of metals and metal oxides nanomaterials have been incorporated in food packaging materials to develop antimicrobial bionanocomposite films. Apart from commercially used antibacterial nanoparticle from Ag and Zn, inorganic metals and metal oxides like silicon dioxide or silica (SiO_2), copper (II) oxide or cupric oxide (CuO), zinc oxide (ZnO), titanium dioxide or titania (TiO_2), magnesium oxide (MgO), and aluminum oxide (Al_2O_3) are also employed in food packaging industries because of their functional benefits. These nanoparticles have gained wide popularity due to their ability to withstand rough processing conditions, strong inhibitory effect against food-borne pathogens, high photo-disinfection activity, improved mechanical strength, and high gas and moisture barrier properties. TiO_2 and ZnO are often used as photocatalysis agent to degrade microorganisms and organic molecules. These nanoparticles generate reactive oxygen species (ROS) and thereby results in cell death by oxidation of cytoplasm of bacterial cells [29]. It was also stated that ZnO is relatively more attractive and efficient than silver due to relatively cheap and less toxicity [30].

10.5.1.1 Zinc

ZnO is a widely used white powder in biological and synthetic polymer to prepare nanocomposites with antimicrobial activity. Food and Drug Administration (FDA) has mentioned ZnO as Generally Recognized as Safe (GRAS), and ZnO nanoparticle with hexagonal quartzite structure is a polar inorganic crystalline component synthesized commercially in a large scale. Sol–gel process, spray pyrolysis, homogeneous precipitation, vapor phase growth, vapor–liquid–solid process, soft chemical method, warm ultrasonic bath, and electrophoretic deposition are commonly used techniques for the preparation of ZnO nanoparticles [31]. From the preparation method, the morphological shape of ZnO nanoparticles can be needles, spherical, nanorods, nanowires, and flowerlike structures, and depending on the morphology of the nanoparticles, the physical and chemical characteristics differ with each structure [31]. The ZnO nanoparticles reinforced nanocomposites possess high photostability and biocompatibility, chemical stability, high catalyst power, high melting temperature, and excellent antimicrobial and UV-blocking properties, which make them as an acceptable ingredient or nanofiller and employed in various applications like pigmentation, cosmetics, medical devices, drug delivery, etc. [32].

Recent studies proved that ZnO nanoparticles reinforcement in food packaging application also exhibited improved mechanical strength and barrier properties to films. Mostly, ZnO nanoparticles have been added to polyolefins like low-density polyethylene (LDPE), HDPE, and PVC and also to biopolymers like poly(lactic acid) (PLA) and poly(butylene adipate-terephthalate) (PBAT). Besides, combination with polymers and ZnO nanoparticles have been used with other antimicrobial components like nisin and allyl isothiocyanate to have synergistic effect on food packaging applications [33]. Pantani et al. [34] prepared a multifunctional biodegradable packaging film with PLA–ZnO nanocomposites by the addition of up to 3 wt% ZnO (silanized nanorods) and resulted in films with improved UV and moisture barrier and antibacterial activity against both Gram-negative and Gram-positive strains. The combined effect of ZnO nanoparticles and MMT to a chitosan matrix was studied by Vaezi et al. [35] and found that the tortuous pathway has improved and thus reduces the permeability of moisture to a great extent. Lately, Roy and Rhim [36] from his experiment on carrageenan-based nanocomposites showed that ZnO nanoparticles capped with melanin had more moisture barrier properties than films reinforced with ZnO nanoparticles alone.

10.5.1.2 Magnesium

MgO is a renewable, colorless, and crystalline material that is GRAS by FDA and economically synthesized on a large scale. MgO is used as a firming agent, anti-caking, pH control agent, flavor enhancer, color retention, and nutrient fortification [37]. The reinforcement with MgO nanoparticles in polymer matrix is known to improve O_2 barrier, thermal stability, and antibacterial activity [3]. Bio-based films were prepared by reinforcement of MgO nanoparticles into PLA matrix by solvent casting method. It was found that incorporation of 2 wt% MgO nanoparticles exhibited higher O_2 barrier and tensile strength when compared with other concentrations of nanoparticles [38].

10.5.1.3 Silica

Silicon containing many hydroxyl group on surface is the most abundant solid metalloid on Earth, which is widely seen in crystalline and amorphous form [3]. Silica (SiO_2) and silicates do not possess any side effects and can be used as a bio-safe food packaging material. SiO_2 have large surface area, low refractive index, high thermal stability, and good mechanical strength. SiO_2 nanoparticles are synthesized mostly by sol–gel process, microemulsion method, ultrasonic spray method, and flame spray pyrolysis process [3]. The addition of SiO_2 nanoparticles over polymer matrix (LDPE, EVOH, and PLA) improves the thermal stability, mechanical strength, and barrier properties.

Xiong et al. [39] reported an improvement in tensile properties and decrease in moisture absorption by SiO_2 nanoparticle-incorporated starch-based nanocomposites. SiO_2 nanoparticles was incorporated by using a compatibilizer, maleic anhydride-grafted polypropylene (PP-g-MA) to an isotactic polypropylene (iPP) substrate. The storage modulus and O_2 barrier was improved thus making the nanocomposites much stiffer and high barrier packaging material [40]. Tang et al. [41] prepared a biodegradable film with the combination of starch/PVOH/SiO_2 nanoparticles. The increase in intermolecular hydrogen bonds and formation of C–O–Si groups, between starch and PVOH with SiO_2 nanoparticles enhanced the compatibility, miscibility, tensile properties, and H_2O resistance. When compared to pure EVOH polymer, 5 wt% SiO_2 nanoparticle-incorporated EVOH–SiO_2 nanocomposites exhibit substantially improved mechanical strength and gas and moisture barrier properties [42]. The blending of ethylene/vinyl acetate (EVA)-modified SiO_2 nanoparticles (0.09 wt%) with PP resulted in polymer nanocomposites with improved tensile strength, reduced gas permeability, and ink solvent adsorption, which are helpful for using this kind of PP layer for printing in food packaging [43]. Another method to use SiO_2 nanoparticles is by using it for coating polymers. A PLA-based transparent biodegradable packaging material was synthesized by sol–gel process using tetraethoxysilane as precursor and 3-isocyanatopropyltriethoxysilane as coupling agent and prepared a PLA-based coating by incorporating SiO_2 nanoparticles. It was found that gas barrier properties increased up to 70% than of pure PLA films [44]. Donglu et al. [43] developed a multifunctional nanocomposite for packaging mushroom from a combination of $SiO_2/TiO_2/Ag$ nanoparticles reinforced into a LDPE matrix, which is able to scavenge ethylene, regulate O_2 and CO_2 levels, and augment the shelf life of the product by inhibiting microbial growth.

10.5.1.4 Titanium

TiO_2 is a naturally occurring bright white chemical that is poorly soluble. Since many years TiO_2 nanoparticles are used as a colorant and anti-caking agent in granular and powdered foods agents in a recommended safe dosage [31]. TiO_2 nanoparticles exists in two different forms: crystalline (photocatalytic active) and amorphous (photocatalytically inactive) forms [3]. TiO_2 nanoparticles are nontoxic valuable metal oxide nanoadditives with acceptable thermostability and inertia. TiO_2 nanoparticles also possess dielectric and catalytic property, photostability, UV protection, and antimicrobial activity [45]. TiO_2 nanoparticles exhibiting different reactivities and properties with respect to their characteristic band gap are prepared in the anatase, rutile, and brookite form by sol–gel method [46]. Many ROS are generated, which can kill microbes, when TiO_2 nanoparticles are irradiated with UV light [45]. Due to their O_2-scavenging property and good antioxidant effect, TiO_2 nanoparticle-based packaging materials are used for O_2-sensitive products. Besides, TiO_2 nanoparticles incorporation into a polymer matrix can improve mechanical strength, UV light barrier, and O_2 and moisture barrier properties.

It was noted that edible films from TiO_2 nanoparticle-incorporated whey protein nanocomposites showed good mechanical strength, moisture barrier, and antimicrobial effects [46]. Nasiri et al. [47] made good O_2 barrier nanocomposites by incorporating TiO_2 nanoparticles (4 wt%) to LDPE matrix and observed that packaging almonds with the developed nanocomposites decreased the rancidity by 78% when compared to control sample. Incorporation of TiO_2 nanoparticles (0.5–2 wt%) significantly ($p < 0.05$) increased the hydrophobicity and water vapor barrier (approx. 35%) of agar films. It was noted that the film solubility in water and mechanical properties are affected by the concentration of nanoadditives used. The addition of TiO_2 nanoparticles (<1 wt%) improved the intermolecular hydrogen bond formation in the film matrix, thereby enhancing the moisture resistance (up to 16%) and tensile strength (up to 96%). But in higher TiO_2 nanoparticles concentration, moisture resistance and tensile strength are reduced due to discontinuity between polymer molecules and reduction in intermolecular forces [31]. The composites are also able to subdue the ethylene production by oxidation of ethylene to H_2O and CO_2 and extend the shelf life of fruits [48]. Shankar et al. [49] demonstrated the use of TiO_2-based nanocomposites as the potential ethylene absorbers because of their capacity to absorb profuse ethylene. Thus TiO_2 nanoparticles can be used to extend the storage period of fruits and vegetables. Li et al. [50] investigated the effect of TiO_2 nanoparticles as antioxidant agent on novel TiO_2 nanoparticles–LDPE packaging material for analyzing quality of strawberries. Similar kind of nanocomposites have been identified to lessen the decay and improve the quality preservation of various food products like cheese [51] and shrimp [52].

10.5.1.5 Copper

Cu nanoparticles and CuO nanoparticles are two extensively studied metal and metal oxide nanoparticles in the fields of H_2O purification, food packaging materials, and polymer nanocomposites [45]. The wider application of CuO when compared with other metal oxide is because of its large surface area, ease of preparation, low cost, and antimicrobial properties [31]. A decrease in water vapor permeability (WVP) and improvement in water contact angle and tensile strength was noticed in Cu nanoparticle-reinforced chitosan/soy protein isolates (SPIs) nanocomposite film [53]. A nanocomposite film as food packaging for an Indian sweet dairy product, Peda, was prepared from LDPE/Cu nanocomposites to augment the shelf life of the product [54]. Roy et al. [55] prepared agar films by incorporating surface-modified Cu nanoparticles by capping agents and sulfidation to form CuS nanoparticles. It was observed that there was significant ($p < 0.05$) improvement in mechanical strength, moisture barrier, and antimicrobial activity in agar films.

10.5.1.6 Aluminum Oxide

Al is a metal that is widely employed for the purpose of protection of food from UV radiation and O_2 by preparing a metalized film in packaged foods. Recently Al_2O_3 films are the mostly preferred coating materials, as they are transparent, lightweight, and microwavable [46]. Al_2O_3 thin films are usually prepared by chemical vapor deposition or physical vapor deposition techniques [56]. Besides, Al_2O_3 thin films have also been prepared by atomic layer deposition technique, which is observed to have excellent barrier properties. This kind of films is greatly preferred for food or biodegradable packaging [57]. It was observed that O_2, moisture, and aroma barrier properties were substantially improved by coating board paper with PLA–Al_2O_3, followed by double coating with alginate–chitosan layer and finally coating again with Al_2O_3 [57, 58]. By double coating of PLA matrix, the WVP of the nanocomposite greatly reduced from 53 to 25 $g/m^2/day$ when compared with pure PLA film [58]. Struller et al. [59] coated Al_2O_3 over PET film and noticed an improvement in the barrier activity of the composite films. The researchers suggested that the composite film can be used as a substitute to presently used metalized retortable packaging materials. However, heterogeneous outcome as results were obtained by coating biaxial-oriented PP, thus recommending a thorough research on Al_2O_3 incorporated nanocomposites.

10.5.2 Nanoclays

Polymer/clay nanocomposites were first commercialized by Toyota in the late 1980s. But it was developed for the food packaging use in late 1990s by the researchers [2].

Nanoclay is an abundant commercially employed nanomaterial, markedly approx. 70% market value all over the world [45]. Nanoclay-reinforced polymer nanocomposite is the first known application of a nanoadditive in food packaging [37]. Nanoclays as filler in the form of plate, spherical, cylindrical, etc. creates a tortuous pathway to the diffusing penetrants. The permeability of nanocomposites with nanoadditives loading percentages were predicted by Nielsen's model [60]. The presence of nanoclays increases the length of diffusive path instead of following straight-line diffusion, thereby reduces the O_2 and moisture flow speed and improves the barrier properties of the polymer. The high aspect ratio and percent volume of nanoclay filler in the nanocomposites determines the length of tortuous path. A new model to predict the polymer nanocomposites permeability was proposed by Beall [61] and provided a correction factor to Nielsen's model. Besides focusing on tortuous path, polymer–clay interface was also considered as the governing factor. Low cost, availability, surface area, high aspect ratio, simple processability, and good performance make nanoclays to compete and win over other nanoadditives used in food packaging industries to enhance polymer performance [2].

Clays are naturally occurring eco-friendly materials, which are made up of complex metallic ores in multiple layers [62]. Nanoclays and layered silicates are basically aluminum silicates that are naturally made up of fine-grained minerals showing sheetlike geometrical structure [63]. The layered silicate that has 1 nm thickness and several microns long is commonly used as a nanoadditive in polymer nanocomposites to augment the tortuosity of the diffusive pathway against penetrants [2]. Presently MMT or bentonite, halloysite, kaolinite, and hectorite are some type of commonly used nanoclays in industrial scale, which are found to have multifarious applications [32]. MMT derived from volcanic ash/rocks is the most recognized exfoliated clay nanoparticle widely used for improving barrier properties in polymer with its large surface area and surface reactivity [45]. MMT is a hydrated aluminosilicate-layered clay comprising edge-shared octahedral sheet of aluminum hydroxide ($Al(OH)_3$) sandwiched between two tetrahedral silica layers [31]. Therefore, many researchers have demonstrated the reinforcement of layered silicate in natural and synthetic polymer to improve the resistance to diffusing particles. The non-permeable inorganic crystal structure of MMT reduced the permeability of O_2 and moisture by inducing longer complex tortuous pathway. Thus varied range of food products like meat, cheese, fruit juices, dairy products, and carbonated beverages use MMT in packaging material to resist the permeation of gases and moisture [37].

Reinforcement of nanoclays (<10 wt%) into polymer substrate can greatly enhance the thermal, mechanical, and barrier properties. When a recycled polymer is incorporated with nanoclay in low concentration, characteristics like the tensile strength, yield stress, hardness, rigidity, and resistance to humidity is increased [32]. Previously, it was commercially claimed that nanoclays could cut permeability up to 75%. But recent studies have proved that almost 100% permeability could be achieved from a transparent clay–polymer matrix [64]. For PET-based nanocomposites, it was observed to exhibit the lowest O_2 permeability (<0.005 cc/(m^2 day atm)) ever declared for a polymer–clay reinforcement. This is achieved by the brick wall nanostructure developed by highly oriented, exfoliated clay platelets and the alternate adsorption of polymeric layers [64]. The poor barrier property, which is the main limitation of biopolymer films, was improved by LbL assembly of branched PEI and thin films of sodium MMT on various substrates [65]. Indeed, numerous research works have been reported to demonstrate the effect of nanoclays in reducing the barrier properties of O_2 [66–71] and moisture [72–74].

Nanoclay is used as O_2 barrier layers in multilayer polymer and paper-based beverage, and it is also utilized as an ethylene glycol copolymer replacement in food packaging [75]. Busolo and Lagaron [76] prepared antioxidant films from HDPE and LLDPE by reinforcing with kaolinite containing metallic nanoparticles and found that nanocomposite scavenges the O_2 molecules and acts as an O_2 barrier. Nalcabasmaz et al. [77] developed a multilayered material and tested it for packaging sliced salami under vacuum and modified atmosphere packaging. The nanocomposite containing 5% poly-beta-pinene (PβP) and 1% nanoclay showed no significant changes in moisture content and hardness during storage for longer days. Junqueira-Goncalves et al. [78] examined the effect of MMT nanoparticles incorporation over a lacto-biopolymer coating and observed that WVTR and OTR improved. In an industrial level, Southern Clay Products, Inc. (Gonzales, TX) and Nanocor Inc. (Arlington Heights, IL), two companies in the United States, currently work on the production of MMT-reinforced nanocomposites. They produce lighter and stronger materials with improved heat resistance and barrier against gases and moisture [2]. A German company, Bayer, incorporated clays from Nanocor Inc. to nylon-6 polymer and developed nanocomposites known as hybrid system and observed that the strength of these nanocomposites has increased two times than the normal nylon-6 polymer product [62].

10.5.3 Carbon-Based Nanomaterials

Carbon-based nanomaterials like graphene, graphite, and CNT have been incorporated into many polymers to develop nanocomposites with enhanced properties. Generally, these nanomaterials have the ability to deactivate bacteria and viruses [28]. Few research finding even demonstrated that carbon-based nanomaterials can be also used with recycled polymers. Graphene is the only known bidimensional elastic material that is flexible and lightweight. Graphene consist of a single layer of sp^2 C-atoms, similar to graphite in a hexagonal network. It is the most resistant material, which is 100 times stronger than steel despite of its thin profile [32]. Graphite is another allotrope of carbon that is added to polymers to exhibit high conductivity for the manufacturing of electronic goods like batteries, electrodes, and solar panels [32]. Because of abundance, cost effectiveness, ease of modification, and layered structure, graphene and GO are employed to prepare thermal-resistant and high barrier food packaging nanocomposites [5]. By graphene nanoplates (0.5 wt%) addition, the gas effective diffusivity of PU matrix reduced by three orders of magnitude, and the glass transition temperature increased by 30 °C with 1–5 wt% graphene nanoplates loading to form nanocomposites with polymethyl methacrylate (PMMA) [79]. Huang et al. [80] developed GO/PVA nanocomposite film and found that by the incorporation of GO (0.72 vol%), the OTR and WVTR decreased by 98% and 68%, respectively. They also suggested that this nanocomposite can be used as antimicrobials, sensors, or carriers of active species in intelligent packaging.

CNTs received great attention since last decade because of its acceptable flexibility, thermal and electrical characteristics, mechanical resistance, barrier, and biocidal properties. CNTs possessing metallic or semiconducting properties are actually rolled graphene sheets, forming a cylindrical piece [32]. CNTs having exceptional elastic modulus and high aspect ratio, usually comprise of one-atom thick single-walled nanotube (SWNT) or multi-walled nanotubes (MWNTs) with number of concentric graphitic rings up to a diameter of 10 nm [2]. Since the early 1990s, MWNTs are commercially produced by companies like Arkema, Hyperion Catalysis, and Nanocyl as a conductive additive for polymeric matrix [81]. Till date, a lot of studies have made reinforcing CNT in a polymer nanocomposite frame. CNTs are widely known to increase the tensile strength/modulus of varied polymers like PP, PVA, PA, etc. The CNTs were surface modified by carboxylic acid groups to improve their intermolecular interactions when incorporated into poly(ethylene-2,6-naphtalene) (PEN) matrix. It was observed that even low concentration of CNTs (0.1 wt%,) greatly enhanced the tensile strength and modulus and thermal stability of the nanocomposites [82].

The motion of water molecules between CNTs is rapid than that of other pore structures of similar size [83]. By disarraying the polymeric chain packing through hollow nanochannels and interspaces, CNT provides a pore network in a polymeric matrix. This generates external nanoscale pathways inside the membrane and thereby aid in the improvement of permeability of matrix [28]. It was found that reinforcement of PLA with CNT exhibited 200% better WVTR and increased modulus and toughness when compared with pure PLA films [84]. By functionalizing CNTs surfaces and tips, the hydrophobic and hydrophilic fouling of polymer membrane can be reduced [28]. The narrow size distribution, atomic scale smoothness, hydrophobic graphitic walls, high aspect ratio, uniformity of the tubes, chemically inert interior walls, and excellent biofouling resistance are the most important characteristics of CNTs, which made it a suitable nanoadditive for improving salt rejection, water flux, and biofouling properties of polymer matrix [28]. Yet, the incorporation of CNTs in polymer nanocomposites till date has been restricted because of high material cost, difficulty in processing, and health concern [5].

10.6 Barrier Improvement of Biopolymers by Reinforcement with Organic Nanoadditives

The concept of biopolymer-based economy is gaining environmental, social, economic, and scientific attention, and the driving force for developing the strategies is mainly by public interest. Polymer waste disposal problem, a crucial environmental issue, can be solved to a great extent by the use of biopolymer-based packaging materials [85]. The main aim in producing biopolymer-based packaging materials is to use renewable resources instead of nonrenewable resources like pure petroleum-derived products [45]. The biopolymers are naturally occurring biomolecules like carbohydrates and proteins that are synthesized physically, chemically, or microbiologically [85]. Mostly used biopolymers include carbohydrates-based materials like cellulose, starch, chitosan, etc. and protein-based materials like zein, gelatin, whey, soy protein, etc. Properties like abundance, biodegradability, low cost, and better film-forming attributes have made protein-based films to receive great research attention. Among available biopolymer for food packaging applications, soy proteins gained popularity due to its promising film-making properties when compared with petroleum-derived polymers.

The processing cost, difficulty in biopolymer preparation methods, and limited applications have made biopolymer to compete with petroleum-derived polymer counterparts to reach wider acceptability in industrial purposes [85]. Addition of nanomaterials to biodegradable polymers like polyhydroxybutyrate (PHB), PLA, and polycaprolactone (PCL), which possess necessary functionalities, is a favorable alternative to currently used polymer food packaging materials [85]. Preparation and incorporation of nanomaterials from biopolymers to synthetic or organic polymer matrix is another method to improve the characteristics of food packaging materials.

10.6.1 Cellulose

Cellulose, the most abundant carbohydrate, is a promising biodegradable polymer seen in almost all plant materials [3]. Cellulose is eco-friendly, needs low energy requirement during manufacturing, possesses remarkable film forming properties, and is easy to recycle by combustion [2]. Cellulose is a structural polysaccharide made up of microfibrils linked together to make cellulose fibers [86]. Apart from its vast advantages, the direct commercial application of cellulose in food packaging is limited by factors such as poor solubility, high crystallinity and high affinity toward moisture [3]. In spite of being a matrix material, it has also gain recognition in nanoscale in commercial level as cellulose-based nanomaterials reinforced in the production of various functional nanocomposites. The cellulose for reinforcement in polymer food packaging is mainly derived from natural sources like cotton, wood, dried hemp, sawdust, and marine animal tunicate [86]. It is also synthesized from recycled cardboard and sugarcane bagasse [32]. The cellulose-based nanoreinforcements comprise of nanomaterials like cellulose microfibrils (or nanofibers) and CNCs (or nanocrystals, nanorods, and rodlike cellulose microcrystals).

Cellulose nanofibers (CNFs) are bundles of elongated molecules that are stabilized by hydrogen bonding. CNFs are materials with length in micrometer range with nanosized diameters (2–20 nm, depending on the origin) [2]. CNF is formed by aggregate elementary fibrils, having amorphous and crystalline parts. CNF is an acceptable class of nanomaterials in paper making and polymer food packaging as high barrier material, because of low cost, high stiffness, low density, high aspect ratio, and low thermal expansion [3]. Cellulose whiskers are rigid rodlike monocrystalline cellulose particles, usually synthesized by acid hydrolysis and have morphology with 1–100 nm diameter and 10–100 nm length [86]. Cellulose whiskers are commonly used as a reinforcing nanomaterial because of its large surface area, high aspect ratio, and high bending strength [2]. CNCs are unique nanomaterials that are mostly widely used as organic nanoadditive due to its exceptional chemical, mechanical, optical, rheological, and barrier properties.

CNCs have been commonly applied in the manufacturing of polymer nanocomposites by adding into polyolefins like PP, PE, PVA, PLA, PHB, etc. Many studies have demonstrated that by cellulose nanoreinforcements the WVP of polymer films has improved [87–89]. Researchers like Dufresne and Vignon [90], Dufresne et al. [91], and Lu et al. [92] proved that increasing cellulose whisker content could reduce the H_2O uptake by starch films. Arun et al. [93] prepared starch–CNC nanocomposites to study the penetration and diffusion of moisture in the film and found that moisture uptake is reduced by hindering movement of H_2O through starch chains due to the presence of CNCs. Other than moisture and gases, the starch-based nanocomposites also demonstrated that permeability of D-limonene was deprived by the presence and high aspect ratio of CNC in film structure [94].

Sanchez-Garcia and Lagaron [95] performed a comparative study by preparing PLA nanocomposites by incorporating CNC synthesized by two methods, namely, freeze-drying and solvent exchange using chloroform. The nanocomposites with freeze-drying CNCs (3 wt%) decreased the O_2 permeability by 90% and WVP by 82%, while the same amount of CNCs prepared by solvent exchange with chloroform decreased the O_2 permeability by 90% and moisture permeability by 49%. This is because of the increase in tortuosity pathway by the crystalline fibers, thus lowering permeability [88]. Fortunati et al. [96] studied the enhancement in barrier properties of pristine CNCs and surface-modified CNCs (by means of a surfactant) reinforcing to the PLA matrix. They found that the OTR values reduced by 43% and 48% for CNCs and surface-modified CNCs, respectively, and the WVTR were reduced by 4% and 15% for CNCs and surface-modified CNCs, respectively. The effect of CNCs from agro-waste, *Phormium tenax* leaves in plasticized PLA in the presence of D-limonene, was also studied. It was observed from the water contact angle measurement that PLA-D-limonene had a surface wettability of 54°, and by the addition of CNCs (3 wt%), the surface wettability changed to 75°, which is due to the alteration in the chemical properties and topography of the system. The hydrophilic CNCs decreased the hydrophobic nature of PLA matrix, by acting as nucleating agent in PLA and affecting the surface properties and interaction with H_2O [97]. The surprising improvement in O_2 barrier property of poly(L-lactic acid) (PLLA) matrix reinforced with polyethylene glycol (PEG)-functionalized CNCs when compared with pristine PLLA films and unmodified CNC-reinforced PLLA nanocomposite was proved by Li et al. [98].

Dhar et al. [99] prepared PHB–CNC nanocomposites and demonstrated the decrease in OTR, diffusion coefficient, and solubility values by augmenting the concentration of CNCs. The OTR values changed from 46% to 65% by the addition of very low loading of CNCs (1 and 2 wt%). The transient percolation network formed by CNCs in the polymer matrix by the hydrogen bonding interactions improved the resistance to O_2 permeation. With CNCs (5 wt%), the diffusivity and solubility reduced from 10.7×10^{-12} m^2/s to 2.2×10^{-12} m^2/s and from 3.22×10^{-6} g/g Pa to 18.43×10^{-6} g/g Pa, respectively, when compared to pure PHB film. The poly(3-hydroxybutyrate-*co*-3-hydroxyvalerate) (PHBV)-based nanocomposites were prepared by the addition of methyl ester-functionalized CNCs by Yu et al. [100]. It was noted that the tortuosity of the transport pathway was increased by the addition of increased amount of functionalized CNCs, which prevent the vapor molecules to permeate through the polymer. The results said that the H_2O uptake and WVP values reduced by 74% and 56% with the incorporation of functionalized CNCs from 1 to 20 wt% concentration. Cano et al. [101] developed starch–PVA nanocomposites incorporated with CNCs (5 wt%) and found that the overall migration levels of food stimulants through the nanocomposites were lowered than the established migration limits (isooctane and ethanol 10% v/v). It was also noted that there was a marginal reduction in WVTR values for the CNCs-reinforced starch–PVA nanocomposites. Nanocomposites using solution casting method was prepared from the blend of starch and carboxymethyl cellulose (CMC) along with addition of CNCs. The study demonstrated a 13% reduction in WVTR in CNCs (5 wt%)-reinforced bionanocomposites when compared with nonreinforced blend. This is because of the interaction of CNCs and polymer matrix by the hydrogen bonding, thus leading to the decrease in the number of free hydroxyl groups, which possess hydrophilic character [102]. Few studies on bionanocomposites prepared with the combination of PLA–PHB reinforced with CNCs and surface-modified CNCs were performed by Arrieta et al. [103–105] to observe the reduction in wettability and O_2 and moisture permeability through the bionanocomposites. The improvement in barrier properties is because of the high number of hydroxyl groups of nanocellulose that aids in the hydrogen bonding between the hydroxyl groups of cellulose and carbonyl groups of biopolymers.

10.6.2 Starch

Starch is a storage polysaccharide that comprises of linear amylose and branched amylopectin chains of size 0.1–1 nm. The major sources of starches are roots, tubes, and cereals. Thermoplastic starches are the type of starch that is thermally and/or mechanically plasticized to substitute the use of synthetic polymers [106]. A nanocomposite film prepared by the reinforcement of CNF to unripe banana starch matrix showed superior optical, mechanical, and moisture barrier properties [107]. Aqlil et al. [108] prepared starch/lignin polymer bionanocomposite by the incorporation of GO and found that the concentration of GO positively influenced the mechanical properties and reduced the WVTR and moisture uptake by the film. MWNTs with or without hydroxylation were incorporated to starch films and observed that hydrophobic character remarkably enhanced by the addition of nanotubes [109]. TiO_2 and MMT nanomaterials were incorporated to potato starch matrix films to identify the change in barrier, mechanical, and thermal properties. It was found that WVTR decreased with increasing amount of nanomaterials [110].

The platelet morphology of starch crystalline particles prepared by extended-time hydrolysis of native starch granules at a temperature below gelatinization temperature showed a thickness of 6–8 nm [111]. It was also noted that incorporation of starch nanocrystals (SNC) to pullulan matrix film improved the tensile strength and reduced the WVTR by incorporation of 20% or more SNC. A comparative study on nanocomposites incorporated with starch crystals and cellulose crystals was conducted by Ali et al. [112]. It was observed that even though starch crystals-reinforced composites had better UV barrier, cellulose crystals-reinforced composites possessed high thermal stability, great mechanical performance, and processability.

10.6.3 Chitosan

The biocompatibility, biodegradability, antimicrobial, and antioxidant characteristics of chitosan make it the mostly accepted biopolymer in food packaging systems in the form of films, blends, coatings, and nanocomposites and in active packaging techniques [3]. The inherent properties of chitosan is from the three functional groups at C2, C3, and C6 positions of the chitosan structure, which are amino group (NH_2), primary, and secondary hydroxyl groups (OH), respectively. Chitosan has positive charge at biological pH, which makes it unique when compared to other hydrocolloids that possess neutral or negative charge [3]. Two research team prepared chitin whiskers from chitin by acid hydrolysis. Addition of chitin whiskers to SPI films [113] and to chitosan films [114] improved the H_2O resistance and tensile properties. De Moura et al. [115] demonstrated that addition of chitosan–tripolyphosphate nanoparticles to hydroxypropyl methylcellulose films enhanced the mechanical and barrier properties. Ultrasonication method was adopted to prepare a PE packaging films by *in situ* deposition of chitosan and antimicrobial vanillin. The microbial spoilage on fresh-cut watermelons was strongly inhibited by increased contact surface of film over the fruit [116]. For instance, incorporation of gallic acid/chitosan nanoparticles (size 80–100 nm) improved the moisture barrier property of nanocomposites made with hydrophilic konjac glucomannan by forming hydrogen bonding between the reinforcement material and the substrate and removing the free volume of hydrophilic film [117].

10.6.4 Zein

Zein comprises a group of prolamines-based biopolymer from corn, which is GRAS food ingredient [45]. Zein is neither fully soluble in 100% water nor alcohol, but it is soluble in specific concentrations of aqueous alcohol solution (50–95% (v/v)) [3]. By the presence of three fractions of zein, specifically α-, β-, and γ-zein, the protein possesses greater solubility [3] and hydrophobicity, thus making it suitable for the production of bio-based films having low WVP [45]. These advantages with other characteristics like glossy appearance, wide isoelectric point range, and antioxidant activity have widened the application of zein in biomedical sectors. The major drawback of zein is its brittle nature, which is overcome by the use of plasticizers, by the production of zein blends, or by the addition of nanoadditives [3].

A PP film coated with zein nanocomposites reinforced with organomodified MMT by Ozcalik and Tihminlioglu [118] was prepared to know the efficiency of nanocomposites. The results revealed that OTR and WVTR of PP films decreased by 25% and 30%, respectively, by coating with nanocomposite film in a 5.9 mm thickness. Moreover, WVTR and OTR decreased by 60% and 65%, respectively, by 5% (w/w) of reinforcement. Addition of silicate nanoparticles (laponite) to zein matrix can form Si—N bond and can improve the surface hydrophobicity and mechanical, thermal, and barrier properties of nanocomposite film [119]. Thymol (THY)/γ-cyclodextrin(γ-CD) inclusion complex (IC)-encapsulated electrospun zein nanofibrous webs (zein-THY/γ-CD-IC-NF) was prepared as a food packaging material and reported that the bacterial growth was significantly inhibited by zein-THY/γ-CD-IC-NF in 2 : 1 ratio [120]. Zn nanoparticles coated with sodium caseinate embedded to whey protein isolate (WPI)-based films by casting method exhibited enhanced mechanical and moisture barrier properties [121].

10.6.5 Gelatin

Gelatin is a water-soluble denatured fibrous protein obtained by the partial or controlled hydrolysis of insoluble protein, collagen [85]. Gelatin is a preferred biopolymer with OH, NH_2, COOH, etc. functional groups that aid in the production of bionanocomposites. Gelatin is a widely accepted protein biopolymer due to its abundance, notable biodegradability, biocompatibility, film forming property, gelling ability, superior gas barrier properties, and cost effectiveness [85]. The hydrophilic amino acid in gelatin makes the developed film poor in mechanical properties and moisture barrier properties [3]. Plasticizers along with suitable nanoadditives are incorporated into gelatin-based films to combat these problems. Byun et al. [122] developed an active film with warm-water fish gelatin by adding O_2 scavenging nanomaterials of iron chloride(II) and α-tocopherol. Ejaz et al. [123] prepared an active packaging nanocomposite for peeled shrimp from bovine skin gelatin. The UV and O_2 barrier, mechanical strength, and antimicrobial properties were improved by the incorporation of clove essential oils and ZnO nanorods. Remarkable changes in O_2 and moisture properties were noticed by the reinforcement of bovine gelatin films with TiO_2 nanorods [124].

10.6.6 Whey Protein Isolates

WPI, which contains highly bio-available essential amino acids, is one of the most abundant by-products from milk industries [125]. WPI-based biopolymers make a flexible transparent film possessing fair functional properties. A WPI-based composite film prepared by sulfuric acid hydrolysis of oat husk nanocellulose (ONC) in different proportion found that film transparency and WVP decreased by the addition of ONC [126]. By the incorporation of TiO_2 NPs to a chitosan/WPI film, WVTR was decreased by 7.60%, whereas the tensile strength and the elongation at break of film augmented by 11.51% and 12.01%, respectively [127]. The WVTR of WPI/pullulan (PUL) films with different amount of nano-SiO_2 prepared by casting method decreased. Thus the H_2O resistance and barrier properties of the nanocomposite films were increased with increasing SiO_2 nanoparticle concentration [125]. Carvalho et al. [128] prepared a WPI-based bionanocomposites incorporated with CNF and observed that maximum moisture barrier property was obtained for the film prepared with 4% (w/w) out of various other concentrations.

10.6.7 Soy Protein Isolates

SPI is a commercially graded by-product of soy bean oil that comprises over 90% protein. SPI is a biopolymer that can be utilized as a potential alternative to petroleum-based polymers. Soy-based biopolymer possess superior mechanical properties because of the polar and functional groups of amino acids that help in inter- and intramolecular interactions [3]. Besides, rigidity and moisture sensitivity are two major characteristics that obstruct the use of SPI in polymer films [3]. Thus certain nanomaterials are added to overcome these issues. For example, Cu and Zn nanoclusters and microcrystalline cellulose together are employed as reinforcement for SPI films to enhance the H_2O resistance and tensile strength [129]. A study on preparation of SPI-based nanocomposites by LbL assembly along with CNCs modified with PEI and graphene exhibited improved tensile strength, contact angle, UV–visible light, and moisture barrier when compared with pure SPI film [130]. For instance, on the preparation of SPI–MMT nanocomposites, the morphological arrangement (level of intercalation and exfoliation) and properties (thermal stability, mechanical strength, and moisture permeability) are influenced by the variation in the type and content of clay, processing techniques, and pH of the filmogenic solutions [131, 132].

10.7 Conclusion

On considering the drawbacks of pure polymeric packaging materials, polymer nanotechnology developed polymer nanocomposites for improving the functionalities, especially high barrier performance against gas and moisture. This way, the use of polymers as packaging material in food industries was able to combat to a great extent with other food packaging materials like glass and metal. The nanomaterial characteristics like size, structure, aspect ratio, chemical composition, surface chemistry, loading concentration, and interfacial properties (degree of adhesion to the biopolymer) considerably affect the extent of barrier property exhibited when such materials are incorporated to polymer matrix. Presently, polymer researchers and technologist in food packaging industries have shown keen interest and put great deal of effort in preparing nanomaterials possessing the ability to reduce gas and moisture permeability along with other functional characteristics. For preparing a high barrier packaging with inorganic and/or organic nanomaterials, some factors like type and morphology of nanoreinforcement, matrix material characteristics, interaction between reinforcement material and matrix, angle of orientation, and presence of other external additives should be taken into consideration. The barrier property of a polymer nanocomposite is also determined by external factors like relative humidity, temperature, and storage period. These are the reasons why many world renowned food packaging companies are actively engaged in exploring the potential of polymer nanotechnology in producing food packaging materials with high barrier packaging by nanomaterial incorporation to maintain the food quality during transport and storage.

References

1 Choudalakis, G. and Gotsis, A.D. (2009). Permeability of polymer/clay nanocomposites: a review. *Eur. Polym. J.* 45 (4): 967–984.

2 De Azeredo, H.M.C. (2009). Nanocomposites for food packaging applications. *Food Res. Int.* 42 (9): 1240–1253.

3 Bharathi, S.K.V., Murugesan, P., Moses, J.A., and Anandharamakrishnan, C. (2020). *Recent Trends in Nanocomposite Packaging Materials.* Elsevier.

4 Yun, X. and Dong, T. (2017). Fabrication of high-barrier plastics and its application in food packaging. In: *Food Packaging* (ed. A.M. Grumezescu), 147–184. Academic Press.

5 Cabedo, L. and Gamez-Pérez, J. (2018). Inorganic-based nanostructures and their use in food packaging. In: *Nanomaterials for Food Packaging*, 13–45. Elsevier.

6 Datta, K.K.R., Achari, A., and Eswaramoorthy, M. (2013). Aminoclay: a functional layered material with multifaceted applications. *J. Mater. Chem. A* 1 (23): 6707–6718.

7 Johnsy, G., Datta, K.K.R., Sajeevkumar, V.A. et al. (2009). Aminoclay: a designer filler for the synthesis of highly ductile polymer-nanocomposite film. *ACS Appl. Mater. Interfaces* 1 (12): 2796–2803.

8 Zhang, S., Xue, W., and Yu, Z. (2015). Moisture barrier evaluation of SiO_x/SiN_x stacks on polyimide substrates using electrical calcium test. *Thin Solid Films* 580: 101–105.

9 Tseng, W.J. and Chang, J.-H. (2014). Preparation of κ-Al_2O_3/resin composite coating on polyethylene terephthalate foil for gas-permeation barrier. *Ceram. Int.* 40 (10): 16779–16784.

10 Kasanen, J., Suvanto, M., and Pakkanen, T.T. (2011). Improved adhesion of TiO_2-based multilayer coating on HDPE and characterization of photocatalysis. *J. Appl. Polym. Sci.* 119 (4): 2235–2245.

11 Maikap, A.N., Jain, S., Singh, H.K. et al. (2015). Effect of primers and surface treatments on the gas barrier, mechanical, optical, printing, and flex-crack properties of the polyvinylidene chloride coated biaxially oriented polyethylene terephthalate films. *J. Plast. Film Sheeting* 31 (2): 158–185.

12 Herrera, M.A., Mathew, A.P., and Oksman, K. (2014). Gas permeability and selectivity of cellulose nanocrystals films (layers) deposited by spin coating. *Carbohydr. Polym.* 112: 494–501.

13 Hagen, D.A., Saucier, L., and Grunlan, J.C. (2014). Controlling effective aspect ratio and packing of clay with pH for improved gas barrier in nanobrick wall thin films. *ACS Appl. Mater. Interfaces* 6 (24): 22914–22919.

14 Yang, Q., Fukuzumi, H., Saito, T. et al. (2011). Transparent cellulose films with high gas barrier properties fabricated from aqueous alkali/urea solutions. *Biomacromolecules* 12 (7): 2766–2771.

15 Shukla, A., Avadhany, S.N., Fang, J.C., and Hammond, P.T. (2010). Tunable vancomycin releasing surfaces for biomedical applications. *Small* 6 (21): 2392–2404.

16 Lee, K.H., Hong, J., Kwak, S.J. et al. (2015). Spin self-assembly of highly ordered multilayers of graphene-oxide sheets for improving oxygen barrier performance of polyolefin films. *Carbon* 83: 40–47.

17 Kim, S.G., Park, O.K., Lee, J.H., and Ku, B.C. (2013). Layer-by-layer assembled graphene oxide films and barrier properties of thermally reduced graphene oxide membranes. *Carbon Lett.* 14 (4): 247–250.

18 Priolo, M.A., Gamboa, D., Holder, K.M., and Grunlan, J.C. (2010). Super gas barrier of transparent polymer-clay multilayer ultrathin films. *Nano Lett.* 10 (12): 4970–4974.

19 Holder, K.M., Priolo, M.A., Secrist, K.E. et al. (2012). Humidity-responsive gas barrier of hydrogen-bonded polymer-clay multilayer thin films. *J. Phys. Chem. C* 116 (37): 19851–19856.

20 Walther, A., Bjurhager, I., Malho, J.M. et al. (2010). Large-area, lightweight and thick biomimetic composites with superior material properties via fast, economic, and green pathways. *Nano Lett.* 10 (8): 2742–2748.

21 Choi, J.H., Park, Y.W., Park, T.H. et al. (2012). Fuzzy nanoassembly of polyelectrolyte and layered clay multicomposite toward a reliable gas barrier. *Langmuir* 28 (17): 6826–6831.

22 Song, E.H., Kang, B.H., Kim, T.Y. et al. (2015). Highly oriented gold/nanoclay-polymer nanocomposites for flexible gas barrier films. *ACS Appl. Mater. Interfaces* 7 (8): 4778–4783.

23 Priolo, M.A., Holder, K.M., Gamboa, D., and Grunlan, J.C. (2011). Influence of clay concentration on the gas barrier of clay-polymer nanobrick wall thin film assemblies. *Langmuir* 27 (19): 12106–12114.

24 Stevens, B.E., Odenborg, P.K., Priolo, M.A., and Grunlan, J.C. (2014). Hydrophobically modified polyelectrolyte for improved oxygen barrier in nanobrick wall multilayer thin films. *J. Polym. Sci. Part B Polym. Phys.* 52 (17): 1153–1156.

25 Tzeng, P., Lugo, E.L., Mai, G.D. et al. (2015). Super hydrogen and helium barrier with polyelectolyte nanobrick wall thin film. *Macromol. Rapid Commun.* 36 (1): 96–101.

26 Xiang, F., Tzeng, P., Sawyer, J.S. et al. (2014). Improving the gas barrier property of clay-polymer multilayer thin films using shorter deposition times. *ACS Appl. Mater. Interfaces* 6 (9): 6040–6048.

27 Jang, W.S., Rawson, I., and Grunlan, J.C. (2008). Layer-by-layer assembly of thin film oxygen barrier. *Thin Solid Films* 516 (15): 4819–4825.

28 Baghbanzadeh, M., Rana, D., Lan, C.Q., and Matsuura, T. (2016). Effects of inorganic nano-additives on properties and performance of polymeric membranes in water treatment. *Sep. Purif. Rev.* 45 (2): 141–167.

29 Bodaghi, H., Mostofi, Y., Oromiehie, A. et al. (2013). Evaluation of the photocatalytic antimicrobial effects of a TiO_2 nanocomposite food packaging film by in vitro and in vivo tests. *LWT Food Sci. Technol.* 50 (2): 702–706.

30 Bumbudsanpharoke, N. and Ko, S. (2015). Nano-food packaging: an overview of market, migration research, and safety regulations. *J. Food Sci.* 80 (5): R910–R923.

31 Mostafavi, F.S. and Zaeim, D. (2020). Agar-based edible films for food packaging applications – a review. *Int. J. Biol. Macromol.*

32 López de Dicastillo, C., Velásquez, E., Rojas, A. et al. (2020). The use of nanoadditives within recycled polymers for food packaging: properties, recyclability, and safety. *Compr. Rev. Food Sci. Food Saf.*: 1–17.

33 Emamifar, A. and Mohammadizadeh, M. (2015). Preparation and application of LDPE/ZnO nanocomposites for extending shelf life of fresh strawberries. *Food Technol. Biotechnol.* 53 (4): 488–495.

34 Pantani, R., Gorrasi, G., Vigliotta, G. et al. (2013). PLA-ZnO nanocomposite films: water vapor barrier properties and specific end-use characteristics. *Eur. Polym. J.* 49 (11): 3471–3482.

35 Vaezi, K., Asadpour, G., and Sharifi, H. (2019). Effect of ZnO nanoparticles on the mechanical, barrier and optical properties of thermoplastic cationic starch/montmorillonite biodegradable films. *Int. J. Biol. Macromol.* 124: 519–529.

36 Roy, S. and Rhim, J.W. (2019). Carrageenan-based antimicrobial bionanocomposite films incorporated with ZnO nanoparticles stabilized by melanin. *Food Hydrocolloids* 90: 500–507.

37 Enescu, D., Cerqueira, M.A., Fucinos, P., and Pastrana, L.M. (2019). Recent advances and challenges on applications of nanotechnology in food packaging. A literature review. *Food Chem. Toxicol.* 134: 110814.

38 Swaroop, C. and Shukla, M. (2018). Nano-magnesium oxide reinforced polylactic acid biofilms for food packaging applications. *Int. J. Biol. Macromol.* 113: 729–736.

39 Xiong, H.G., Tang, S.W., Tang, H.L., and Zou, P. (2008). The structure and properties of a starch-based biodegradable film. *Carbohydr. Polym.* 71 (2): 263–268.

40 Vladimirov, V., Betchev, C., Vassiliou, A. et al. (2006). Dynamic mechanical and morphological studies of isotactic polypropylene/fumed silica nanocomposites with enhanced gas barrier properties. *Compos. Sci. Technol.* 66 (15): 2935–2944.

41 Tang, S., Zou, P., Xiong, H., and Tang, H. (2008). Effect of nano-SiO_2 on the performance of starch/polyvinyl alcohol blend films. *Carbohydr. Polym.* 72 (3): 521–526.

42 Liu, Y., Liu, Y., and Wei, S. (2010). Processing technologies of EVOH/nano-SiO_2 high-barrier packaging composites. *Sci. Res. Publ, Inc.-Srp. Irvin.*

43 Donglu, F., Wenjian, Y., Kimatu, B.M. et al. (2016). Effect of nanocomposite-based packaging on storage stability of mushrooms (*Flammulina velutipes*). *Innovative Food Sci. Emerg. Technol.* 33: 489–497.

44 Bang, G. and Kim, S.W. (2012). Biodegradable poly(lactic acid)-based hybrid coating materials for food packaging films with gas barrier properties. *J. Ind. Eng. Chem.* 18 (3): 1063–1068.

45 Huang, Y., Mei, L., Chen, X., and Wang, Q. (2018). Recent developments in food packaging based on nanomaterials. *Nanomaterials* 8 (10): 1–29.

46 Garcia, C.V., Shin, G.H., and Kim, J.T. (2018). Metal oxide-based nanocomposites in food packaging: applications, migration, and regulations. *Trends Food Sci. Technol.* 82: 21–31.

47 Nasiri, A., Shariaty-Niasar, M., and Akbari, Z. (2012). Synthesis of LDPE/Nano TiO_2 nanocomposite for packaging applications. *Int. J. Nanosci. Nanotechnol.* 8 (3): 165–170.

48 Yildirim, S., Röcker, B., Pettersen, M.K. et al. (2018). Active packaging applications for food. *Compr. Rev. Food Sci. Food Saf.* 17 (1): 165–199.

49 Shankar, S., Jaiswal, L., and Rhim, J.W. (2016). Gelatin-based nanocomposite films: potential use in antimicrobial active packaging. In: *Antimicrobial Food Packaging* (ed. J.B. Velázquez), 339–348. Academic Press.

50 Li, D., Ye, Q., Lei, J., and Zisheng, L. (2016). Effects of nano-TiO_2-LDPE packaging on postharvest quality and antioxidant capacity of strawberry (Fragaria ananassa Duch.) stored at refrigeration temperature. *J. Sci. Food Agric.* 97 (4): 1116–1123.

51 Gumiero, M., Peressini, D., Pizzariello, A. et al. (2013). Effect of TiO_2 photocatalytic activity in a HDPE-based food packaging on the structural and microbiological stability of a short-ripened cheese. *Food Chem.* 138 (2–3): 1633–1640.

52 Luo, Z., Qin, Y., and Ye, Q. (2015). Effect of nano-TiO_2-LDPE packaging on microbiological and physicochemical quality of Pacific white shrimp during chilled storage. *Int. J. Food Sci. Technol.* 50 (7): 1567–1573.

53 Li, K., Jin, S., Liu, X. et al. (2017). Preparation and characterization of chitosan/soy protein isolate nanocomposite film reinforced by Cu nanoclusters. *Polymers* 9 (7): 247.

54 Lomate, G.B., Dandi, B., and Mishra, S. (2018). Development of antimicrobial LDPE/Cu nanocomposite food packaging film for extended shelf life of peda. *Food Packag. Shelf Life* 16: 211–219.

55 Roy, S., Rhim, J.-W., and Jaiswal, L. (2019). Bioactive agar-based functional composite film incorporated with copper sulfide nanoparticles. *Food Hydrocolloids* 93: 156–166.

56 Cibert, C., Hidalgo, H., Champeaux, C. et al. (2008). Properties of aluminum oxide thin films deposited by pulsed laser deposition and plasma enhanced chemical vapor deposition. *Thin Solid Films* 516 (6): 1290–1296.

57 Hirvikorpi, T., Vähä-Nissi, M., Mustonen, T. et al. (2010). Atomic layer deposited aluminum oxide barrier coatings for packaging materials. *Thin Solid Films* 518 (10): 2654–2658.

58 Hirvikorpi, T., Vähä-Nissi, M., Harlin, A. et al. (2011). Enhanced water vapor barrier properties for biopolymer films by polyelectrolyte multilayer and atomic layer deposited Al_2O_3 double-coating. *Appl. Surf. Sci.* 257 (22): 9451–9454.

59 Struller, C.F., Kelly, P.J., and Copeland, N.J. (2014). Aluminum oxide barrier coatings on polymer films for food packaging applications. *Surf. Coat. Technol.* 241: 130–137.

60 Nielsen, L.E. (1967). Models for the permeability of filled polymer systems. *J. Macromol. Sci.* 1 (5): 929–942.

61 Beall, G.W. (2000). New conceptual model for interpreting nanocomposite behavior. In: *Polymer–Clay Nanocomposites* (eds. T.J. Pinnavaia and G.W. Beall), 267–279.

62 Cho, S.I., Kim, Y.R., Lee, J.W. et al. (2010). A review on the application of nanotechnology in food processing and packaging. *Food Eng. Prog.* 14 (4): 283–291.

63 Majeed, K., Jawaid, M., Hassan, A. et al. (2013). Potential materials for food packaging from nanoclay/natural fibres filled hybrid composites. *Mater. Des.* 46: 391–410.

64 Silvestre, C., Duraccio, D., and Cimmino, S. (2011). Food packaging based on polymer nanomaterials. *Prog. Polym. Sci.* 36 (12): 1766–1782.

65 Priolo, M.A., Gamboa, D., and Grunlan, J.C. (2010). Transparent clay-polymer nano brick wall assemblies with tailorable oxygen barrier. *ACS Appl. Mater. Interfaces* 2 (1): 312–320.

66 Bharadwaj, R.K., Mehrabi, A.R., Hamilton, C. et al. (2002). Structure-property relationships in cross-linked polyester-clay nanocomposites. *Polymer* 43 (13): 3699–3705.

67 Cabedo, L., Giménez, E., Lagaron, J.M. et al. (2004). Development of EVOH-kaolinite nanocomposites. *Polymer* 45 (15): 5233–5238.

68 Lagarón, J.M., Cabedo, L., Cava, D. et al. (2005). Improving packaged food quality and safety. Part 2: Nanocomposites. *Food Addit. Contam.* 22 (10): 994–998.

69 Cava, D., Giménez, E., Gavara, R., and Lagaron, J.M. (2006). Comparative performance and barrier properties of biodegradable thermoplastics and nanobiocomposites versus PET for food packaging applications. *J. Plast. Film Sheeting* 22 (4): 265–274.

70 Petersson, L. and Oksman, K. (2006). Biopolymer based nanocomposites: comparing layered silicates and microcrystalline cellulose as nanoreinforcement. *Compos. Sci. Technol.* 66 (13): 2187–2196.

71 Koh, H.C., Park, J.S., Jeong, M.A. et al. (2008). Preparation and gas permeation properties of biodegradable polymer/layered silicate nanocomposite membranes. *Desalination* 233 (1–3): 201–209.

72 Bharadwaj, R.K. (2001). Modeling the barrier properties of polymer-layered silicate nanocomposites. *Macromolecules* 34 (26): 9189–9192.

73 Jawahar, P. and Balasubramanian, M. (2006). Preparation and properties of polyester-based nanocomposite gel coat system. *J. Nanomater.*: 1–7.

74 Lotti, C., Isaac, C.S., Branciforti, M.C. et al. (2008). Rheological, mechanical and transport properties of blown films of high density polyethylene nanocomposites. *Eur. Polym. J.* 44 (5): 1346–1357.

75 Julkapli, N.M. and Bagheri, S. (2016). Developments in nano-additives for paper industry. *J. Wood Sci.* 62 (2): 117–130.
76 Busolo, M.A. and Lagaron, J.M. (2012). Oxygen scavenging polyolefin nanocomposite films containing an iron modified kaolinite of interest in active food packaging applications. *Innovative Food Sci. Emerg. Technol.* 16: 211–217.
77 Nalçabasmaz, S., Ayhan, Z., Cimmino, S. et al. (2017). Effects of PP-based nanopackaging on the overall quality and shelf life of ready-to-eat salami. *Packag. Technol. Sci.* 30 (10): 663–679.
78 Junqueira-Gonçalves, M.P., Salinas, G.E., Bruna, J.E., and Niranjan, K. (2017). An assessment of lactobiopolymer-montmorillonite (MMT) composites for dip coating applications on fresh strawberries. *J. Sci. Food Agric.* 97 (6): 1846–1853.
79 Xiang, C., Cox, P.J., Kukovecz, A. et al. (2013). Functionalized low defect graphene nanoribbons and polyurethane composite film for improved gas barrier and mechanical performances. *ACS Nano.* 7 (11): 10380–10386.
80 Huang, H.D., Ren, P.G., Chen, J. et al. (2012). High barrier graphene oxide nanosheet/poly(vinyl alcohol) nanocomposite films. *J. Membr. Sci.* 409: 156–163.
81 Markarian, J. (2005). Automotive and packaging offer growth opportunities for nanocomposites. *Plast. Addit. Compd.* 7 (6): 18–21.
82 Kim, J.Y., Han, S. II, and Hong, S. (2008). Effect of modified carbon nanotube on the properties of aromatic polyester nanocomposites. *Polymer* 49 (15): 3335–3345.
83 Kar, S., Bindal, R.C., and Tewari, P.K. (2012). Carbon nanotube membranes for desalination and water purification: challenges and opportunities. *Nano Today* 7 (5): 385–389.
84 Brody, A.L. (2006). Nano and food packaging technologies converge. *Food Technol.* 60 (3): 92–94.
85 Shankar, S. and Rhim, J.W. (2018). Bionanocomposite films for food packaging applications. In: *Reference Module in Food Science*, 1–10. Elsevier.
86 Arrieta, M.P., Peponi, L., López, D. et al. (2017). An overview of nanoparticles role in the improvement of barrier properties of bioplastics for food packaging applications. In: *Food Packaging* (ed. A.M. Grumezescu), 391–424. Academic Press.
87 Svagan, A.J., Hedenqvist, M.S., and Berglund, L. (2009). Reduced water vapour sorption in cellulose nanocomposites with starch matrix. *Compos. Sci. Technol.* 69 (3–4): 500–506.
88 Sanchez-Garcia, M.D., Gimenez, E., and Lagaron, J.M. (2008). Morphology and barrier properties of solvent cast composites of thermoplastic biopolymers and purified cellulose fibers. *Carbohydr. Polym.* 71 (2): 235–244.
89 Paralikar, S.A., Simonsen, J., and Lombardi, J. (2008). Poly(vinyl alcohol)/cellulose nanocrystal barrier membranes. *J. Membr. Sci.* 320 (1–2): 248–258.
90 Dufresne, A. and Vignon, M.R. (1998). Improvement of starch film performances using cellulose microfibrils. *Macromolecules* 31 (8): 2693–2696.
91 Dufresne, A., Dupeyre, D., and Vignon, M.R. (2000). Cellulose microfibrils from potato tuber cells: processing and characterization of starch–cellulose microfibril composites. *J. Appl. Polym. Sci.* 76 (14): 2080–2092.

92 Lu, Y., Weng, L., and Cao, X. (2005). Biocomposites of plasticized starch reinforced with cellulose crystallites from cottonseed linter. *Macromol. Biosci.* 5 (11): 1101–1107.

93 Arun, S., Kumar, K.A.A., and Sreekala, M.S. (2012). Fully biodegradable potato starch composites: effect of macro and nano fiber reinforcement on mechanical, thermal and water-sorption characteristics. *Int. J. Plast. Technol.* 16 (1): 50–66.

94 Liu, S., Li, X., Chen, L. et al. (2018). Tunable D-limonene permeability in starch-based nanocomposite films reinforced by cellulose nanocrystals. *J. Agric. Food. Chem.* 66 (4): 979–987.

95 Sanchez-Garcia, M.D. and Lagaron, J.M. (2010). On the use of plant cellulose nanowhiskers to enhance the barrier properties of polylactic acid. *Cellulose* 17 (5): 987–1004.

96 Fortunati, E., Peltzer, M., Armentano, I. et al. (2012). Effects of modified cellulose nanocrystals on the barrier and migration properties of PLA nano-biocomposites. *Carbohydr. Polym.* 90 (2): 948–956.

97 Fortunati, E., Luzi, F., Puglia, D. et al. (2014). Investigation of thermo-mechanical, chemical and degradative properties of PLA-limonene films reinforced with cellulose nanocrystals extracted from Phormium tenax leaves. *Eur. Polym. J.* 56 (1): 77–91.

98 Li, L., Bao, R.Y., Gao, T. et al. (2019). Dopamine-induced functionalization of cellulose nanocrystals with polyethylene glycol towards poly(L-lactic acid) bionanocomposites for green packaging. *Carbohydr. Polym.* 203: 275–284.

99 Dhar, P., Bhardwaj, U., Kumar, A., and Katiyar, V. (2015). Poly(3-hydroxybutyrate)/cellulose nanocrystal films for food packaging applications: barrier and migration studies. *Polym. Eng. Sci.* 55 (10): 2388–2395.

100 Yu, H., Yan, C., and Yao, J. (2014). Fully biodegradable food packaging materials based on functionalized cellulose nanocrystals/poly(3-hydroxybutyrate-co-3-hydroxyvalerate) nanocomposites. *RSC Adv.* 4 (104): 59792–59802.

101 Cano, A., Fortunati, E., Cháfer, M. et al. (2015). Effect of cellulose nanocrystals on the properties of pea starch–poly(vinyl alcohol) blend films. *J. Mater. Sci.* 50 (21): 6979–6992.

102 El Miri, N., Abdelouahdi, K., Barakat, A. et al. (2015). Bio-nanocomposite films reinforced with cellulose nanocrystals: rheology of film-forming solutions, transparency, water vapor barrier and tensile properties of films. *Carbohydr. Polym.* 129: 156–167.

103 Arrieta, M.P., Samper, M.D., López, J., and Jiménez, A. (2014). Combined effect of poly(hydroxybutyrate) and plasticizers on polylactic acid properties for film intended for food packaging. *J. Polym. Environ.* 22 (4): 460–470.

104 Arrieta, M.P., Fortunati, E., Dominici, F. et al. (2014). PLA-PHB/cellulose based films: mechanical, barrier and disintegration properties. *Polym. Degrad. Stab.* 107: 139–149.

105 Arrieta, M.P., Fortunati, E., Dominici, F. et al. (2015). Bionanocomposite films based on plasticized PLA-PHB/cellulose nanocrystal blends. *Carbohydr. Polym.* 121: 265–275.

106 Peelman, N., Ragaert, P., De Meulenaer, B. et al. (2013). Application of bioplastics for food packaging. *Trends Food Sci. Technol.* 32 (2): 128–141.

107 Pelissari, F.M., Sobral, P.J.D.A., and Menegalli, F.C. (2014). Isolation and characterization of cellulose nanofibers from banana peels. *Cellulose* 21 (1): 417–432.

108 Aqlil, M., Moussemba Nzenguet, A., Essamlali, Y. et al. (2017). Graphene oxide filled lignin/starch polymer bionanocomposite: structural, physical, and mechanical studies. *J. Agric. Food. Chem.* 65 (48): 10571–10581.

109 Shahbazi, M., Rajabzadeh, G., and Sotoodeh, S. (2017). Functional characteristics, wettability properties and cytotoxic effect of starch film incorporated with multi-walled and hydroxylated multi-walled carbon nanotubes. *Int. J. Biol. Macromol.* 104: 597–605.

110 Oleyaei, S.A., Zahedi, Y., Ghanbarzadeh, B., and Moayedi, A.A. (2016). Modification of physicochemical and thermal properties of starch films by incorporation of TiO_2 nanoparticles. *Int. J. Biol. Macromol.* 89: 256–264.

111 Kristo, E. and Biliaderis, C.G. (2007). Physical properties of starch nanocrystal-reinforced pullulan films. *Carbohydr. Polym.* 68 (1): 146–158.

112 Ali, A., Xie, F., Yu, L. et al. (2018). Preparation and characterization of starch-based composite films reinfoced by polysaccharide-based crystals. *Compos. Part B Eng.* 133: 122–128.

113 Lu, Y., Weng, L., and Zhang, L. (2004). Morphology and properties of soy protein isolate thermoplastics reinforced with chitin whiskers. *Biomacromolecules* 5 (3): 1046–1051.

114 Sriupayo, J., Supaphol, P., Blackwell, J., and Rujiravanit, R. (2005). Preparation and characterization of α-chitin whisker-reinforced chitosan nanocomposite films with or without heat treatment. *Carbohydr. Polym.* 62 (2): 130–136.

115 De Moura, M.R., Aouada, F.A., Avena-Bustillos, R.J. et al. (2009). Improved barrier and mechanical properties of novel hydroxypropyl methylcellulose edible films with chitosan/tripolyphosphate nanoparticles. *J. Food Eng.* 92 (4): 448–453.

116 Buslovich, A., Horev, B., Rodov, V. et al. (2017). One-step surface grafting of organic nanoparticles: in situ deposition of antimicrobial agents vanillin and chitosan on polyethylene packaging films. *J. Mater. Chem. B* 5 (14): 2655–2661.

117 Wu, C., Li, Y., Du, Y. et al. (2019). Preparation and characterization of konjac glucomannan-based bionanocomposite film for active food packaging. *Food Hydrocolloids* 89: 682–690.

118 Ozcalik, O. and Tihminlioglu, F. (2013). Barrier properties of corn zein nanocomposite coated polypropylene films for food packaging applications. *J. Food Eng.* 114 (4): 505–513.

119 Rouf, T.B., Schmidt, G., and Kokini, J.L. (2018). Zein–Laponite nanocomposites with improved mechanical, thermal and barrier properties. *J. Mater. Sci.* 53 (10): 7387–7402.

120 Aytac, Z., Ipek, S., Durgun, E. et al. (2017). Antibacterial electrospun zein nanofibrous web encapsulating thymol/cyclodextrin-inclusion complex for food packaging. *Food Chem.* 233: 117–124.

121 Oymaci, P. and Altinkaya, S.A. (2016). Improvement of barrier and mechanical properties of whey protein isolate based food packaging films by incorporation of zein nanoparticles as a novel bionanocomposite. *Food Hydrocolloids* 54: 1–9.

122 Byun, Y., Bae, H.J., and Whiteside, S. (2012). Active warm-water fish gelatin film containing oxygen scavenging system. *Food Hydrocolloids* 27 (1): 250–255.

123 Ejaz, M., Arfat, Y.A., Mulla, M., and Ahmed, J. (2018). Zinc oxide nanorods/clove essential oil incorporated type B gelatin composite films and its applicability for shrimp packaging. *Food Packag. Shelf Life* 15: 113–121.

124 Nassiri, R. and Mohammadi, N.A. (2013). Antimicrobial and barrier properties of bovine gelatin films reinforced by nano TiO_2. *J. Chem. Heal. Risks* 3 (3): 21–28.

125 Hassannia-Kolaee, M., Khodaiyan, F., Pourahmad, R., and Shahabi-Ghahfarrokhi, I. (2016). Development of ecofriendly bionanocomposite: Whey protein isolate/pullulan films with nano-SiO_2. *Int. J. Biol. Macromol.* 86: 139–144.

126 Qazanfarzadeh, Z. and Kadivar, M. (2016). Properties of whey protein isolate nanocomposite films reinforced with nanocellulose isolated from oat husk. *Int. J. Biol. Macromol.* 91: 1134–1140.

127 Zhang, W., Chen, J., Chen, Y. et al. (2016). Enhanced physicochemical properties of chitosan/whey protein isolate composite film by sodium laurate-modified TiO_2 nanoparticles. *Carbohydr. Polym.* 138: 59–65.

128 Carvalho, R.A., Santos, T.A., De Azevedo, V.M. et al. (2018). Bio-nanocomposites for food packaging applications: effect of cellulose nanofibers on morphological, mechanical, optical and barrier properties. *Polym. Int.* 67 (4): 386–392.

129 Li, K., Jin, S., Chen, H. et al. (2017). A high-performance soy protein isolate-based nanocomposite film modified with microcrystalline cellulose and Cu and Zn nanoclusters. *Polymers* 9 (5): 167.

130 Li, K., Jin, S., Han, Y. et al. (2017). Improvement in functional properties of soy protein isolate-based film by cellulose nanocrystal-graphene artificial nacre nanocomposite. *Polymers* 9 (8): 321.

131 Kumar, P., Sandeep, K.P., Alavi, S. et al. (2010). Preparation and characterization of bio-nanocomposite films based on soy protein isolate and montmorillonite using melt extrusion. *J. Food Eng.* 100 (3): 480–489.

132 Kumar, P., Sandeep, K.P., Alavi, S. et al. (2010). Effect of type and content of modified montmorillonite on the structure and properties of bio-nanocomposite films based on soy protein isolate and montmorillonite. *J. Food Sci.* 75 (5): 46–56.

11

Nanostructure-Based Multilayer Food Packaging Films

Shiji Mathew

Mahatma Gandhi University, School of Biosciences, Priyadarshini Hills, Kottayam 686 560, Kerala, India

11.1 Introduction

Plastics or polymers have extensively replaced the use of metals, glass, and papers as important food packaging materials due to their exclusive properties such as lightweight, ease in processing, and low-cost. However, the weak mechanical and moisture and oxygen barrier properties of polymers are considered as an important drawback, which limits its wider application. In this context, researchers have found better alternative strategies to improve the weaker properties of polymer-based packages. These include the use of (i) polymer blends, where two or more polymers are blended together to achieve improved properties; (ii) development of polymer composites, where suitable fillers are added to polymer matrix; (iii) chemical modification of polymers; and (iv) formation of multilayer films [1]. Among these approaches, multilayer films, as a single unit, are capable of satisfying all the properties essential for an ideal packaging. Multilayer structures used for packaging purpose can be in the form of coatings or interlayers.

Multilayer films are systems that combine two or more polymers and their functions, in a single structure. The preparation of multilayer films is a feasible and promising approach where two or more materials are combined to form multiple layers to develop highly improved food packaging material. Multilayer packagings are found to possess a blend of advantages such as water resistance, reduced oxygen permeability, mechanical stability, thermal resistance, and aroma barrier capability when compared with monolayer packages. However, as these structures are composed of different polymers, the compatibility between the materials used for layer formation has to be assessed. In the case of chemically dissimilar layers, an adhesive layer is often needed to tie them together to prevent their separation during layer formation. So while designing a multilayer structure, various factors related to polymers such as mechanical strength, heat sealing ability, barrier properties, and also formability and recyclability should be taken into account.

The recent advancements of nanotechnology in food packaging industry have demonstrated that the incorporation of nanostructures in one or more inner

Nanotechnology-Enhanced Food Packaging, First Edition. Edited by Jyotishkumar Parameswaranpillai, Radhakrishnan Edayileveettil Krishnankutty, Aswathy Jayakumar, Sanjay Mavinkere Rangappa, and Suchart Siengchin.

layers of multilayer films can result in development of film with high barrier and mechanical properties for improved durability and protection [2]. Accordingly, the incorporation of nanostructures into the inner layers of multilayer structure has shown to further improve the functionalities of the film for active and intelligent packaging performance. Moreover, this technology also omits the use of the additional adhesive layers, which is needed to hold different layers together.

This chapter provides a brief description on the general requirements of a food packaging material. The following sections then discuss about the typical structure and functions of multilayer films and their advantages over other packaging systems. The next section focuses on the recent advances of nanostructure-based multilayer films in food packaging sector and their main preparative methodologies. Then the practical applications of multilayer films for packaging of food samples will be discussed in the subsequent sections.

11.2 Requirements of Food Packaging Systems

Packaging serves an important role in protecting the food against the external barriers and microbial contamination. It also enables effective distribution of food and facilitates the end use convenience and communication at consumer levels [3]. But now, the advancements in material science and technology, as well as changing demands of modern society and industrial production, have resulted in a consistent evolution of food packaging industry [4]. Today, food packaging materials have evolved from simple containment and preservation systems to aspects of safety, material reduction, marketing appraisal, convenience, environmental concerns, and tamperproofing [5]. Commercialization of minimal processed food has increased as the consumers show increased interest to ready-to-eat food stuffs [6]. As these minimal processed foods possess high risk of moisture loss and browning/oxidation, they are also prone to contamination by microorganisms. Hence, the packaging of such foods should have an efficient protective barrier. An ideal packaging material should possess basic requirements in terms of physical, mechanical, thermal, gas/water vapor barrier, and optical properties to protect the food during transportation and storage.

Advancements in packaging industry have led to the invention of improved, smart/intelligent packaging and active food packaging materials, which are superior over the conventional packaging materials due to their specific functions. Packaging materials with improved mechanical, gas/moisture barrier, and temperature stability due to the incorporation of filler particles are termed "improved packaging." Intelligent/smart package can provide the traceability and feedback about the quality of food. Here, indicators/sensors are incorporated onto packages, which can interact with the food as well as external environment. With the incorporation of active agents into packaging materials, development of active package is possible, which possesses high antimicrobial, biocatalytic, or antioxidant properties, along with improved mechanical and barrier properties. These active packaging materials can extend the shelf life of food, and ensure the quality and safety of food [7]. The

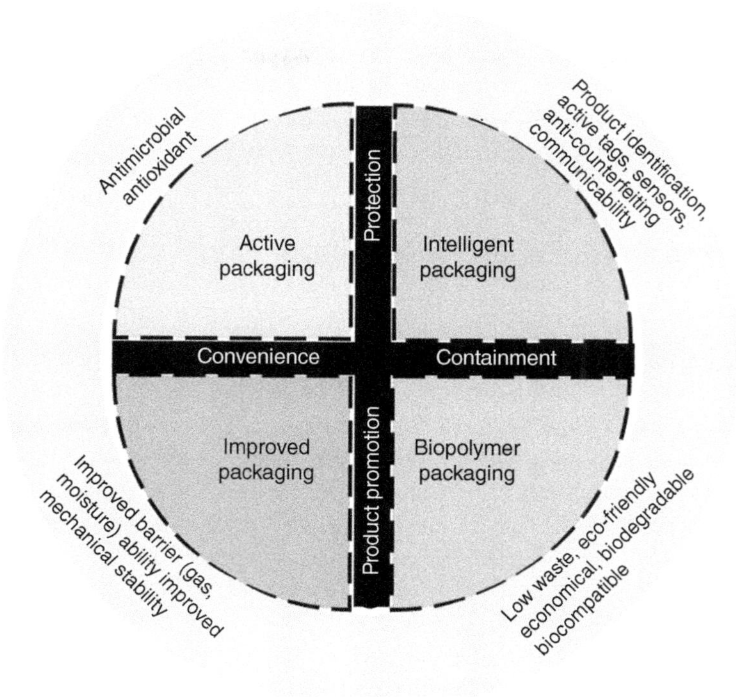

Figure 11.1 Functions of different types of food packaging materials.

essential functions of different kinds of food packaging films are summarized in Figure 11.1.

11.3 Multilayer Packaging Films

It has been found that in most of the cases, no single polymer layer is able to meet all the requirement of a food packaging material. Multilayer films are promising systems that combine various polymers in a single structure with the intension of developing an ideal packaging with multiple superior properties. Here, the assembled polymer layers are intended to perform different functions such as providing mechanical stability (tensile strength, rigidity, and durability), barrier properties (gas, vapor, flavor, and odor), promotion, containment, and convenience, at the same time ensuring cost-effectiveness, and fulfilling other essential requirements essential for a proper packaging [8]. Multilayer packages can also be designed to complement and improve the properties of active and smart packaging materials. In addition, multilayer packages also offer reduction in the overall material used as it provides lightweight, thinner, and compact packaging materials [9].

One of the major issues related to multilayer films is their degradability and recyclability, as they are composed of different polymeric layers of varied properties. Currently, most of the materials used for multilayer preparation are of synthetic origin, which are nondegradable. In this regard, materials of low life cycle and minimum influence on the environment are of great interest for developing multilayer films. So while designing multilayer films, it would be highly advantageous to include biodegradable polymers or biopolymers as an important component in multilayer films, which may alleviate this problem to a great extent.

11.4 Structure and Functions of Multilayer Film Packaging

Multilayer packages for food packaging purpose usually consist of three to seven layers [9]. Based on the degree of flexibility, multilayer packages are classified as semirigid and flexible packages. Beverage cartons and multilayer bottles are examples of semirigid multilayer packages, whereas flexible packages are films, which are mainly composed of polymers [10]. Flexible multilayer packagings are further categorized into five types as (i) flexible packaging without barrier layer, (ii) flexible packaging with a barrier layer, (iii) flexible packagings with metal oxide coatings such as aluminum or silicon oxide, (iv) thermoformed plastic packagings, and (v) plastic packaging with aluminum foil [11].

Each component layer of multilayer flexible packaging is intended to perform or impart essential functions to the overall architecture. Figure 11.2 illustrates the typical structure and functions of each component of a multilayer flexible packaging.

The outer layer of a typical multilayer packaging is strong and provides a printing surface and also protects the food from outside moisture. Owing to the flexibility, abundance, versatility, ease of processing, chemical inertness, excellent vapor barrier properties, polyethylene (PE), polypropylene (PP), and polyethylene terephthalate (PET) are the most commonly used constituents of the outer layer of multilayer films. This layer can be printed using dyes or pigments. An optional thin layer called the coating layer can be sometimes included outside the outer layer, which is intended to protect the printable surface. It can be composed of any type of special polymers. Next comes the structural layer that imparts shape to the packaging and protects it from puncturing and tearing. Usually used polymers

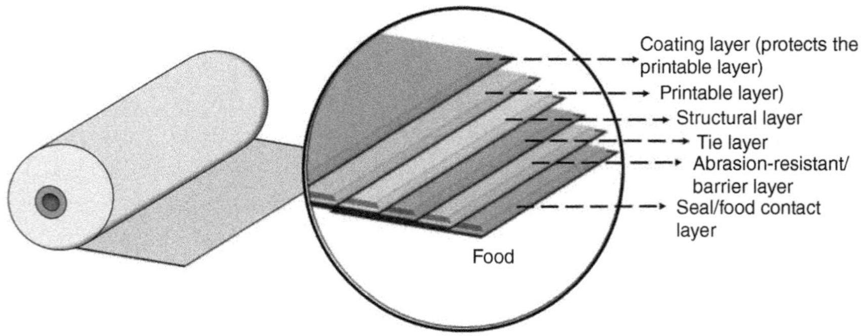

Figure 11.2 Typical layers and functions of a multilayer food packaging.

in this layer are PE and PET. A tie or adhesive layer is introduced to enhance the adhesion and impart strength to layer bonding between two chemically dissimilar polymer layers such as nylon and PE, which tend to separate [12]. Examples of tie layers include functionalized polyolefins. The next layer will be the barrier layer, which primarily inhibits O_2 infiltration into the package. In addition, it also offers light, odor, and flavor barrier capability to the package. Examples of such barrier layers include polyamide resins and poly(ethylene-*co*-vinyl alcohol) (EVOH). The last or inner layer is the food contact layer or seal layer, which is generally strong, flexible, and usually of low melting point so that it can be easily heat sealed. As the inner comes in direct contact with the food, it should be made of an inert polymer, which do not react with any food constituents. PE layer is commonly used for this purpose [13]. Hence each layer of a multilayer packaging has to be pre-designed and chosen based on the final desired properties.

11.5 Nanotechnology-Based Multilayer Films

Nanotechnological interventions have created great breakthrough in food packaging industry with the development of polymer nanocomposites, which incorporates nanostructures in polymeric matrices for the generation of highly improved and functional food packagings. The nanostructures included in polymer composites can be metal or metal oxide nanoparticles, polymeric nanoparticles, or nanofillers such as clay nanoparticles. The inclusion of each of these nanomaterials can bring desired functionality in the packaging material.

Recently, nanotechnology has further enabled the development of highly improved multilayer films/coatings with nanometer range thickness by the incorporation of nanostructures within the inner layers. Layered silicates or clay nanoparticles are usually incorporated in multilayer film so as to provide a high barrier layer. These inorganic nanofillers being impermeable to molecules, such as gas and liquid as well as to light, create a homogeneous layer in matrix, leading to a tortuous pathway in the film. As a result, these molecules cannot follow a straight line, but must diffuse around the nanofillers, culminating in increased path length for diffusion. On the other hand, in pristine polymers, due to its high permeable nature, the diffusion of molecules and gases occurs in a straight path perpendicular to the film surface [2, 14]. It has been observed that the barrier property of multilayer films is superior even when compared with polymer blends and polymer nanocomposite films (Figure 11.3).

11.6 Preparation of Nano-Based Multilayer Films

Nanostructure-based multilayer films can be prepared by a number of ways. This section discusses in brief the details of the most widely used methodologies with suitable examples. The important methods used for fabrication of nanostructure-based multilayer films for food packaging purposes include the layer-by-layer (LbL) nanoassembly, the electrohydrodynamic processing (EHDP) and microlayer coextrusion.

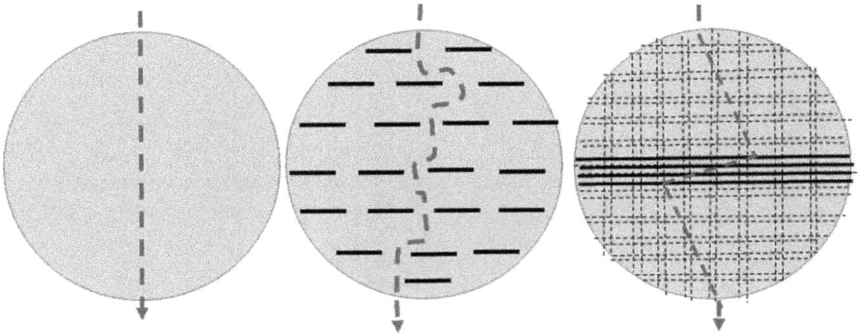

Figure 11.3 Diagram showing the pathway of molecule (gas and water) transport through pristine polymer, a nanocomposite film, and a multilayer film.

11.6.1 Layer-by-Layer (LbL) Nanoassembly

A highly versatile technology for fabricating multilayer thin films, called the LbL method, was first proposed by Decher et al. in 1991 [15]. LbL self-assembly is a method of constructing multilayer thin films by exposing a substrate alternatively to aqueous solution containing cationic and anionic particles [16]. Using this technique, electrically charged molecules are found to get adsorbed from the aqueous solution to a solid support substrate, which is formed by successive adsorption of oppositely charged particles such as nanoparticles or polymers, resulting in formation of multiple layers of nanometer range [17]. Figure 11.4 depicts the diagrammatic representation of the LbL technique for nano-multilayer film formation. Secondary forces such as hydrogen bonding are found to be uniquely responsible for the formation of these multilayered structures. The functionality and properties of the resulting multilayer films will mainly depend upon certain factors like the type of adsorbing molecules used in each layer, number and sequence of layers incorporated, and the preparation conditions [19].

Only those polymeric materials that fulfil the Generally Recognized as Safe (GRAS) status and approved food contact materials should be used for preparing food packaging multilayer films. In this context, various polymers, biopolymers, and edible polymers have been used for the formation of multilayer films through the LbL technique. In initial attempts, numerous studies have employed LbL technique for developing food packaging materials with improved gas and vapor barrier properties. Later on, LbL technology was also exploited for developing packaging materials with better mechanical and thermal properties, active films with antimicrobial and antioxidant activity, and also for intelligent packaging with chemical-sensing ability [20, 21]. Sánchez-Valdes et al. in 2008 prepared multilayer PE–silver nanocomposite film employing various methods such as laminating, casting, and spraying for the nanocomposite layer deposition [22]. The role of silver nanoparticles and the deposition methods on the barrier, antimicrobial, optical, and mechanical properties of the multilayer films was evaluated. It was observed

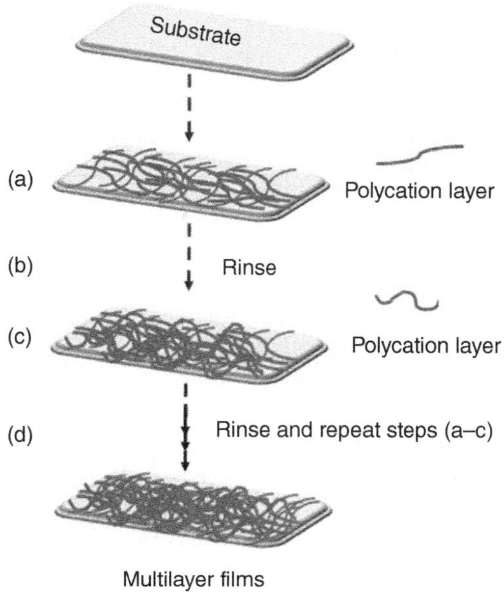

Figure 11.4 Schematic of layer-by-layer (LbL) technique for preparing multilayer films. Source: Modified from Merlin L. Bruening and Maneesh Adusumilli, https://www.sigmaaldrich.com/technical-documents/articles/material-matters/polyelectrolyte-multilayer-films.html [18].

that spraying method was highly effective in enabling antimicrobial activity to the Ag ions by creating tiny voids in the layers, thereby facilitating Ag+ release rate.

The most common processes for attaining self-assembly of multilayers by LbL method are by spraying and dipping. Dipping method possesses potential advantages such as easiness, flexibility, process adaptability along with the capacity of using materials of various geometric shapes, sizes and diverse kinds of polyelectrolytes as well as the opportunity of using various active components. At the same time, the disadvantages of the same method include the longer duration taken for the complete adsorption of layers and relatively less area coated during the film formation. Spraying methods are found to overcome these limitations by providing faster layer formation within lesser time and with better film quality [23]. In 2006, Lu et al. reported the formation of Au-NPs (gold nanoparticles) and polycation nitrodiazoresin containing multilayer thin films by spray-assisted LbL technique [24]. The mechanical properties of this multilayer ultrathin film were found to be controlled by the presence of Au nanoparticles within the film layers. In a recent study, multilayered graphene oxide (GO) film was developed for better oxygen barrier properties using LbL self-assembly method. Here spraying method was adopted to deposit GO+/GO− sheets on flexible PE substrate. The oxygen transmittance rate of the modified PE film was found to be decreased from 3511.5 cc/m^2 d to 1091 cc/m^2 d after depositing a GO layer of 60 nm thickness itself [25]. Table 11.1 gives a list of nanostructure-based multilayer packagings developed by LbL method and their functionalities.

11.6.2 Electrohydrodynamic Processing (EHDP)

Multilayer structures such as interlayers or polymer coatings can be prepared using another simple, straightforward, low-cost, and versatile technique termed EHDP.

Table 11.1 Functions of nanostructure-based multilayer food packaging films prepared by LbL method.

Substrate	Polyelectrolytes		Functions performed	References
	Anion	Cation		
Polylactic acid (PLA)	Nanofibrillated cellulose (NFC) and carboxymethyl cellulose (CMC)	Polyethyleneimine (PEI)	Improved oxygen and vapor barrier properties	[26]
Polyethylene terephthalate (PET)	Sodium montmorillonite (MMT) clay	Polyacrylamide (PAM)	Oxygen barrier	[27]
Polyurethane foam and polylactic acid film	Na^+ MMT nanoclay	Chitosan	Oxygen barrier and flame retardant	[28]
PET	MMT nanoclay	Chitosan and λ-carrageenan	Oxygen barrier	[29]
–	Guar gum	Cellulose nanofibrils	Oil, gas, and water vapor barrier	[30]
PET, PS, fused quartz, silicon, Ti/Au crystals	MMT nanoclay	Branched polyethyleneimine (PEI)	Oxygen barrier	[31]
LDPE	Silver nanoparticles	Chitosan	Antibacterial active films	[32]
LDPE	Polyacrylic acid (PAA) and silver nanoparticles	Polyethyleneimine (PEI)	Antibacterial active films	[33]
–	Poly(L-glutamic acid) (PLGA)	Hen egg white lysozyme (HEWL)	Antibacterial and edible films	[34]
PET	Sodium alginate	Chitosan, carvacrol, zein nanocapsules	Antifungal activity	[35]
PET sheets, quartz slides	Alginate	Chitosan	Edible packaging film	[36]
Poly(ethylene-*co*-vinyl alcohol) (EVOH)	Reduced graphene oxide (rGO)	[2-(Methacryloyloxy) ethyl] trimethylammonium chloride (MTAC)	Antibacterial films	[37]

As the name suggests, it uses electrical and hydrodynamic forces with external high voltage electric field to break polymeric solutions into fine jets [38]. The application of high voltage to the polymeric solution leads to elongation of a jet, resulting in the formation of fiber (electrospinning) or droplet (electrospraying) [39].

Electrospinning technique is used to produce continuous nonwoven mats of polymer nanofibers. Here, under the influence of high voltage electrostatic field, the polymeric solution is extruded from a capillary tube. This results in the formation of a conical-shaped Taylor cone at the tip of the capillary. As the electric field increases, accumulation of positive charges occurs on the Taylor cone, which finally cause fluid ejection from it. As the spinning continues, simultaneous stretching of the polymer fluid and evaporation of solvent leads to generation of continuous nanofibers [40]. Electrospinning techniques have different variants such as needle, wire, vibration, bubble, edge, and roller electrospinning methods [11]. On the other hand, electrospraying or electrohydrodynamic atomization (EHDA) is employed for producing noncontinuous structures like nanobeads or capsules. The suitability of this technique to form nanocapsules of active and bioactive agents provides its great application in food packaging materials. Liu et al. [41] used electrospraying process to develop a multilayer coating of chitosan on a polyvinyl alcohol (PVA)/chitosan blend film prepared by solvent casting method. This multilayer film possessed good antibacterial activity, low oxygen permeability, and high water barrier properties compared with neat PVA films.

Using this technique, nanofibers can be generated in large scale, and the materials will possess vital intrinsic characteristics such as fine size and surface morphology, high surface-to-area ratio, superior mechanical properties, porosity, and the chance to encapsulate bioactive molecules and mediate their controlled release [42, 43]. Figure 11.5 shows an illustration of the electrospraying and electrospinning processes. Recently, both electrospinning and electrospraying techniques are widely used for the development of multilayer food packaging materials with improved barrier properties as well as active (antimicrobial and antioxidant) and intelligent performances. Using EHDP methods, nanostructure-based multilayer films with improved barrier properties can be easily generated by the reinforcement of nanofillers such as nanoclays, nanowhiskers, or nanokeratin into the electrospun fibers or coatings [44].

In a study, montmorillonite nylon-6 clay (MMT-N6) nanofibers were deposited on PP films by electrospinning method. The developed electrospun nanocomposite fiber when applied for packaging potato chips and bread showed improved oxygen permeability and moisture barrier properties [45]. In a similar attempt, Fabra et al. developed a multilayer nano-biocomposite food packaging by electrospinning process. Here polyhydroxy butyrate/bacterial cellulose nanowhisker (PHB/BCNW) fiber mats were directly electrospun onto the surface of thermoplastic corn starch (TPCS) films. These electrospun fibers showed better oxygen permeability and water barrier properties compared to the neat TPCS film [46]. In an interesting work, superhydrophobic multilayer coatings on PET substrate were made by a single step, co-continuous bilayer coating process. Here continuous coating of ultrathin polylactide fibers were deposited on PET by electrospinning process, and nanostructured silica (SiO_2) microparticles were electrosprayed onto the PET/PLA films. Results showed that the electrospun polylactic acid (PLA) nanofibers increased only the water contact angle of PET films, whereas an increase in hydrophobicity was detected after SiO_2 was electrosprayed into the multilayer films [47].

EHDP method can also be applied for the development of active fiber-based multilayer packaging with antimicrobial ability. This can be achieved by two strategies;

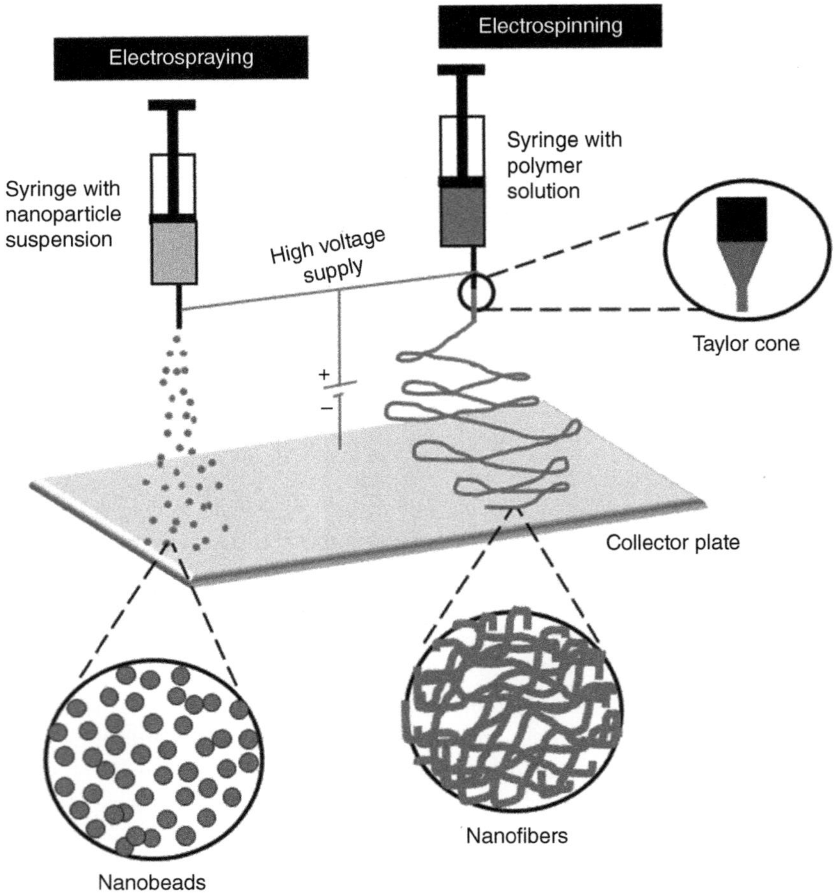

Figure 11.5 Schematic representation of EHDP method for preparation of nanostructured multilayer films.

first involves the direct use of polymers with inherent antimicrobial properties (e.g. chitosan), and the other strategy makes use of introducing external antimicrobial substances into the nanofiber or nanobeads [48]. Using EHDP, nanoparticle agglomeration does not occur, and thus the developed packaging material will be of improved antimicrobial activity, when compared to other preparation techniques. In addition, EHDP can also be applied for the development of active antioxidant packaging through the encapsulation of antioxidants in the inner and outer layers along with their controlled release to aid in preserving the food quality. Liu et al. 2018 reported the fabrication of four types of antimicrobial electrospun PLA/tea polyphenol (TP) nanofibers with PLA/TP in the ratio 5 : 1, 4 : 1, 3 : 1, and 2 : 1. Results indicated that the addition of TP determined the mechanical stability, hydrophilic–hydrophobic behavior, and antioxidant property of the films. Moreover, the electrospun film showed better antibacterial properties against *Escherichia coli* and *Staphylococcus aureus* [49]. In a recent study, chitosan/poly(ε-caprolactone) blend antimicrobial fibrous mats embedded with oregano essential oil were developed by electrospinning. Besides possessing excellent antibacterial property, these hybrid mats also showed improved mechanical and vapor barrier properties [50].

For performing intelligent/smart role in food preservation, multilayer coatings or interlayer packagings with heat management capacity can be developed using EHDP method by the incorporation of phase change materials (PCMs). Such nanolayered structures can easily detect temperature variations (due to chemical reactions and microbial growth) occurring during the transport and storage of food [21]. For example, in a study, a PCM named RT5, a commercial blend of paraffins with a transition temperature of 5 °C, was used as a temperature indicator for the development of polycaprolactone/PCM fibrous mats, which was then electrospun directly onto polystyrene (PS) films. This multilayer film was then used to coat on food packaging polymers, which showed its potential as an effective temperature buffering system with promising application as smart food packaging [51]. In a recent study, for the interest of developing intelligent food packaging materials, graphene nanoplatelets (GNPs) were embedded in EVOH fibers by electrospinning process. The presence of GNPs provided high electrical conductivity to the developed nanofibrous film that could be used as "smart" labels or tags [52].

11.6.3 Multilayer Coextrusion Technique

Multilayer coextrusion is defined as a method of forming an extrudate of more than one polymer melts. It is an advanced and emerging technique that allows the formation of thousands of layers of alternating polymers in nanometer or micrometer range thickness without the involvement of any solvents [21]. In conventional coextrusion process, two or more polymers in granule or pellet forms are melted separately. The apparatus requires separate extruders for each individual polymer, and the melt streams are collected together to develop single package with multiple layers [53]. Based on the point at which the two polymer melts are combined together, two variants of coextrusion process exist: feed block and die extrusion. In the former process, the separate melt streams are collected as a single laminar stream in a feed block positioned just above the extrusion die, whereas in the latter method, the two melt streams are brought together where the die exists [54, 55]. In multilayer coextrusion, these two processes are merged together where two to five feed blocks are followed by sequential layer multiplication dies. The number of multiplication dies added in between the feed block and the film determines the total layers of the multilayer structure. Figure 11.6 shows an illustration of multilayer coextrusion process.

Nanostructure-based multilayer packaging developed by coextrusion method is found to possess significant barrier properties (gas and vapor), mechanical, and optical properties. In a recent study, multilayer coextrusion method was utilized for preparing biodegradable polyester multilayer films with more than 2000 alternating layers of amorphous poly(lactic) (PLA) and semi-crystalline poly(butylene succinate) (PBS). This multilayer film showed improved gas and water barrier properties compared with control films [57]. Zia and colleagues developed a LPDE/curcumin

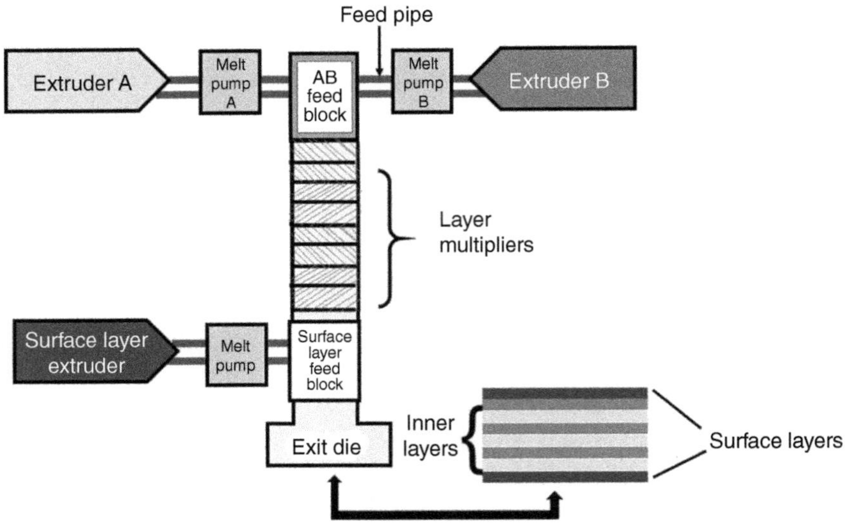

Figure 11.6 Multilayer coextrusion method: a diagrammatic representation. Source: Modified from Zhang et al. [56].

multilayer film through extrusion technique and investigated its thermal, water barrier, and antioxidant properties. The results showed promises of the developed multilayer film as an active food packaging with improved thermal and excellent water vapor barrier ability [58].

11.7 Practical Applications of Multilayer Films/Coatings for Packaging of Food

In recent years many nanostructure-based multilayer packages and coatings composed of synthetic, biopolymeric-based, and a combination of both have been intensively promoted and investigated for the practical packaging application of highly perishable food items such as milk, milk products, meat and fish, fresh fruits, and vegetables and also minimally processed foods. Nowadays, many natural extracts and essential oils are used as important additive to multilayer films, which play an essential role in imparting antimicrobial, antifungal, antioxidant, and smart packaging capability. Table 11.2 provides a list of various multilayer films, their method of preparation, and their application for food packaging purpose.

11.8 Conclusion and Future Outlook

Generally, the many essential requirements of an ideal packaging cannot be provided by a single monolayer film. In this context, a cost-effective approach of combining thin layers of different polymers to form a single ultrathin multilayer

Table 11.2 Food packaging applications of nano-based multilayered structures.

Multilayer structure	Preparation method	Food sample tested	Function performed	References
Gallic acid encapsulated lentil flour/polyethylene oxide nanofibers	Electrospinning	Walnut	Improved oxidative stability	[59]
Poly(ethylene-*co*-vinyl alcohol)/MMT multilayer film with mulberry anthocyanin coating	LbL	Shrimp	Freshness monitoring	[60]
PP/PET/LDPE films with star anise essential oil and thymol coating	Bar coating and adhesive lamination	Sliced wheat bread	Insect repellent and antimicrobial	[61]
ε-Poly(L-lysine) and carboxymethyl cellulose (CMC) was deposited on cellulose paper	LbL	Meat	Antibacterial activity against *E. coli* and *S. aureus*	[62]
ε-Polylysine and chitosan nanofibers	Electrospinning	Chicken	Antibacterial activity against *S. typhimurium* and *S. enteritidis*	[63]
Moringa oil-loaded chitosan nanoparticles embedded in gelatin nanofibers	Electrospinning	Cheese	Antibacterial activity against *L. monocytogenes* and *S. aureus*	[64]
Nanoclay-modified polyamide 6 layered on LDPE	Coextrusion process	Dry cured ham	High barrier property	[65]
Cinnamon essential oil nanophytosome embedded into cross-linked PVA nanofiber	Electrospinning	Shrimp	Extended shelf life of prawns and antibacterial activity against *E. coli* and *S. aureus*	[66]
Polyamide 6 dispersed with nanoclay and layered on LDPE	Coextrusion	Beef loins	Oxidative stability of meat sample	[67]
LDPE blended with PLA	Blown film coextrusion process	Durian	Aroma barrier property	[68]
Molecularly imprinted polymer (MIP) incorporated with graphene–gold nanoparticles		Milk	Electrochemical sensing property for detection of 4-nitrophenol	[69]

Table 11.2 (Continued)

Multilayer structure	Preparation method	Food sample tested	Function performed	References
Zein nanofiber/chitosan bilayer film	Electrospinning	Cut apple slices	Edible film, anti-browning effect due to oxygen barrier	[70]
LDPE/PET films with Algerian sage and bay leaf extract	Lamination	Fried potatoes	Active packaging increased oxidative stability	[71]
Chitosan/sodium alginate multilayer films incorporated with carboxymethyl chitosan ZnO nanoparticles	Solvent casting	Red grapes	Extended shelf life of grapes, antibacterial activity against *E. coli* and *S. aureus*	[72]
PET/PP food grade film incorporated with methyl gallate from mango kernels	Supercritical solvent impregnation	Lettuce and tangerine	Antioxidant capacity, which increased food shelf life	[73]
LDPE films with cinnamaldehyde	Extrusion	Pasteurized milk	Active film	[74]

packaging structure is highly remarkable and attractive. Recently, researchers have exploited nanostructure-based multilayer films for providing more functionalities in food packaging. Multilayer systems comprising of nanostructures layers can improve not only the material properties and barrier function but also can promote active and intelligent packaging performances. Nano-multilayer film structures can be prepared in a scale through simple and feasible methods like LbL and EHDP technology.

Numerous researches are underway to develop multilayer film packaging for food protection and preservation purposes. Even though multilayer films have great scope in packaging industry, the challenges that arise from various issues such as its biodegradability and environmental concerns, health and safety matters, and migration-related problems have to be well addressed in future studies for its wider acceptance.

References

1 Lagarón, J.-M. (2011). Multifunctional and nanoreinforced polymers for food packaging. In: *Multifunctional and Nanoreinforced Polymers for Food Packaging* (ed. J.-M. Lagarón), 1–28. Elsevier. https://linkinghub.elsevier.com/retrieve/pii/B978184569738950001X.

2 Müller, K., Bugnicourt, E., Latorre, M. et al. (2017). Review on the processing and properties of polymer nanocomposites and nanocoatings and their

applications in the packaging, automotive and solar energy fields. *Nanomaterials* 7 (4): 74.

3 Kotzekidou, P. (2016). Factors influencing microbial safety of ready-to-eat foods. In: *Food Hygiene and Toxicology in Ready-to-Eat Foods* (ed. P. Kotzekidou), 33–50. Elsevier. https://linkinghub.elsevier.com/retrieve/pii/B9780128019160000030.

4 Mihindukulasuriya, S.D.F. and Lim, L.-T. (2014). Nanotechnology development in food packaging: A review. *Trends Food Sci. Technol.* 40 (2): 149–167.

5 Bouwmeester, H., Brandhoff, P., Marvin, H.J.P. et al. (2014). State of the safety assessment and current use of nanomaterials in food and food production. *Trends Food Sci. Technol.* 40 (2): 200–210.

6 Llorens, A., Lloret, E., Picouet, P.A. et al. (2012). Metallic-based micro and nanocomposites in food contact materials and active food packaging. *Trends Food Sci. Technol.* 24 (1): 19–29.

7 Silvestre, C., Duraccio, D., and Cimmino, S. (2011). Food packaging based on polymer nanomaterials. *Prog. Polym. Sci.* 36 (12): 1766–1782.

8 Kaiser, K., Schmid, M., and Schlummer, M. (2017). Recycling of polymer-based multilayer packaging: a review. *Recycling* 3 (1): 1.

9 Butler, T.I. and Morris, B.A. (2016). PE-based multilayer film structures. In: *Multilayer Flexible Packaging* (ed. J.R. Wagner Jr.), 281–310. Elsevier. https://linkinghub.elsevier.com/retrieve/pii/B978032337100100017X.

10 Ramos, M., Valdés, A., Mellinas, A., and Garrigós, M. (2015). New trends in beverage packaging systems: a review. *Beverages* 1 (4): 248–272.

11 Anukiruthika, T., Sethupathy, P., Wilson, A. et al. (2020). Multilayer packaging: advances in preparation techniques and emerging food applications. *Compr. Rev. Food Sci. Food Saf.* 19 (3): 1156–1186.

12 Maes, C., Luyten, W., Herremans, G. et al. (2018). Recent updates on the barrier properties of ethylene vinyl alcohol copolymer (EVOH): a review. *Polym. Rev.* 58 (2): 209–246.

13 Morris, B.A. (2017). *The Science and Technology of Flexible Packaging*. Elsevier. https://linkinghub.elsevier.com/retrieve/pii/C20130005063.

14 Youssef, A.M. (2013). Polymer nanocomposites as a new trend for packaging applications. *Polym. Plast. Technol. Eng.* 52 (7): 635–660.

15 Decher, G. and Hong, J.-D. (1991). Buildup of ultrathin multilayer films by a self-assembly process, 1 consecutive adsorption of anionic and cationic bipolar amphiphiles on charged surfaces. *Makromol. Chem. Macromol. Symp.* 46 (1): 321–327.

16 Ariga, K., Hill, J.P., and Ji, Q. (2007). Layer-by-layer assembly as a versatile bottom-up nanofabrication technique for exploratory research and realistic application. *Phys. Chem. Chem. Phys.* 9 (19): 2319.

17 Decher, G. (1997). Fuzzy nanoassemblies: toward layered polymeric multicomposites. *Science* 277 (5330): 1232–1237.

18 Bruening, M.L. and Adusumilli, M. (2011). *Polyelectrolyte Multilayer Films and Membrane Functionalization*. Sigma-Aldrich. https://www.sigmaaldrich.com/

technical-documents/articles/material-matters/polyelectrolyte-multilayer-films .html.
19 Porcel, C., Lavalle, P., Ball, V. et al. (2006). From exponential to linear growth in polyelectrolyte multilayers. *Langmuir* 22 (9): 4376–4383.
20 Humood, M., Chowdhury, S., Song, Y. et al. (2016). Nanomechanical behavior of high gas barrier multilayer thin films. *ACS Appl. Mater. Interfaces* 8 (17): 11128–11138.
21 Cerqueira, M.A., Torres-Giner, S., and Lagaron, J.M. (2018). Nanostructured multilayer films. In: *Nanomaterials for Food Packaging* (eds. M.A. Cerqueira, J.M. Lagaron, L.M. Castro and A.A., de Oliverira Soares Vincente), 147–171. Elsevier. https://linkinghub.elsevier.com/retrieve/pii/B9780323512718000061.
22 Sánchez-Valdes, S., Ortega-Ortiz, H., Ramos-de Valle, L.F. et al. (2008). Mechanical and antimicrobial properties of multilayer films with a polyethylene/silver nanocomposite layer. *J. Appl. Polym. Sci.* 111: 953–962.
23 Izquierdo, A., Ono, S.S., Voegel, J.-C. et al. (2005). Dipping versus spraying: exploring the deposition conditions for speeding up layer-by-layer assembly. *Langmuir* 21 (16): 7558–7567.
24 Lu, C., Dönch, I., Nolte, M., and Fery, A. (2006). Au nanoparticle-based multilayer ultrathin films with covalently linked nanostructures: spraying layer-by-layer assembly and mechanical property characterization. *Chem. Mater.* 18 (26): 6204–6210.
25 Heo, J., Choi, M., and Hong, J. (2019). Facile surface modification of polyethylene film via spray-assisted layer-by-layer self-assembly of graphene oxide for oxygen barrier properties. *Sci. Rep.* 9 (1). http://www.nature.com/articles/s41598-019-39285-0.
26 Aulin, C., Karabulut, E., Tran, A. et al. (2013). Transparent nanocellulosic multilayer thin films on polylactic acid with tunable gas barrier properties. *ACS Appl. Mater. Interfaces* 5 (15): 7352–7359.
27 Jang, W.-S., Rawson, I., and Grunlan, J.C. (2008). Layer-by-layer assembly of thin film oxygen barrier. *Thin Solid Films* 516 (15): 4819–4825.
28 Laufer, G., Kirkland, C., Cain, A.A., and Grunlan, J.C. (2012). Clay–chitosan nanobrick walls: completely renewable gas barrier and flame-retardant nanocoatings. *ACS Appl. Mater. Interfaces* 4 (3): 1643–1649.
29 Laufer, G., Kirkland, C., Cain, A.A., and Grunlan, J.C. (2013 Jun). Oxygen barrier of multilayer thin films comprised of polysaccharides and clay. *Carbohydr. Polym.* 95 (1): 299–302.
30 Dai, L., Long, Z., Chen, J. et al. (2017). Robust guar gum/cellulose nanofibrils multilayer films with good barrier properties. *ACS Appl. Mater. Interfaces* 9 (6): 5477–5485.
31 Priolo, M.A., Gamboa, D., and Grunlan, J.C. (2010). Transparent clay–polymer nano brick wall assemblies with tailorable oxygen barrier. *ACS Appl. Mater. Interfaces* 2 (1): 312–320.
32 Plácido, A., de Oliveira Farias, E.A., Marani, M.M. et al. (2016). Layer-by-layer films containing peptides of the Cry1Ab16 toxin from *Bacillus thuringiensis* for potential biotechnological applications. *Mater. Sci. Eng., C* 61: 832–841.

33 Azlin-Hasim, S., Cruz-Romero, M.C., Cummins, E. et al. (2016). The potential use of a layer-by-layer strategy to develop LDPE antimicrobial films coated with silver nanoparticles for packaging applications. *J. Colloid Interface Sci.* 461: 239–248.

34 Rudra, J.S., Dave, K., and Haynie, D.T. (2006 Jan). Antimicrobial polypeptide multilayer nanocoatings. *J. Biomater. Sci., Polym. Ed.* 17 (11): 1301–1315.

35 Fabra, M.J., Flores-López, M.L., Cerqueira, M.A. et al. (2016). Layer-by-layer technique to developing functional nanolaminate films with antifungal activity. *Food Bioprocess Technol.* 9 (3): 471–480.

36 Acevedo-Fani, A., Salvia-Trujillo, L., Soliva-Fortuny, R., and Martín-Belloso, O. (2015). Modulating biopolymer electrical charge to optimize the assembly of edible multilayer nanofilms by the layer-by-layer technique. *Biomacromolecules* 16 (9): 2895–2903.

37 Wang, H., Chen, M., Jin, C. et al. (2018). Antibacterial [2-(methacryloyloxy) ethyl] trimethylammonium chloride functionalized reduced graphene oxide/poly(ethylene-*co*-vinyl alcohol) multilayer barrier film for food packaging. *J. Agric. Food. Chem.* 66 (3): 732–739.

38 Li, D. and Xia, Y. (2004). Electrospinning of nanofibers: reinventing the wheel? *Adv. Mater.* 16 (14): 1151–1170.

39 Echegoyen, Y., Fabra, M.J., Castro-Mayorga, J.L. et al. (2017 Feb). High throughput electro-hydrodynamic processing in food encapsulation and food packaging applications: Viewpoint. *Trends Food Sci. Technol.* 60: 71–79.

40 Huang, Z.-M., Zhang, Y.-Z., Kotaki, M., and Ramakrishna, S. (2003 Nov). A review on polymer nanofibers by electrospinning and their applications in nanocomposites. *Compos. Sci. Technol.* 63 (15): 2223–2253.

41 Liu, Y., Wang, S., and Lan, W. (2018). Fabrication of antibacterial chitosan-PVA blended film using electrospray technique for food packaging applications. *Int. J. Biol. Macromol.* 107: 848–854.

42 Kim, J. and Reneker, D.H. (1999 Feb). Mechanical properties of composites using ultrafine electrospun fibers. *Polym. Compos.* 20 (1): 124–131.

43 Frenot, A. and Chronakis, I.S. (2003). Polymer nanofibers assembled by electrospinning. *Curr. Opin. Colloid Interface Sci.* 8 (1): 64–75.

44 Fabra, M.J., Pardo, P., Martínez-Sanz, M. et al. (2016). Combining polyhydroxyalkanoates with nanokeratin to develop novel biopackaging structures. *J. Appl. Polym. Sci.* 133 (2): 42695.

45 Agarwal, A., Raheja, A., Natarajan, T.S., and Chandra, T.S. (2014). Effect of electrospun montmorillonite-nylon 6 nanofibrous membrane coated packaging on potato chips and bread. *Innovative Food Sci. Emerg. Technol.* 26: 424–430.

46 Fabra, M.J., López-Rubio, A., Ambrosio-Martín, J., and Lagaron, J.M. (2016). Improving the barrier properties of thermoplastic corn starch-based films containing bacterial cellulose nanowhiskers by means of PHA electrospun coatings of interest in food packaging. *Food Hydrocolloids* 61: 261–268.

47 Pardo-Figuerez, M., López-Córdoba, A., Torres-Giner, S., and Lagaron, J. (2018). Superhydrophobic bio-coating made by co-continuous electrospinning and electrospraying on polyethylene terephthalate films proposed as easy emptying transparent food packaging. *Coatings* 8 (10): 364.

48 Torres-Giner, S. (2011). Novel antimicrobials obtained by electrospinning methods. In: *Antimicrobial Polymers* (eds. J.M. Lagarón, M.J. Ocio and A. López-Rubio), 261–285. Hoboken, NJ: Wiley https://doi.org/10.1002/9781118150887.ch10.

49 Liu, Y., Liang, X., Wang, S. et al. (2018). Electrospun antimicrobial polylactic acid/tea polyphenol nanofibers for food-packaging applications. *Polymers* 10 (5): 561.

50 Hasanpour Ardekani-Zadeh, A. and Hosseini, S.F. (2019). Electrospun essential oil-doped chitosan/poly(ε-caprolactone) hybrid nanofibrous mats for antimicrobial food biopackaging exploits. *Carbohydr. Polym.* 223: 115108.

51 Chalco-Sandoval, W., JoséFabra, M., AmparoLópez-Rubio, M., and Lagaron, J. (2015). Development of polystyrene-based films with temperature buffering capacity for smart food packaging. *J. Food Eng.* 164: 55–62.

52 Torres-Giner, S., Echegoyen, Y., Teruel-Juanes, R. et al. (2018). Electrospun poly(ethylene-co-vinyl alcohol)/graphene nanoplatelets composites of interest in intelligent food packaging applications. *Nanomaterials* 8 (10): 745.

53 Ajitha, A.R., Aswathi, M.K., Maria, H.J. et al. (2016). Multilayer Polymer Films. In: *Multicomponent Polymeric Materials*, Springer Series in Materials Science, vol. 223 (eds. J.K. Kim, S. Thomas and P. Saha), 229–258. Dordrecht: Springer Netherlands. http://link.springer.com/10.1007/978-94-017-7324-9_8.

54 Ebnesajjad, S. and Khaladkar, P.R. (2018). *Fluoropolymer Applications in the Chemical Processing Industries*. Elsevier. https://linkinghub.elsevier.com/retrieve/pii/C20150044197.

55 Drobny, J.G. (2014). Processing methods applicable to thermoplastic elastomers. In: *Handbook of Thermoplastic Elastomers*, 33–173. Elsevier. https://linkinghub.elsevier.com/retrieve/pii/B9780323221368000041.

56 Zhang, X., Xu, Y., Zhang, X. et al. (2019). Progress on the layer-by-layer assembly of multilayered polymer composites: strategy, structural control and applications. *Prog. Polym. Sci.* 89: 76–107.

57 Messin, T., Marais, S., Follain, N. et al. (2020). Biodegradable PLA/PBS multinanolayer membrane with enhanced barrier performances. *J. Membr. Sci.* 598: 117777.

58 Zia, J., Paul, U.C., Heredia-Guerrero, J.A. et al. (2019). Low-density polyethylene/curcumin melt extruded composites with enhanced water vapor barrier and antioxidant properties for active food packaging. *Polymer* 175: 137–145.

59 Aydogdu, A., Yildiz, E., Aydogdu, Y. et al. (2019). Enhancing oxidative stability of walnuts by using gallic acid loaded lentil flour based electrospun nanofibers as active packaging material. *Food Hydrocolloids* 95: 245–255.

60 Kang, S., Wang, H., Guo, M. et al. (2018). Ethylene-vinyl alcohol copolymer–montmorillonite multilayer barrier film coated with mulberry anthocyanin for freshness monitoring. *J. Agric. Food. Chem.* 66 (50): 13268–13276.

61 Lee, J., Park, M.A., Yoon, C.S. et al. (2019). Characterization and preservation performance of multilayer film with insect repellent and antimicrobial activities for sliced wheat bread packaging. *J. Food Sci.* 84 (11): 3194–3203.

62 Li, H., Cui, R., Peng, L. et al. (2017). Preparation of antibacterial cellulose paper using layer-by-layer assembly for cooked beef preservation at ambient temperature. *Polymers* 10 (1): 15.

63 Lin, L., Xue, L., Duraiarasan, S., and Haiying, C. (2018). Preparation of ε-polylysine/chitosan nanofibers for food packaging against Salmonella on chicken. *Food Packag. Shelf Life* 17: 134–141.

64 Lin, L., Gu, Y., and Cui, H. (2019). Moringa oil/chitosan nanoparticles embedded gelatin nanofibers for food packaging against *Listeria monocytogenes* and *Staphylococcus aureus* on cheese. *Food Packag. Shelf Life* 19: 86–93.

65 Lloret, E., Fernandez, A., Trbojevich, R. et al. (2016). Relevance of nanocomposite packaging on the stability of vacuum-packed dry cured ham. *Meat Sci.* 118: 8–14.

66 Nazari, M., Majdi, H., Milani, M. et al. (2019). Cinnamon nanophytosomes embedded electrospun nanofiber: its effects on microbial quality and shelf-life of shrimp as a novel packaging. *Food Packag. Shelf Life* 21: 100349.

67 Picouet, P.A., Fernandez, A., Realini, C.E., and Lloret, E. (2014). Influence of PA6 nanocomposite films on the stability of vacuum-aged beef loins during storage in modified atmospheres. *Meat Sci.* 96 (1): 574–580.

68 Winotapun, C., Phattarateera, S., Aontee, A. et al. (2019). Development of multilayer films with improved aroma barrier properties for durian packaging application. *Packag. Technol. Sci.* 32 (8): 405–418.

69 Zheng, L., Zhang, C., Ma, J. et al. (2018). Fabrication of a highly sensitive electrochemical sensor based on electropolymerized molecularly imprinted polymer hybrid nanocomposites for the determination of 4-nonylphenol in packaged milk samples. *Anal. Biochem.* 559: 44–50.

70 Bharathi, S.K.V., Leena, M.M., Moses, J.A., and Anandharamakrishnan, C. (2020). Nanofibre-based bilayer biopolymer films: enhancement of antioxidant activity and potential for food packaging application. *Int. J. Food Sci. Technol.* 55 (4): 1477–1484.

71 Oudjedi, K., Manso, S., Nerin, C. et al. (2019). New active antioxidant multilayer food packaging films containing Algerian Sage and Bay leaves extracts and their application for oxidative stability of fried potatoes. *Food Control* 98: 216–226.

72 Wang, H., Gong, X., Miao, Y. et al. (2019 Jun). Preparation and characterization of multilayer films composed of chitosan, sodium alginate and carboxymethyl chitosan-ZnO nanoparticles. *Food Chem.* 283: 397–403.

73 Belizón, M., Fernández-Ponce, M.T., Casas, L. et al. (2018). Supercritical impregnation of antioxidant mango polyphenols into a multilayer PET/PP food-grade film. *J. CO2 Util.* 25: 56–67.

74 Haghighi-Manesh, S. and Azizi, M.H. (2017). Production and evaluation of type of multi-layer active film for packaging of pasteurized milk: multi-layer active film. *J. Food Process Eng* 40 (3): e12442.

12

Characterization Techniques for Nanostructures in Food Packaging

Ashitha Jose[1], *R. Aswani*[1], *and Radhakrishnan E. Krishnankutty*[2]

[1] *School of Biosciences, Mahatma Gandhi University, PD Hills (PO), Kottayam 686560, India*
[2] *Research Scholar, School of Biosciences, Mahatma Gandhi University, PD Hills (PO), Kottayam 686560, India*

12.1 Introduction

The major objective of a packaging material is to protect the food from adverse environmental conditions by acting as a barrier to gases and moisture. This can also help to preserve the quality and improve the shelf life of the food. However, an ideal packaging material should remain chemically inert to the food packed within. The diverse consumer demands and the negative environmental impact of the synthetic polymers have led the research toward the development of novel packaging materials with biodegradable and environment-friendly nature. However, the natural polymers are inadequate in acting as a stand-alone material for the same, owing to its insufficient material properties. This has put forward the concept of composite materials, wherein the material properties can be tailor-made to meet the requirements through the simultaneous utilization of multiple polymers. The addition of various filler molecules to the composite matrix is aimed at imparting both mechanochemical properties and specified functions dependent on the filler–matrix interactions.

The concept of ideal food packaging material has progressed over the time. The current research, however, focuses on the development of materials that are capable of ensuring shelf life extension as well as monitoring the safety and quality of the packed food [1]. The paradigm shift in the packaging industry toward the development of active packaging is aimed at controlling the growth of undesirable microorganisms on the food through the incorporation of a range of antimicrobial substances. Recent advancements have led to the development of active packaging that can act as a sensor for detecting the freshness of the food, microbial contamination, and microbial load along with the improvement of the shelf life through maintaining the atmosphere of the material packed within. The primary source of microbial contamination in packed food is the inappropriate post-processing and handling, which can result in the migration of the microorganisms into the food along with its rapid multiplication. Here comes the importance of novel packaging materials with tailor-made properties. The active functionalization of these materials can improve the consumers' health and safety through the

Nanotechnology-Enhanced Food Packaging, First Edition. Edited by Jyotishkumar Parameswaranpillai, Radhakrishnan Edayileveettil Krishnankutty, Aswathy Jayakumar, Sanjay Mavinkere Rangappa, and Suchart Siengchin.

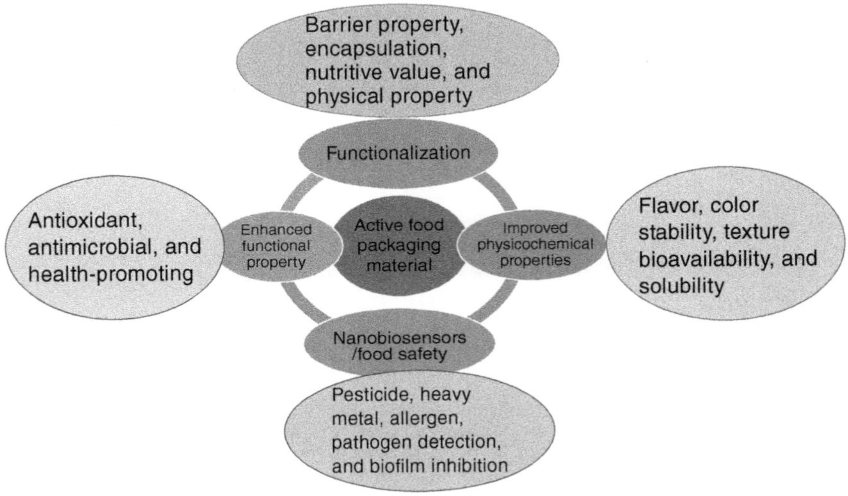

Figure 12.1 Functionalities of active packaging.

modulation of properties including the extended shelf life and reduced release of the matrix components into the food along with much advanced requirements such as recording, tracking, and providing substantial information on the quality of the packed food [2]. Figure 12.1 depicts the various functionalities of the active packaging. However, for the implementation of such active packaging materials, the incorporation of bioactive molecules and various active nanoparticles (NPs) is of high significance.

A good packaging material must be able to protect the food from dirt/dust, light, atmospheric gases, moisture, microorganisms, and other deleterious contaminants. The components used must be safe for use, inert, cheap, easy to reuse, and biodegradable and should be capable of withstanding extreme conditions of processing. The recent adaptations in the packaging industry aimed at the development of active packaging, having inbuilt antimicrobial, sensory, or shelf life extension applications through the incorporation of various nanofillers into the matrix (Figure 12.2). These nanofillers can in turn modulate the material properties of the developed material ensuring their superior nature. However, the individual components being exploited as a food contact agent – either in the form of packaging or coating – must be recognized under the Generally Recognized as Safe (GRAS) category of chemicals.

12.2 Nanoparticles

The word “nano” is derived from the Greek word nanos, which means dwarf, and the nanoparticles (NPs) can be defined as the particles having a dimension of 1–100 nm in size [3]. There are different groups of nanoparticles, and these include fullerenes, metal NPs, ceramic NPs, and polymeric NPs, and all of them are known to have unique physical and chemical properties due to their high

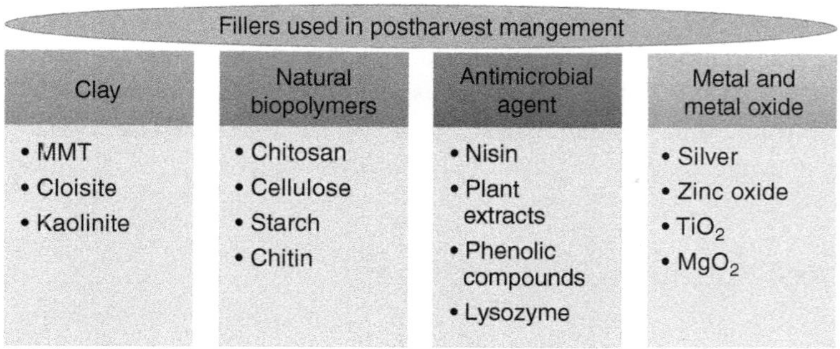

Figure 12.2 Nanofillers used in active packaging.

surface area and nanoscale size. The reactivity, toughness, and other properties of nanoparticles are also dependent on their unique size, shape, and structure. Due to these characteristics, the nanoparticles are expected to be used as a suitable candidate for various commercial and domestic applications [4].

12.3 Role of Nanoparticles in Packaging Applications

The nanoparticles have been well thought out as the fifth state of matter having lesser atoms with multi-beneficial properties than the bulk crystals. Nanoparticles adopt shape, structure, and properties differently from their bulk counterparts. Various methods can be employed for the synthesis of nanoparticles and are broadly divided into two main classes: (i) bottom-up approach and (ii) top-down approach. The bottom-up approach is considered as the constructive method where the nanoparticles are formed from relatively simpler substances. While the top-down approach is destructive in nature where the nanoparticles are formed from larger molecule by decomposing them into smaller units, which is then converted into suitable nanoparticles [5]. The synthesized nanoparticles can exhibit a range of ideal properties such as near-ideal strength, enhanced catalytic properties, and detached energy levels, which help in tailoring the electrical properties to explore the application of these nanoparticles in various fields. They find applications in optical data storage, sensor development, and conversion of solar energy. This in turn makes the analysis of its shape and size an important criterion in determining various nanotechnological advancements for its characterization. The surface properties like surface tension play a major role in defining its distinguishing properties [6].

Even though various synthesis routes are employed, the minor variations in the handling, synthesis, and processing of the same can alter the properties of nanoparticles. The nanoparticles are considered to be dynamic in nature as they are capable of shape shifting, subjected to time, proximity to other particles or surfaces, and the environment. The dynamic nature of both natural and engineered nanoparticles, especially in smaller size range, can be confirmed through various analyses including the electron microscopy. The energy bands of the nanoparticles have many local minima configurations conforming to various structures. Hence a small variation is adequate enough to induce greater transformations in the material. The growth or solubility of the particles is capable of lowering its surface energy. The growth in the particle size can result in the formation of soft agglomerates that are relatively easy to break apart or aggregates that are difficult to be broken apart [6, 7].

12.4 Nanocomposite in Food Packaging

Nanocomposite is considered as a good alternative for the regular conventional food packaging [8]. Generally, composites are composed of a polymer matrix and a filler [9]. Nanocomposites are made by the incorporation of nanofillers that have nanoscale measurements in at least one dimension. The incorporation of nanoparticles is aimed at enhancing the mechanical and thermal properties of the developed nanocomposite, improving the barrier property against oxygen, carbon dioxide, and flavor compounds by increasing the path length for the gas diffusion, modulating the surface wettability or the hydrophobicity of the packed material, and imparting antimicrobial potential to the developed material, oxygen scavenging activity, enzyme immobilization, biosensing, etc. The development of nanocomposites can improve the sensor properties such as flavor and better maintenance of color and texture and also improve the shelf life and thus ensure lesser food loss by spoilage. The nanocomposite development must assure good dispersion of the nanofillers in the polymer matrix to attain intercalated or exfoliated structure. The exfoliated structures can be either ordered or disordered in nature. The partial exfoliation is considered to have an intermediate morphology between the exfoliation and intercalation. This variation in the morphology can result in considerable variation in the physical and mechanical properties of the developed polymer. Significant changes in the properties can be seen even at lower concentrations such as <5% of the nanoparticle loads [10].

The thickness and mechanical properties of the packaging films prepared can be modified based on the incorporated fillers. Studies have shown that the incorporation of oregano essential oil was able to enhance the thickness of the films. However, the tensile strength of the same was found to be reduced at higher concentrations of the same filler. Here the essential oils perform the function of a plasticizer, increasing the flexibility of the polymer chains and reducing the tensile strength resulting in an improved elongation at break. On the other hand, the addition of graphene oxide nanoparticles, which possess high tensile properties as such was found to enhance the mechanical properties of the developed films [11].

12.5 Methods for the Development of Nanocomposites

The nanocomposite development can follow many routes; among which the commonly exploited ones are solvent casting, extrusion or melt compounding, sol–gel processing, electrospinning/electrospraying, etc. In solvent casting technique, the polymer and the active components are dissolved in a suitable solvent through the vigorous stirring or ultrasonication of the components. The film-forming solution thus developed is then casted onto an inert surface, and then the complete evaporation of the solvent phase is ensured at the optimum temperature, which can result in the development of thin films having desired functionality. The intercalation of the polymer and the filler components is subjected to the polarity of the solvent used. The emulsion polymerization can result in the homogeneous distribution of the nanoparticles into the polymer matrix. This laboratory scale technique can be translated into the industrial scale through its efficient coupling with a wet-coating system and a roll-to-roll system. However, the technique is limited by the fact that the solubility of certain components can occur only at higher temperature and requires certain organic solvents. This in turn can affect the stability and effectiveness of many active/bioactive compounds of interest. Hence the technique can be applied only to biodegradable polymers, which can be dissolved at milder temperature. Thus, the common polymers that can be developed using the solvent casting technique include poly(lactic acid) (PLA), poly(butylene adipate-*co*-terephthalate), cellulose and its derivatives, polyvinyl alcohol (PVA), etc.

The extrusion or melt compounding is an industrial approach where the compounding of the polymer and the nanofiller is performed using a single or twin-screw extruder where the components are in a melt stage. The inherent lack of thermal stability is a disadvantage in this case. This can in turn result in the loss of active ingredients either through the degradation or evaporation. However, the release of the active compounds is controlled by the characteristic properties of the packaging material including the affinity of the filler with the matrix.

The sol–gel processing has been widely employed to obtain the nano-based coatings, where the viscous colloidal sol is applied to the surface through the dip/spray/spin coating techniques, and the resulting layers have a thickness in between 0.5 and 3 μm. The major advantages of the process include the lower temperature requirement and lower energy consumption along with short time span required for its development. Moreover, the structure and functionality of the formed complex is not affected.

Electrospinning/electrospraying is a versatile technique employed in the development of a biopolymer. Submicron to nano-range particle diameter can be obtained in the technique by altering the applied voltage. Here one or more bioactive compounds can be deposited onto the active surface [10]. In the electrospinning technique, the polymers are dissolved in the desired solvent and then subjected to appropriate electrostatic force, which is essential in producing fibers having submicrometric diameter. The solvents used can cause instability under the electric field known as whipping, which leads to the unpredictable deposition of the polymer, leading to development of randomized structures [12].

12.6 Various Nanoparticles Employed in Packaging

The most commonly used nanoparticles as nanofillers or reinforcement agents include clay (montmorillonite [MMT] and kaolinite), silica nanoplates and nanoparticles, carbon nanotubes, graphene, TiO_2, silver, zinc oxide, copper and copper oxides, starch, cellulose nanofibers, cellulose nanocrystals, chitosan, chitin, etc.

12.6.1 Nanoclay

The nanoclays have found its application in both food and drug packaging. They are capable of preventing oxidative damage in the food and hence reduce the food loss by extending the shelf life. Nanoclays have shown performance equivalent or better to that of other nanofillers such as nanosilica, calcium carbonate, and nanocellulose. Nanoclay can also impart better mechanical properties to the film matrix. They possess a platelet form with soft structure, lower specific gravity, and higher aspect ratio. Among the various nanoclays, MMT clay and organic modified montmorillonite (OMMT), which are miscible with hydrophilic polymers, have gained immense advantages in the packaging industry. Since the packaging industry exploits hydrophobic materials for the film formation, the nanoclay is chemically modified using ion exchange approaches with organic cations such as ammonium salt. The quaternary ammonium groups on the surface of the modified clay can render them to have antibacterial potential. The impregnation of various phytochemical components onto the nanoclay develops them as efficient indicator systems also. The addition of OMMT has been confirmed to accelerate the biodegradability of the packaging material by improving the oxidation of the polymer matrix. Nanoclay-incorporated bottled beverages and plastic bags that come in direct contact with the food are readily available in the current market. The nanoclays account for lower migration rate from the packaging material under various conditions on comparison with silver nanoparticles at same conditions. MMT, accounting for the largest proportion of the clay market, belongs to the smectite group of clay minerals and is widely exploited in the packaging industry. Its wide acceptance can be attributed to the natural abundance, environmental friendliness, and safety. Clays are generally composed of aluminosilicates with neutral or negatively charged layers. The layers form a stacked arrangement and are held together by the electrostatic force [4, 6]. The probing of nanoclays as a reinforcement material in the composite matrix has shown to reduce the moisture absorbance and brittleness of the prepared films. The nanoencapsulation of anthocyanin was carried out with the help of nanoclays. It was aimed at minimizing the damage incorporated in the phytochemical due to the higher extrusion temperature [1].

12.6.2 Titanium Dioxide

Titanium dioxide (TiO_2), in the micro and nanoform, is categorized under the food grade and has been widely accepted in the consumer products. It is commonly known as E171 in the European Union and is considered as a common food additive. Inductively coupled-mass spectrometry (ICPMS) and single particle inductively coupled-mass spectrometry (spICPMS) are used for the detection and quantification of TiO_2 particles. E171 is being incorporated in the manufacturing of candies, chocolate, sauces, and chewing gums [13]. The TiO_2 nanoparticles possess low cytotoxicity and high antimicrobial activity and block the UV light efficiently. The TiO_2 nanoparticles are thermostable, inert materials capable of modifying the material properties of biodegradable films. They form a class of superior photocatalyst materials having antimicrobial applications. However, their tendency to aggregate is a limiting factor in determining its functionality. Tailoring the functional properties of these nanomaterials as to obtain synergistic effect has enhanced the application of TiO_2 nanoparticles. Their incorporation was found to enhance the optical and tensile properties of the prepared films [14].

12.6.3 Zinc Oxide Nanoparticles

Among the wide range of nanomaterials used in industry, ZnO NPs are considered as a multifunctional one because of their high near-UV emission, antimicrobial efficacy, optical transparency, electrical conductivity, and piezoelectricity [15]. ZnO NPs are also used in various fields including cosmetics, drug delivery systems, biosensors, and food packaging materials due to their excellent biocompatibility. The zinc oxide nanoparticles also exhibit superior UV blocking ability in addition to the antioxidation and antimicrobial property of the developed films. As the ZnO NP has been considered under the GRAS category by the US Food and Drug Administration (21 CFR 182.8991), these materials are considered to be for the use in food industry. Therefore, it could be frequently used in food packaging materials as an active antimicrobial agent. Physical vapor synthesis or mechanochemical processing is exploited in the synthesis of zinc oxide nanoparticles (ZnO NP). By contacting or penetrating into the bacterial cells, the ZnO NP can exert their antimicrobial mechanism [8]. The antibacterial activity of ZnO NP involves the destruction of bacterial cell integrity, release of antimicrobial Zn^{2+} ions, and generation of ROS when it comes in direct contact with the microbial cell wall [16, 17]. The ZnO NP has many hanging bonds on the surface, making it highly reactive and thus leads to the formation of aggregates. The superimposition of the regular crystalline lattice on an amorphous background was confirmed in the TEM analysis. Also the particles were well distributed on the surface and had a hexagonal structure [18].

12.6.4 Graphene-Based Nanomaterials

Graphene is a two-dimensional carbon crystal existing in the sp^2 hybridization possessing excellent material properties. The graphene family of nanomaterials includes the graphite, graphite oxide, graphene oxide, expanded graphite, etc. The addition of high modulus graphene in to the low modulus polymer matrix can also exhibit significant reinforcing effects. Delayed volatilization of gases and improvement of thermal stability can be ensured through the homogeneous dispersion of graphene oxide sheets. Increment in the nanoparticle content can delay the onset of degradation temperature of the matrix along with improving the thermal stability. The addition of graphene oxide nanoparticle in to the film preparation has led to the formation of coarse structure of the developed film as confirmed through the surface micrographs. The GO-based composites can also act as an efficient antibacterial agents with better barrier properties and lower toxicity [19].

12.6.5 Silver Nanoparticles

Silver nanoparticles have emerged as a potent candidate material in the food packaging, owing to its antimicrobial mechanism. They are synthesized by the reduction of silver salts such as $AgNO_3$ to produce colloidal silver with particle diameter of several nanometers. While the physical reduction can present with regular-shaped particles capable of better reduction, the chemical reduction leads to the formation of agglomerated particles. The silver nanoparticles exhibit a broad-spectrum antimicrobial activity against Gram-positive, Gram-negative, and multidrug-resistant organisms. The immobilization of AgNPs to the polymer matrix can ensure the even dispersion of the nanoparticles [8]. AgNP is applied in the food packaging industry to absorb moisture and other exudates of the packed food, leaving the food fresh over a longer time period. The uniform dispersion of the silver nanoparticles in the film matrix can be confirmed through the SEM micrographs where the nanoparticles are seen as minute white dots. The incorporation of clay nanoparticles along with silver nanoparticles has shown to induce a synergistic effect through their effective homogeneous dispersion. Furthermore, the incorporation of clay nanoparticles has been found to enhance the dispersion of silver nanoparticles along with the modification of the surface polarity of the prepared films [20]. The silver nanomaterials exhibit strong surface plasmon resonance, larger surface-area-to-volume ratio, and enhanced antimicrobial and catalytic activity. They are found to have low volatility and are highly stable in nature. These enhanced properties of the nanoparticles have led to its exploitation as an efficient coating material for the food containers, medical devices, wound dressing material, etc. in addition to its packaging applications [21]. The polymers such as polyethylene (PE), polyvinyl chloride, and polyvinylpyrrolidone are the commonly used matrices to host the silver nanoparticles for its packaging applications. A larger proportion of silver nanoparticles can be easily coated onto the hydrophilic matrixes. However, the addition of AgNPs to the biodegradable films has been found to result in the reduction of mechanical resistance and thermal stability of the novel developed film.

12.7 Issues Associated with the Nanoparticle Incorporation

The inadequate characterization data of the nanoparticle under consideration remains the major limiting factor in determining the functionality of the concerned nanoparticles. This in turn limits the application of nanotechnology in food packaging. The dynamic nature of the nanoparticle can thus determine the substantial properties of the developed materials subjected to its handling process. These challenges can limit both the technological advancements and the scientific understanding through the exploitation of the nanoparticle and at the same time pose questions in regard to the product shelf life and health concern. This has brought forward the importance of the nanoparticle characterization approaches, which in turn helps to determine the variations in the material properties subjected to environmental impact. The traditional surface sensitivity analysis includes the X-ray photoelectron spectroscopy (XPS), electron microscopic technique, mass spectroscopy, etc. in addition to other morphological characterizations. The surface analysis can be used either to recognize the presence or absence of impurities on the surface of the nanomaterials or to confirm the chemical state of the elements that make up the nanoparticles. Majority of atoms present in the nanoparticles can be considered as surface or next to the surface atoms. Hence the nanoparticle cannot be defined without proper surface data of the same. However, the surface data of the nanomaterials cannot be considered as the sole characterization data as the results obtained can be affected by various factors. The larger fraction of atoms or molecules in association with the surface interfaces, impurities, enrichment or depletion, and contamination act as the prime factors modulating the obtained results. These in turn can dominate the material properties of the obtained nanoparticles falsifying the data procured. Hence, the electrical properties of an nanoparticle can be determined by the surface area, and the environment of the developed material is highly significant in defining the properties of the nanomaterials. Moreover, the spontaneous adsorption of the nanoparticles can reduce its free energy [7].

12.8 Characterization of Nanoparticles in the Packaging Materials

Nanomaterials have attracted huge interest over the years due to its wide range of industrial and pharmaceutical applications. There are many techniques used to characterize the nanoparticles, and it is based on the size, structure, elemental composition, and other physical properties (Figure 12.3). However, the limitations of a single technique in determining the nature and properties of the nanoparticle lead to the exploitation of combinatorial characterization approach utilizing multiple platforms. Hence the current chapter discusses the main characterization techniques for various nanoparticle samples both in the dry state and in solution. The techniques used in the quantification of nanoparticles in various matrices should be extremely sensitive as the net concentration of the nanomaterial will be too low to detect. However, it is essential to determine the concentration, composition, and physicochemical properties of the nanomaterials to identify their unique features. Moreover, a single method might be insufficient in providing a feasible data on the quantification of the nanoparticles present. The nanomaterials can be characterized by various physicochemical characters such as the size, shape, structure, dispersion, and coagulation of the nanoparticles in the matrix by microscopic techniques, and the functional group analysis can be done by Fourier transform infrared (FTIR). The electron microscopic techniques such as atomic force microscopy (AFM), scanning electron microscopy (SEM), and transmission electron microscopy (TEM) are used widely for both structural and quantitative analysis, owing to their higher magnification and resolution. These advanced techniques allow for a resolution as low as 0.1–1 nm [8].

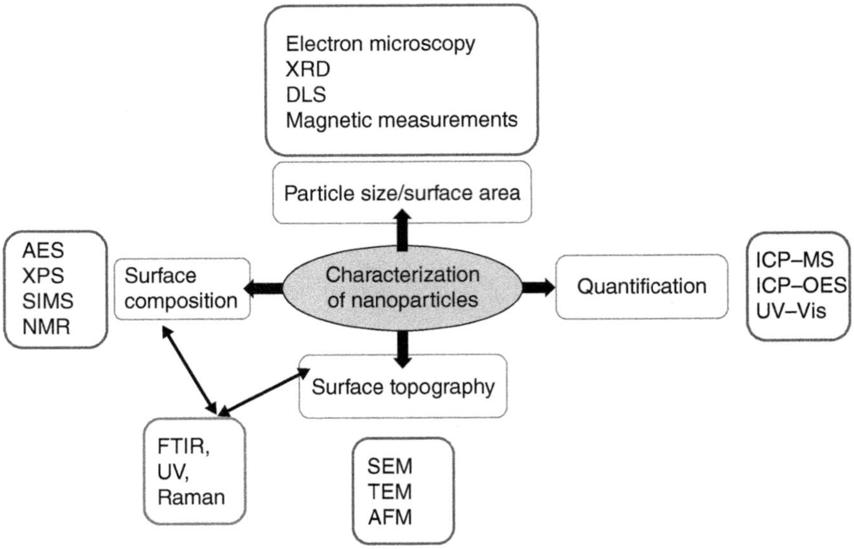

Figure 12.3 Characterization techniques of nanoparticles.

12.8.1 FTIR

The possibility of exploiting the technique to gather information from virtually all sample types despite its shape and form, along with the ease of identification of various in situ processes, makes FTIR a versatile technique. It can thus be used in the antimicrobial investigation of polymers such as chitosan. While the other common techniques give an insight into the release profile of the active component emphasizing on the sensitivity of the techniques to trace levels of components present, the vibrational pattern formed provides a direct information on the chemistry of attachment, antimicrobial activity, release profile, chemical changes, and kinetics [22].

FTIR spectroscopic technique can be successfully utilized for the analysis and identification of the microorganisms. The analysis develops a fingerprint spectrum of the microorganisms, allowing the rapid microbial strain characterization. Moreover, the attenuated total reflection (ATR) can be exploited for the identification of biofilm formation on the ATR crystal interfaces. The spectra can be attained on a nondestructive mode in real time. The Diffuse Reflectance Infrared Fourier Transform (DRIFT) helps in the probing of the reflecting surfaces such as metals or smaller samples. The molecules that are irradiated by a spectrum of infrared radiation absorb the energy and move to an excited state. This can result in complex rotational and vibrational movements of the molecules. These vibrations can in turn modulate the molecular dipole moment. The vibrations thus formed can be discriminated on the basis of the angles and binding forces of the atoms in the molecules, resulting in a complex internal vibration spectrum for a similarly complex molecule. The frequency of the absorption band is in proportion to the force constant and inversely proportional to the atomic mass. The specific vibrational bands produced can thus be correlated to the single bonds or functional groups of a molecule acting as a specific barcode for the molecular/atomic identification. The position of the absorption bands formed is expressed in wavenumbers (cm^{-1}) and is directly proportional to the absorbed energy. In a broader aspect, the functional groups give away an absorption peak at a range of 4000 to 1500 cm^{-1}, while those seen below the 1500 cutoff marks the significant deformations, bending, and ring vibrations. This region being specific for the substances or the substitutions made is known as the "fingerprint region."

The working principle of the FTIR technique involves the simultaneous recording of the spectroscopic radiation over the entire wavelength range with the help of an interferometric modulation, and the resulting interferogram produced contains the sample information under the entire wavelength range. This can be further employed in the development of a simple spectrum with the help of mathematical algorithms. Attenuated total reflectance is an analytical tool that offers the possibility of direct investigation of the chemical configuration of the smooth surfaces of materials. Sample preparation is indeed not required in the ATR; hence the sample can be analyzed under a reasonably uninterrupted state and thus reused. This technique can be used in the biofilm research wherein the in situ analysis is made nondestructively possible and in real time at the substratum/liquid interface. In the ATR technique, the sample is kept in such a way as to maintain the contact with the internal reflection element (IRE) having a higher refractive index when compared with the sample and the medium surrounding it. The existing electromagnetic field shows the occurrence of the inward light, while the amplitude diminishes exponentially with the distance from the surface. The depth of the penetration is limited by the increasing incident angle along with the higher refractive index. These parameters in turn help in determining the depth of the samples such as the biofilm. Elements with multiple internal reflection can be utilized to increase the sensitivity of the analysis. The background spectrum, obtained at the start of an experimentation, helps in determining the sample spectra [23].

The technique can be exploited in microbiological analysis, owing to the fact that a specific spectrum is generated by the bacteria under consideration. However, due to the complex structure of these microorganisms, a range of signals are produced as a result of the molecular vibrations produced by the extracellular polymeric substances (EPSs), cell wall, cytoplasm, etc. This in turn results in the broadening of the bands due to the overlapping spectrum formed, and the complete breakdown of the so formed spectra still remains a herculean task despite the existing resolution enhancement techniques. The fingerprint region, falling between 1500 and 500 cm^{-1}, can be considered further for the characterization of the microorganisms. Even then, a highly complex structure is obtained due to the vast majority of molecules present within the organism. The variations produced in the quantity and the distribution of the various functional groups among the bacterial strains can bring about a minor variation in the formed spectra. Establishment of spectral library and the development of mathematical and statistical modeling can help in the further investigation of the bacterial FTIR spectrum [23, 24]. While analyzing the film-forming potential of chitosan, an evolution of the spectra was seen over the time subjected to the drying of the film-forming solution. A strong absorption spectrum at 1638 and 1706 cm^{-1}, associated with the in-plane OH group and the bends in free acetic acid, was not perceived later in the solid state once the sample was completely dry. Similarly, upon allowing the prepared films to swell, the water diffused rapidly into the matrix of the film, and a strong in-plane OH band was evident at 1638 cm^{-1}. Also as moisture penetrates the chitosan film, it has been reported to release a significant fraction of the carboxylate group, which diffused out into the upper liquid phase and that in turn leads to the reduction of the corresponding peak at 1405 cm^{-1}. It has also been found that the heavily protonated amine group dissolves quickly into the aqueous phase, indicating the exitance of a molecular heterogeneity in its protonation pattern [22].

Recent studies on the biofilm formation using the ATR-FTIR analysis have shown that the ratio of the integrated intensities of the amide II bands varied among the exponential and stationary phases of the bacterial growth. It was in correlation with the amount of nucleic acid synthesized during the log phase, where rapid cell division occurs. The nucleic acid molecules exhibited characteristics IR spectra at 1713, 1238, 1220, and 1080 cm^{-1}, and the phospholipids had specific peaks at 2923, 2853, 1732, 1238, 1220, and 1080 cm^{-1}. Similarly, CH_2 stretching bands at positions 2923 and 2853 cm^{-1} are attributed to the lipids on the bacterial periphery, while the CH_2 bending band formed at 1450 cm^{-1} might be contributed by the CH_2 bending bands of other bacterial biomolecules. Variations in this profile can be confirmed with the growth stages of the bacteria wherein the metabolism rate varies significantly as a subjected to time. The thin depth of analysis (1–2 μm) requires the ATR-FTIR technique to be a valuable tool in monitoring the changes associated with the biofilm formation in situ on a real-time basis [24]. The FTIR technique has a higher detection limit, and it is an efficient tool in scanning the transfer properties of the film, which is responsible in determining the release of components in a nondestructive manner.

The FTIR spectra of pure ZnO nanopowder showed major spectral bands at 3439 cm^{-1} because the OH stretching and the Zn–O stretching was observed at 430 cm^{-1}. However, a weak band was observed in the 500–700 range, and this could be accredited to the stretching of ZnO NP. Moreover, the bending of OH in the range of 1631 cm^{-1} could be attributed to the nanoparticle polymer interaction [18].

12.8.2 Electron Microscopic Techniques

The resolution power of human eye is 0.2 mm. Various lenses and microscopes have been developed with the aim of magnifying the objects. The better the resolution power, better will be the magnification of the system used. While visible light is exploited as the light source in optical microscope, electron microscope makes use of short wavelength (0.859–0.037 Å) electron beams for the surface analysis of the samples. The short wavelength electron beams can cause diffraction in the samples, which can be of nanometer scale.

12.8.2.1 Scanning Electron Microscopy

SEM is a versatile technique, which enables the imaging of the sample surface through the detection of secondary electrons emitted during the interaction of samples with the impinging electron beam. Compared with the TEM, SEM exploits lower beam energies for sample imaging, which has resulted in a limited penetration depth of beam and provides images solely to the specimen surface. But this shallow interaction implies to get more data about the morphological, topographical, crystallographic information, and composition. Morphology specifies the form and size, while topography designates the surface characteristics like its texture, evenness, or coarseness. Crystallography indicates the atomic arrangement within the materials as well as the composition determining the information of elements and compounds that constitute the material. SEM is an important tool capable of achieving a thorough visual image of a particle with high-quality and spatial resolution of 1 nm in size. The surface microstructural analysis of the developed nanocomposite film could be done using the SEM technique. Here, a randomly chosen small fragment of the dry film will be fixed using conductive adhesive material and then will be sprayed by gold prior to analysis on being subjected to an accelerated voltage [11].

The SEM analysis of the cryogenic fractured surfaces of the films indicated that the nanofillers were contained in the developed bionanocomposite films. The surface adhesion between the matrix and the nanofillers could be identified along with the granular nature of the particles using the SEM analysis. The orientation of the nanofillers with respect to each other and their degree of exfoliation can be confirmed through the SEM analysis [1]. The SEM micrographs of the developed films indicated the surface roughness along with the presence of small aggregates, which was not present in the native PVA films. Furthermore, the plain PVC films had a uniform surface morphology and a smooth surface. The nanoparticles were shown to be distributed evenly as bright spots in the micrograph [18].

The SEM analysis of chitosan/PVA films having zinc oxide nanoparticles as fillers showed the uniform dispersion of the nanoparticle in the matrix, which has caused the appearance of the pore-like structure having a diameter of approximately 8 μm. This pore formation could be attributed to the formation of hydrogen peroxide at the metal surface. However, the porosity was found to be in correlation with the concentration of tween 80 used as a plasticizer. This in turn has led to the reduction of interatomic distance in the matrix, which resulted in the reduced tensile strength [25].

12.8.2.2 Transmission Electron Microscopy

The transmission electron microscope was discovered by Knoll and Ruska in 1931 and is one of the most widely used technique for the nanoparticle characterization. Micrographs of nanoscale materials with high lateral spatial resolution are obtained by focusing an electron beam onto the thin sample [26]. The better resolution of electron microscope even at 0.05–0.1 nm size of nanoparticles is an added advantage in the proper understanding of the nanoparticle used. TEM analysis enables the formation of crystalline structure of the particles by spatially confining and focusing the impinging beam and through the detection of the electron diffraction pattern. Here, an intense ray is targeted to expose the sample, and, therefore, the interactions among the electrons and atoms of the sample form a high-resolution image. Considering the miniscule nature of the electrons and the possibility of easy distraction of the electrons from their path by impurities and gases, it is essential to ensure a vacuum source. Various vacuum pumps are used to establish vacuum in TEM. Rotary pump or roughing pump is used to set a low vacuum range, and it is also exploited as a backing pump for other ultrahigh vacuum pumps such as turbo or cryopump, diffusion pump, etc. Similarly, the electron gun used for extracting and regulating the electron beam are of two types: (i) field emission and (ii) thermionic emission. However, the field emission source is generally preferred for obtaining images having higher resolution, which is formed by focusing the electron beams on to the sample with the help of various lenses; briefly, once the electron is extracted from the source, this beam is focused on to the sample with the help of lens, while the extra electrons are filtered out by apertures. This beam is then focused onto the test sample, illuminating it, with the help of a condenser lens, while the image formation is supported by the objective lens. A CCD camera attached with the microscope helps to capture and store the images.

The electron microscopy makes use of the magnetic or electrostatic lens instead of optical lens. It works under a high vacuum condition where the magnification can be improved to 90–800 000×, and the resolving power can be as low as 3 Å. Electron microscopy helps in confirming the composition and crystallographic nature of the samples. The contrast generation occurs in the SEM through the scattered absorption, while that in TEM occurs through diffraction/transmission. While the SEM deals with the surface morphology, TEM helps in analyzing the internal composition of the material. However, both these electron microscopic techniques require a conductive layer of metal or carbon on which the samples are placed. Moreover, the sample preparation must be carried out carefully in TEM so as to obtain a thin layer (less than 100 nm) of the sample, while such a prerequisite is not essential in case of SEM. The higher resolution of the TEM microscopy enables it to obtain high magnification images as when compared with the SEM [27].

TEM is a beam characterization technique in which the specimen is illuminated and magnified by a lens system situated below the sample. SEM can be used for the surface analysis of a sample, which confirms the surface morphology. The research advancements in the electron microscopy have led to the development of scanning–transmission electron microscopy (STEM), which combines both the operation modes simultaneously. The focusing of a convergent electron beam onto the sample, which then propagates through the sample, can lead to the matter–electron interactions. As a result, there occurs a deviation in the pathway of the electron, leading to the generation of wide range of signals. The probe dimensions determine the resolution of the obtained image. These scattered signals can be detected as a function of the probe positioning. The transmitted electrons will be detected in the STEM imaging. The electron–optical parameters of STEM are similar to the well-acknowledged HRTEM wherein the accelerating voltage can be of 100–300 kV and the spherical aberration of the objective lens can be 1.0 mm. This in turn helps in obtaining images having atomic resolution. The use of a geometrically large annular detector in the optical far field is known as the annular dark-field imaging. Here, the intensity detected over the whole detector is recorded and later displayed as a function of the illuminating probes position.

Signal will be received only in the presence of the sample, leaving the vacuum dark. Moreover, larger atoms can lead to greater scattering, which in turn leads to the atomic number contrast (*Z*). This contrast produces images that vary strongly with the atomic number, while the chemical sensitivity is limited as well [28, 29].

12.8.3 Thermal Analysis of the Packaging Material Containing Nanoparticles

The thermal analysis helps in determining the thermal weight loss and stability variations of the samples as subjected to a temperature range of 25–600 °C. The temperature gradually increases with an increment of 10 °C/min and is done in a nitrogen atmosphere (N_2). The weighed samples will be mounted onto an aluminum plate, and liquid nitrogen acts as the coolant for the analysis. Moreover, an empty plate will be used as a reference during the entire analysis [11]. In the thermal stability analysis, the obvious weight loss in the initial temperatures could be attributed to the evaporation of free water from the samples. This is then followed by the thermal decomposition stage in which the decomposition or degradation of the compounds of the film-forming solution can result in the weight loss.

The TGA analysis of the thermoplastic starch (TPS)-based thin films indicated that the degradation process occurs in three stages. The primary stage occurs at a temperature range of 50–100 °C, which is related to the evaporation of both water present in the environment and the free water contained within the material. The evaporation of glycerol rich phases occurs at a temperature of 160–290 °C where a part of starch is also evaporated. In the third phase of the thermal degradation, starch-rich phase is disintegrated at a temperature of 300–320 °C. The elimination of starch poly hydroxyl groups along with depolymerization and decomposition of the TPS matrix occurs in this temperature, leading to the production of carbon. However, the residual mass content of the bionanocomposite film containing nanoclay was found to be more than the plain TPS film. This could be attributed to the fact that the nanoclay used (MMT) does not decompose at the temperature used in the analysis. The study also confirmed that, among the various nanoclays employed, those having a greater organic content within the layers resulted in the lower residual mass following the thermal degradation.

Moreover, an increased thermal resistance was observed in the glycerol rich phase of the bionanocomposite films confirming the positive interactions among the nanofillers and TPS with glycerol, which prevents the evaporation. This in turn indicated that the nanofiller served as a core for the recrystallization of the starch. Also, the clay nanofillers has been suggested to prevent the phase separation due to the greater interlayer spacing they offer [1]. The TGA analysis of nanoclay indicates three steps. First step can last up to a temperature of 150 °C where the loss of physically adsorbed water occurs. The water loss from the interlayers can be confirmed at a temperature range of 150–450 °C. Beyond 650 °C, the dehydroxylation of the structural OH groups of the clay occurs [30]. The application of ZnO NP has shown to improve the glass transition temperature of the PVC matrix under study [18]. The addition of the nanoparticles resulted in a higher glass transition temperature of the nanocomposite films. The specific heat and the thermal stability of the nanocomposites were also found to be enhanced through the addition of the nanoparticles [18]. Addition of ZnO NP improved the thermal stability of the chitosan/PVA films [25]. Again, during the application of ZnO NP as a filler in the pea starch matrix, where glycerol is used as a plasticizer, the low filler concentration has been found to improve the glass transition temperature of the developed nanocomposite. This can be attributed to the strong filler matrix interaction in the nanocomposite [31].

12.8.4 X-Ray Photoelectron Spectroscopy

XPS/electron spectroscopy for chemical analysis is a surface analysis technique to obtain the chemical state information from the material surface. It detects the photoelectrons emitted from the sample by single-energy X-ray photons. The average depth of the analysis can be about 10 nm. It can be used to determine the elemental composition of the material surface and binding state of the elements. It is an ultrahigh vacuum technique and requires the sample evacuation prior to use [32].

The short distance traveled by the excited photoelectrons prior to losing its energy determines the surface and structural sensitivity in XPS. The XPS analysis can provide surface composition and the chemical state of the particles deposited on a substrate/support along with the thickness of the coating or contamination of up to 10 nm. Enrichment/exhaustion of surface elements is possible through this analysis. Surface functionalization and electrical properties along with the surface defects can be identified through the analysis. It can identify the oxidation state as well. However, proper sample preparation and handling is a prerequisite in the analysis [7]. The surface chemical composition of the copper nanoparticles done using the XPS analysis confirmed the presence of copper, carbon, and oxygen on the surface of the material; among which, the latter two could be adventitious surface contaminations [33].

12.8.5 XRD

The characterization of crystal structure determined by the arrangement of atoms can be done with the help of X-ray diffraction (XRD) technique. It is also used to analyze the structural adaptation of the nanofillers in the matrix. The interlayer spacing of the nanofillers within the matrix could be observed from the X-ray diffractograms at region between 2 and 6°. The total exfoliation of the nanofillers within the TPS matrix indicated strong interactions between the nanofiller and the matrix. The confirmation for the same could be obtained through the XRD analysis, wherein the peak between the regions 2 and 6° was almost nonexistent. The interlayer spacing of the nanofillers was found to be increased following the extrusion, which led to the exfoliation of the nanofillers in the matrix [1]. In the XRD analysis, the distance between the planes of the crystals could be calculated from the diffraction angle as obtained from the X-ray patterns [30].

The XRD analysis of the developed ZnO/PVC nanocomposite films (having either 10% or 20% ZnO nanopowder) resulted in diffractograms having peaks nearly similar to that of plain zinc oxide nanopowder. This confirmed that the crystal nature of the nanoparticle was not altered in the presence of PVC. However, reduction of the peak intensity at 2θ value of 12.92° was confirmed with increasing concentration of ZnO nano powder. This confirmed the decrease in the degree of crystallization with increase in the amorphous region. This further increased the conductivity of the nanocomposite films. From the analysis, it can be confirmed that the complexation among the nanofiller and the polymer matrix occurs at the amorphous region rather than the crystalline region [18].

12.8.6 ICPMS

The novel physical and chemical characteristics of engineered nanomaterials with wide range of biological applications demands for the need for an accurate, rapid, and sensitive technique for characterization and quantification of nanoparticles in a wide range of sample types. Pulsed laser ablation microsampling, along with the analysis of the ablation products using inductively coupled plasma spectrometry (ICPMS), has found enumerable applications over the recent years.

Inductive coupled plasma mass spectroscopy (ICPMS), a sensitive detection technique, is a promising tool for the detection and quantification of nanoparticles. The temporal resolution helps in the detection of individual nanoparticles. The presence of the nanoparticles can be detected as signal spikes on the time trace if the dissolved background is low and the mass of the particle under consideration is high enough. The advancements in the instrumentation have allowed to quantify nanoparticles having mass down to single digit attogram (10^{-18} g). The similarities in the temporal duration and sensitivity with the microdroplets and nanoparticles have allowed the usage of microdroplets as nanoparticle proxies [34].

The sample in the aqueous form is introduced continually into the system, which acquires data at high time resolution. The nanoparticles enter the plasma once it is nebulized. Then the sample is vaporized, and the ionization of the individual atom results in a cloud of ions, which is further sampled by mass spectrometer (MS), and the signal pulse is detected by the detector. The typical run time can be 60 seconds. The number of pulses detected per second is proportional to the analyte concentration in the sample. Also, the signal pulse obtained is directly proportional to the mass of the detected compound. The sizing and quantification of nanosilver in the meat samples could be done using ICPMS technique. Here, the desired sample is processed and then subjected to the following enzymatic digestion, which is characterized using ICPMS.

In an analysis on the quantification of nanosilver present in meat samples, 5–25 mg/kg of 60 nm silver gave adequate results using the technique confirming its application at lower concentrations. The results obtained are linear, proportional to the concentration of the nanosilver in the sample [35]. ICP-MS is a powerful tool exploited in the detection of large number of elements including the titanium. Among the five naturally occurring isotopes (^{46}Ti, ^{47}Ti, ^{48}Ti, ^{49}Ti, ^{50}Ti), ^{48}Ti is the most abundant form (73.72%). The spICPMS technique monitors the ^{47}Ti as the common form; ^{48}Ti can have isobaric interference due to the ^{48}Ca [13].

12.8.7 Raman Spectroscopy

Raman spectroscopy, a sophisticated spectroscopic analysis, helps in obtaining the molecular fingerprints of biological samples and hence find a wide range of applications. The technique is based on the vibrational transitions and thus provides information about the molecular structure, conformation, dynamics and interactions, etc. However, the higher sensitivity of the instrument and the Brownian motion of the smaller particles is the real limiting factor for the applications [36].

12.9 Conclusions

Food packages and the packaging material have played a vital role in maintaining the food quality and shelf life. The packaging material can primarily modulate the barrier properties to form a desirable environment, which in turn can delay the processes such as ripening, oxidation, and degradation. The advancements in nanotechnology have further opened up new avenues in the food packaging industry. The incorporation of nanoparticles to the polymer matrix fabricates active packaging materials with enhanced mechanical, thermal, and barrier properties, which ensures longer shelf life. As the nano-packaging system has the potential to serve as an effective alternative for the conventional packaging system, the different nanomaterials and their characterization techniques described in this chapter can aid in designing a novel and effective food packaging material.

References

1 Toro-Márquez, L.A., Merino, D., and Gutiérrez, T.J. (2018). Bionanocomposite films prepared from corn starch with and without nanopackaged Jamaica (*Hibiscus sabdariffa*) flower extract. *Food Bioprocess Technol.* 11 (11): 1955–1973. https://doi.org/10.1007/s11947-018-2160-z.

2 Bumbudsanpharoke, N. and Ko, S. (2019). Nanoclays in food and beverage packaging. *J. Nanomater.* 2019 https://doi.org/10.1155/2019/8927167.

3 Auffan, M., Rose, J., Bottero, J.-Y. et al. (2009). Towards a definition of inorganic nanoparticles from an environmental, health and safety perspective. *Nat. Nanotechnol.* 4 (10): 634–641. https://doi.org/10.1038/nnano.2009.242.

4 Khan, I., Saeed, K., and Khan, I. (2019). Nanoparticles: properties, applications and toxicities. *Arabian J. Chem.* 12 (7): 908–931. https://doi.org/10.1016/j.arabjc.2017.05.011.

5 Thakur, A. and Reddy, S.G. (2016). Biological materials: green source towards silver nanomaterials synthesis. *Orient. J. Chem.* 32 (3): 1465–1472. https://doi.org/10.13005/ojc/320320.

6 Yacamán, M.J., Ascencio, J.A., Liu, H.B., and Gardea-Torresdey, J. (2001). Structure shape and stability of nanometric sized particles. *J. Vac. Sci. Technol., B* 19 (4): 1091. https://doi.org/10.1116/1.1387089.

7 Baer, D.R., Engelhard, M.H., Johnson, G.E. et al. (2013). Surface characterization of nanomaterials and nanoparticles: important needs and challenging opportunities. *J. Vac. Sci. Technol., A* 31 (5): 050820. https://doi.org/10.1116/1.4818423.

8 Huang, J.Y., Li, X., and Zhou, W. (2015). Safety assessment of nanocomposite for food packaging application. *Trends Food Sci. Technol.* 45 (2): 187–199. https://doi.org/10.1016/j.tifs.2015.07.002.

9 Honarvar, Z., Hadian, Z., and Mashayekh, M. (2016). Nanocomposites in food packaging applications and their risk assessment for health. *Electron. Physician* 8 (6): 2531–2538. https://doi.org/10.19082/2531.

10 Vasile, C. (2018). Polymeric nanocomposites and nanocoatings for food packaging: a review. *Materials (Basel)* 11 (10) https://doi.org/10.3390/ma11101834.

11 Lin, D., Wu, Z., Huang, Y. et al. (2020). Physical, mechanical, structural and antibacterial properties of polyvinyl alcohol/oregano oil/graphene oxide composite films. *J. Polym. Environ.* 28 (2): 638–646. https://doi.org/10.1007/s10924-019-01627-4.

12 Wunner, F.M., Wille, M.-L., Noonan, T.G. et al. (2018). Melt electrospinning writing of highly ordered large volume scaffold architectures. *Adv. Mater.* 30 (20): 1706570. https://doi.org/10.1002/adma.201706570.

13 Candás-Zapico, S., Kutscher, D.J., Montes-Bayón, M., and Bettmer, J. (2018). Single particle analysis of TiO_2 in candy products using triple quadrupole ICP-MS. *Talanta*, 180. https://doi.org/10.1016/j.talanta.2017.12.041.

14 Huang, Y., Mei, L., Chen, X., and Wang, Q. (2018). Recent developments in food packaging based on nanomaterials. *Nanomaterials* 8 (10): 830. https://doi.org/10.3390/nano8100830.

15 Mihindukulasuriya, S.D.F. and Lim, L.-T. (2014). Nanotechnology development in food packaging: a review. *Trends Food Sci. Technol.* 40 (2): 149–167. https://doi.org/10.1016/j.tifs.2014.09.009.

16 Jones, N., Ray, B., Ranjit, K.T., and Manna, A.C. (2008). Antibacterial activity of ZnO nanoparticle suspensions on a broad spectrum of microorganisms. *FEMS Microbiol. Lett.* 279 (1): 71–76. https://doi.org/10.1111/j.1574-6968.2007.01012.x.

17 Yamamoto, O. (2001). Influence of particle size on the antibacterial activity of zinc oxide. *Int. J. Inorg. Mater.* 3 (7): 643–646. https://doi.org/10.1016/S1466-6049(01)00197-0.

18 Elashmawi, I.S., Hakeem, N.A., Marei, L.K., and Hanna, F.F. (2010). Structure and performance of ZnO/PVC nanocomposites. *Physica B* 405 (19): 4163–4169. https://doi.org/10.1016/j.physb.2010.07.006.

19 Ahmed, J. (2019). Use of Graphene/Graphene oxide in food packaging materials: thermomechanical, structural and barrier properties. In: *Reference Module in Food Science*. Elsevier. Doi: https://doi.org/10.1016/B978-0-08-100596-5.22499-2.

20 Abreu, A.S., Oliveira, M., De Sá, A. et al. (2015). Antimicrobial nanostructured starch based films for packaging. *Carbohydr. Polym.* 129: 127–134. https://doi.org/10.1016/j.carbpol.2015.04.021.

21 Mei, L. and Wang, Q. (2020). Advances in using nanotechnology structuring approaches for improving food packaging. *Annu. Rev. Food Sci. Technol.* 11: 339–364. https://doi.org/10.1146/annurev-food-032519-051804.

22 Lagaron, J.M., Fernandez-Saiz, P., and Ocio, M.J. (2007). Using ATR-FTIR spectroscopy to design active antimicrobial food packaging structures based on high molecular weight chitosan polysaccharide. *J. Agric. Food. Chem.* 55 (7): 2554–2562. https://doi.org/10.1021/jf063110j.

23 Schmitt, J. and Flemming, H.-C. (1996). FTIR Spectroscopy. In: *Microbially Influenced Corrosion of Materials*, 143–157. Berlin, Heidelberg: Springer Berlin Heidelberg.

24 Quilès, F., Humbert, F., and Delille, A. (2010). Analysis of changes in attenuated total reflection FTIR fingerprints of *Pseudomonas* fluorescens from planktonic state to nascent biofilm state. *Spectrochim. Acta, Part A*, Editors: Heitz, Ewald, Flemming, Hans-Curt, Sand, Wolfgang. 75 (2): 610–616. https://doi.org/10.1016/j.saa.2009.11.026.

25 Vicentini, D.S., Smania, A., and Laranjeira, M.C.M. (2010). Chitosan/poly(vinyl alcohol) films containing ZnO nanoparticles and plasticizers. *Mater. Sci. Eng., C* 30 (4): 503–508. https://doi.org/10.1016/j.msec.2009.01.026.

26 Surrey, A., Pohl, D., Schultz, L., and Rellinghaus, B. (2012). Quantitative measurement of the surface self-diffusion on Au nanoparticles by aberration-corrected transmission electron microscopy. *Nano Lett.* 12 (12): 6071–6077. https://doi.org/10.1021/nl302280x.

27 S. K. Sharma, D. S. Verma, L. U. Khan, S. Kumar, and S. B. Khan, *Handbook of Materials Characterization*. 2018. Springer Berlin Heidelberg.

28 Nellist, P.D. and Pennycook, S.J. (2000). The principles and interpretation of annular dark-field Z-contrast imaging. In: *Advances In Imaging And Electron Physics Vol. 113* (ed. Peter W. Hawkes), 147–203. Elsevier.

29 Pennycook, S.J. (1989). Z-contrast stem for materials science. *Ultramicroscopy* 30 (1–2): 58–69. https://doi.org/10.1016/0304-3991(89)90173-3.

30 Gutiérrez, T.J., Ponce, A.G., and Alvarez, V.A. (2017). Nano-clays from natural and modified montmorillonite with and without added blueberry extract for active and intelligent food nanopackaging materials. *Mater. Chem. Phys.* 194: 283–292. https://doi.org/10.1016/j.matchemphys.2017.03.052.

31 Yu, J., Yang, J., Liu, B., and Ma, X. (2009). Preparation and characterization of glycerol plasticized-pea starch/ZnO-carboxymethylcellulose sodium nanocomposites. *Bioresour. Technol.* 100 (11): 2832–2841. https://doi.org/10.1016/j.biortech.2008.12.045.

32 Mather, R.R. (2009). Surface modification of textiles by plasma treatments. In: Edited by: Q. Wei *Surface Modification of Textiles*, 296–317. Elsevier.

33 Longano, D., Ditaranto, N., Cioffi, N. et al. (2012). Analytical characterization of laser-generated copper nanoparticles for antibacterial composite food packaging. *Anal. Bioanal.Chem.* 403 (4): 1179–1186. https://doi.org/10.1007/s00216-011-5689-5.

34 Gundlach-Graham, A. and Mehrabi, K. (2020). Monodisperse microdroplets: a tool that advances single-particle ICP-MS measurements. *J. Anal. At. Spectrom.* 35 (9): 1727–1739. https://doi.org/10.1039/D0JA00213E.

35 Peters, R.J.B., Rivera, Z.H., van Bemmel, G. et al. (2014). Development and validation of single particle ICP-MS for sizing and quantitative determination of nano-silver in chicken meat. *Anal. Bioanal.Chem.* https://doi.org/10.1007/s00216-013-7571-0.

36 Barkur, S., Lukose, J., and Chidangil, S. (2020). Probing nanoparticle-cell interaction using micro-Raman spectroscopy: silver and gold nanoparticle-induced stress effects on optically trapped live red blood cells. *ACS Omega* 5 (3): 1439–1447. https://doi.org/10.1021/acsomega.9b02988.

13

Biodegradability Assessment of Biopolymer-Based Films

Andrelina Maria Pinheiro Santos[1], Betty Del Carmen Jarma Arroyo[2], Luana de Souza Cavalcante Carnaval[2], and Enayde de Almeida Melo[2]

[1] *Universidade Federal de Pernambuco, Bioprocess Laboratory, Department of Chemical Engineering, Av. Professor Moraes Rego, 1235, Cidade Universitária, Recife, 50670-901, Brazil*
[2] *Universidade Federal Rural de Pernambuco, Consumer Science Department, Rua Dom Manoel de Medeiros s/n, Dois Irmãos, Recife, 52171-900, Brazil*

13.1 Introduction

Synthetic polymers, for example, polystyrene (PS), polyamides (PAs), polyethylene terephthalate (PET), and polyethylene (PE), have been widely used in packaging because they present large availability [1]. These materials are used in the food packing sector because of their mechanical and barrier properties and relatively low cost. However, the problem is that these materials are not derived from sustainable materials nor are compostable, fully recyclable, or biodegradable; they cause serious environmental problems. Measures are proposed to reduce the environmental impact as polymers from natural sources to develop food packaging. Several renewable biopolymers as polysaccharides, proteins, natural gums, and their composites have the property of forming a cohesive and continuous matrix. This biomaterial has been investigated for use in biodegradable packaging materials to replace their nonbiodegradable material [1–5].

The biodegradable films, which may be inedible or edible, are polymers obtained by natural or microbial sources, which generate carbon dioxide, water, methane, and biomass in their biodegradation process. These polymers still have limited use by industries in packaging development due to their lower mechanical properties, barriers, and cost when compared with traditional polymers. In the last decades, researches have been developed to improve these properties and reduce cost through property modification. One of them is incorporating nanoparticles, promising route to enhance mechanical and barrier properties and biodegradable. Nevertheless, it is necessary to clarify the influence of nanoparticles on the biodegradability of polymers and how biodegradability could be affected by the type of polymer and the presence of additives [3, 6, 7].

By definition, biodegradation is the chemical degradation of materials by the action of microorganisms like bacteria, fungi, and algae. In nature, biotic and abiotic

Nanotechnology-Enhanced Food Packaging, First Edition. Edited by Jyotishkumar Parameswaranpillai, Radhakrishnan Edayileveettil Krishnankutty, Aswathy Jayakumar, Sanjay Mavinkere Rangappa, and Suchart Siengchin.

factors act synergistically to decompose organic matter. The biodegradation occurs in two steps and is carried out by enzymes and/or chemical deterioration associated with microorganisms. In the first step, the molecular weight of polymers is reduced to lower molecular mass by oxidation, enzyme degradations, action microorganisms, photodegradation, or hydrolysis. In the second step, bio-assimilation of the polymer occurs by microorganisms and their mineralization [2, 8–12].

Biodegradation is a complex natural phenomenon. However, biopolymers' biodegradability assessment must be implemented to estimate the environmental impact of biopolymer-based films [1]. Therefore, this chapter contains an overview of the major biodegradable polymers used in films and their degradability, as well as the influence of the incorporation of nanoparticles in natural polymers and their biodegradability.

13.2 Commercial and Renewable Biodegradable Polymers and Plasticizers

Plastic has many different advantages such as flexibility, elasticity, low density, and also its facility to work and shape. Unfortunately, these same advantages make it the number one ally for pollution once it comes from oil, a limited nonrenewable resource, and it generates environmental impact since only a small part of plastics is recycled.

Due to the increasing awareness and attention to the widespread use of synthetic and nonbiodegradable polymer packaging, the demand for biodegradable and renewable materials not just in packaging applications is raising more and more [13, 14]; thus, the needs are focused on looking for materials that are eco-friendly and could be generated through sustainable physicochemical processes, to obtain packaging that can compete especially with the highly known polyethylene [11, 15, 16].

In this sense, bioplastics can either replace or at least reduce the dependence on petrochemicals through the use of biological resources [17]. Although packaging industries lead the development and use of biopolymers, mainly from those of renewable organic resources since it must search for green substitutes of nonbiodegradable plastic materials in the future, their use could potentially expand to other areas such as medicine and textile industry, among others [18].

Considering the previously mentioned aspects, then this is worth asking: What requirements must a material meet to be considered ideal for food packaging? The answer is simple, product safety is a primary concern. The material must be capable of protecting the food quality over time and able to withstand environmental, chemical, physical, and microbiological elements, among other aspects that must be considered. Priyadarshi and Rhim complement the concept by defining it also as *inexpensive, renewable, or biodegradable, resulting in no municipal solid waste accumulation*. So far, no ideal material has been produced. Each material has its advantages and disadvantages [19, 20].

The combined action of ultraviolet radiation (photodegradable), heat, and atmospheric oxygen breaks through the chemical bonds of biopolymers (especially the metabolism of microorganisms) [7, 21]. Films and coatings made of biopolymers can be produced with an intention beyond having mechanical barrier and biodegradability characteristics, which interact with food when there are antimicrobial or antioxidant properties in them, and suitable for manufacture active packaging materials, which can be beneficial in enhancing the shelf life of food products [19, 22–24].

Different types of biopolymer classifications can be found in the literature; however, it is important to understand the sources. According to Vieira et al., biodegradable polymers are classified into five categories depending on their fabrication methods as follows [1, 8, 25]:

(i) From plants (e.g. starch, cellulose, zein, pectin, and lipid).
(ii) From animals and dairy products (e.g. casein, whey protein, gelatin, chitosan, and lipid).
(iii) From fermenting products or by-products (e.g. polyhydroxy butyrate [PHB] and polyhydroxyalkanoates [PHAs]).
(iv) From chemical synthesis of naturally derived substances (e.g. polylactic acid [PLA]).
(v) From chemical synthesis of fossil resources (e.g. polycaprolactones [PCLs] and polyesteramides [PEAs]).

In this chapter, the three most used biopolymers will be discussed, including the food packaging industry.

13.2.1 Thermoplastic Starch (TPS)

Starch is a promising polysaccharide biopolymer for producing biocomposite materials because it is renewable and abundant in nature, making it a low-cost material as can be found in several species [26, 27]. Also, starch can be easily modified physically and chemically [28]. However, it has low gas permeability, poor water vapor barrier, and mechanical properties due to its hydrophilic nature [29], besides low tensile strength and severe deformations, which limit its application in films or food packaging [27].

To successfully improve its flexibility, raw starches are normally mixed with other additives and plasticizers to improve their processability and the properties of final products, converting starch into thermoplastic starch (TPS) by applying heat and shear force during the extrusion process, which allows it to show similar behaviors as conventional thermoplastic polymers [13, 30–34]. Without plasticizers, the application of heat and shear force, starch cannot be transformed into a thermoplastic material [30, 35, 36].

TPSs are based on the combination of different starches and plasticizers. However, the final physical properties are highly influenced by the amount of plasticizer present [1, 37–39]; they are commonly processed by conventional techniques such as extrusion, compression molding, and injection molding to obtain ductile materials with improved water vapor barrier and mechanical properties [40–44].

Depending on the plasticizer/starch (suitable), TPS exhibits a wide range of properties due to the gelatinization of molecules that must be under heat and high shear force [40]. Plasticizers such as glycerol, glycol, xylitol, sorbitol, sugars, and ethanolamine go through the starch granules causing a structural breakdown (modification), and the action of shear force and temperature in the presence of plasticizers that do not evaporate easily during processing builds a matrix that does not exhibit diffraction anymore [27, 45].

Blends with other mineral-based particles or biodegradable polymers such as PHAs, PLA, and PCL can improve TPS properties and also maintain its biodegradability [2, 40, 46–48]. Due to its potential, TPS could become a promising material for applications in the manufacturing of packaging materials [45].

13.2.2 Polylactic Acid (PLA)

PLA is a thermoplastic, aliphatic, and compostable polyester of natural origin obtained from lactic acid during the fermentation of carbohydrates or through chemical synthesis of direct condensation of lactic acid (2-hydroxy propionic acid) as well as ring-opening polymerization (ROP) of cyclic lactide dimer [3, 13, 14, 20, 49–51].

PLA has raised great interest in recent years, thanks to its biocompatibility, processability, and good mechanical and chemical barrier properties when compared to standard thermoplastic materials [22, 52]. It can be processed, like most thermoplastics, using the conventional process (extrusion and film blowing, among others) [13, 20, 41, 50, 51, 53]. However, some negative aspects about PLA are its sensitivity to moisture, as it degrades rapidly above ±90 °C in high humidity conditions, and also its low flexibility and toughness. As an alternative to enhance physical properties and molecular characteristics, PLA can be blended with other copolymers or polymers such as thermoplastic, which results in obtaining materials with improved mechanical strength [13, 14, 50, 51, 53–55].

13.2.3 Polyhydroxyalkanoates (PHAs)

As previous biopolymers mentioned, PHAs never rely on fossil fuel as raw material, instead they are synthesized by microorganisms through the bacterial fermentation of lipids or sugar, which common source is corn crops. PHA is recognized by its thermoplastic properties like polypropylene and polyethylene, and its forms range from highly elastic and crystalline materials to opaque solids with high material strength [56–58].

Although there are several varieties of PHAs, they can be categorized in the short-chain and medium-chain, the first one generally composed of between 3 and 5 carbon atoms and the second containing between 6 and 14 atoms [59]. PHAs can be obtained efficiently from genetically modified plants; however, for both cases, the extraction process is through the use of solvents [57, 60, 61].

PHAs can be elaborated by injection molded or extruded; some known PHAs include poly-(3-hydroxybutyrate), poly-(4-hydroxybutyrate), poly-(3-hydroxybutyrate-4-hydroxybutyrate), and poly-(β-malic acid) [55, 59]. Despite the many advantages of PHAs, the cost of its production is high; however, the two most common and marketed classes exhibit properties that resemble PP (used in packaging and coatings) such as polyhydroxybutyrate homopolymer (PHB) and the polyhydroxybutyrate–polyhydroxyvalerate copolymer known as polyhydroxybutyrate–valerate (PHBV) [61, 62].

The reason for the high cost of its production is due to the energy required for sterilization to avoid risks of contamination. The final structure and properties of PHAs vary and depend on the carbon source, the producer of microorganisms (pure and mixed cultures), and the environmental conditions of biosynthesis [63].

In general, the research, production, and use of biopolymers result from the need to search for alternatives to reduce the environmental footprint. Through the life cycle analysis (LCA) technique, it is possible to evaluate potential impacts associated with a product since the obtaining of the raw material for its fabrications to its disposal [64].

13.2.4 Plasticizers

The use of plasticizers in biopolymer processing is important since they are considered as a possible solution to improve the mechanical properties of biopolymers; however, if there is no compatibility between plasticizer and biopolymer, the final product will remain brittle and rigid. Factors like solubility, polarity, dielectric constant, and hydrogen bonding must be considered at the time of selection, which generally acts by reducing viscosity, decreasing the glass transition temperature and Young's modulus, and improving the elastic behavior [18, 50, 65, 66].

According to Vieira et al., *plasticizers can be classified as primary and secondary;* the first one is when they are highly compatible with polymers and can be added in large quantities, and the second one is when it has limited compatibility with the polymer and/or high volatility, so they are usually blended with primary plasticizers for property improvement and cost and viscosity reduction, among other uses [25, 32].

Glycerol, for example, draws a lot of attention as it is presented as an attractive alternative because of its high compatibility, its nontoxic nature, and its low cost, being the most widely used plasticizer for bio-based polymers in research [37, 67, 68]. This plasticizer can cause structural changes in the formation of the biopolymer, leading to higher crystallinity, softness, and flexibility, enabling various uses in the packaging and manufacturing industry [69].

The selection of the plasticizer depends on the biopolymer and what is sought in the final product. Glycerol is no exception to the rule; normally in the plasticization process, the minimum amount of glycerol used is N25% [68–71]. Although some starch polymer films can show unattractive or unlikable flexibility due to the use of a single plasticizer that prevents its use for food packaging, co-plasticizers are being applied into those matrices to enhance their properties [72, 73].

Migration of plasticizers from the material or exudation can be produced when there exists an incompatibility between the polymer and plasticizer or when working with hydrophilic plasticizers at high relative humidity environments, which is the importance of the right selection for each application [74, 75]. According to Ma et al., "*urea and ethanolamine can form more stable and stronger hydrogen bonds with starch molecules than conventional plasticizers.*" Khan et al. [76] suggested that "*boric acid and glycerol can act as a co-plasticizer to form thermoplastic starch films*" [28, 75].

13.3 Biodegradation Mechanism

To comprehend the evolution of new biomaterials, it is important to evaluate the role of packaging for society [3, 11]. If we look closely at the evolution of humanity, we can see how packaging has been influenced by these social changes. The importance of packaging was realized by its practicality and safety, for example, in transport and food, and the evolution of materials used in its manufacture trails from the use of straw, ceramics, glass, and paper to the synthetic polymers that have dominated the market. However, this growth has a price for society and the environment. Although plastics produce packaging with high practicality, easy molding, differentiated systems, and easier access to the product, the disadvantages came with the great environmental impact. Making changes in concepts, applications, and new materials is necessary. Thus, new materials have been appearing with natural polymers. Not only in the design as attractive and decisive in the sale but also in the definition of security.

Before understanding the process of biodegradation of biopolymers, it is important to know the following terms and definitions [49, 64]:

- *Biopolymers/bioplastics*: They are polymers structurally classified as polysaccharides, polyesters, or PAs. The main raw material for obtaining these is a renewable carbon source, which can be a carbohydrate derived from sugar cane, corn, beet, or a vegetable oil extracted from soy, sunflower, palm, or other oilseed.
- *Composting*: It is a biodegradation process that results in organic compounds, with the final destination of the soil. Remembering that these compounds have to be following the laws of each country.
- *Biodegradation*: Biodegradation is nothing more than a degradation caused by microorganisms present in the environment, requiring the presence of oxygen, light, heat, and humidity.
- *Additives*: Substances that are added to plastic to modify a certain property or set of properties.
- *Microbial biodegradation*: This type of biodegradation can be classified as aerobic (13.1) and anaerobic (13.2), differentiated by the concentration of oxygen present during the process.

$$C_{\text{Polymer}} + O_2 \rightarrow CO_2 + H_2O + C_{\text{residual}} \tag{13.1}$$

$$C_{\text{Polymer}} \rightarrow CO_2 + CH_4 + H_2O + C_{\text{residual}} \tag{13.2}$$

The study of the biodegradability of biopolymers becomes essential for the characterization as biodegradable. According to the North American (ASTM) and European (ISO/CEN) standards, a biopolymer will be classified as biodegradable if, after undergoing the tests under the ideal conditions defined by the protocol that defines the biomaterial as biodegradable, it presents a conversion of 60% of the carbon present in its composition in CO_2, in 45 days of study. It is worth mentioning that other parameters can also significantly affect the efficiency of the biodegradation process, for example, the chemical composition of the polymer, conditions of the medium, and presence of additives [36, 60, 67, 77]. Considering that, biodegradation, by definition, anaerobically or aerobically occurs in the presence of microorganisms, more specifically fungi and bacteria [2].

Expanding the discussion about the importance of knowledge of the capacity and understanding of the biodegradation process mechanism of biopolymers, it is important to emphasize that it is not only due to the application in food packaging but also pharmaceutical, medical, and cosmetics packaging, considering the attractive properties of these biopolymers such as biocompatibility, availability, and biodegradation [9, 36, 43, 58].

The understanding of biodegradation mechanism is important because food packaging application has become crucial due to the growth of research involving new materials from synthetic and natural polymers, with or without additives, nanocomposites. The incorporation of nanocomposites aims to improve mechanical and barrier properties and antioxidant and antimicrobial activity [6, 14, 22, 50, 65, 73, 78, 79]. Studies incorporating extracts, essential oils [11, 21, 23, 80–82], and nanocomposites [6, 22, 34, 58, 83–85] have also been developed.

The use of nanotechnology in the food packaging aimed to develop and improve new polymers from natural or synthetic source to obtain nanostructured films [26, 63, 86] and edible coatings [22, 87–89]. The application of nanotechnology in food packaging has been showing improvements in mechanical and barrier properties, increasing resistance temperature and humidity. Also silver, zinc oxide, copper, and clay nanoparticles [34, 52, 58, 90–92] of essential oils extracts rich in phenolic compounds [80, 82, 93–96] are enabled with antimicrobial and antioxidant activity to develop food active and smart packaging. There is another interesting type of smart packaging. In the polymer, sensors are added that can indicate the maturity and quality of the product to the consumer [57, 97–99] by color change due to changes in pH and temperature. Regarding food safety, it was evaluated by cytotoxicity assessment, studying the migration of these composites that are incorporated into the polymer [12, 57, 100–102].

Commission Regulation (EU) No. 10/2011, about material plastic in direct or indirect contact with food, determines an overall migration limit of 10 mg/cm^2 surface area [11]. It is important because of the incorporated composite in micro- or nanoscale in polymers for food packaging. In the case of 1 cubic liter of a package containing 1 kg of food, the migration permitted by law is 60 mg of the component for a kg of food [100, 103]. In the case of nanomaterial for films and/or edible coating, studies of migration concentration, cytotoxity, and biodegradability are essential to evaluate safety not only for human health but also for animals and the environment [20, 102]. It is important to highlight that studies have to evaluate nanoparticle toxicity in a biological system, aiming at safety for human health, animals, and the environment [12, 78, 104, 105].

13.4 Biodegradation of Biopolymers with Additives

Bio-based polymers have become a potential strategy to reduce environmental impact. Biopolymers are biodegradable and renewable sources for the development of new materials in many applications [25, 106]. However, for several applications some properties of biopolymers, as mechanical, need to be improved to make them more effective. Blends with other biopolymers and incorporation with additives are relevant options. For example, PLA is a bio-based polymer widely investigated in biomedical, bioengineering, drug delivery, and pharmaceutical area. To make it suitable for these applications, PLA has received reinforcement, for example, with small amounts of nanofillers [33, 59, 107].

Kan et al. [66] developed active food packaging based on chitosan/gelatin blend films incorporated with Chinese hawthorn (*Crataegus pinnatifida*) fruit extract [2]. Through the extract, incorporation was possible to improve mechanical and barrier properties. The inner microstructure blend films became more compact; thickness, tensile strength, and elongation at break increased; and water vapor permeability was significantly reduced. Also by the addition of extract, the light barrier property of these films was enhanced (decrease in light transmittance) [1, 25, 56, 66, 78]. Amjadi et al. [108] prepared a gelatin-based nanocomposite containing chitosan nanofiber and ZnO nanoparticles [2]. Good compatibility was demonstrated between these components, and the nanocomposite showed high mechanical and water barrier properties and a high dense and less permeable structure [38, 88, 108]. Thus, research has been carried out to improve the physical properties of composites by incorporating additives and often to reach active characteristics [11, 35, 80, 99, 109].

As biopolymers are promising substitutes for petroleum-based plastics and due to the need for additives to improve their properties, researchers have been studying the effect of this incorporation in the biodegradation process of these materials because of concerns about their disposal [52, 55, 106, 110]. Studies about composites of PLA and graphene nanoplatelets are conducted for future application in medical devices [8], and results have shown that addition of graphene nanoplatelets to PLA produces composites with improved mechanical properties and the nanofillers demonstrated not to impair the polymer degradation process (hydrolytic degradation for six months in phosphate-buffered saline [PBS] at 37 ° C). Besides that the composites, comparing with unfilled PLA, after six months of degradation in physiological conditions, do not release toxic products toward human fibroblasts [1, 17, 45, 96, 111].

Jumaidin et al. developed a biodegradable thermoplastic composite from sugar palm starch and agar, with the addition of *Eucheuma cottonii* seaweed waste as a biofilter. The incorporation of seaweed (0–40 wt%) has brought improvements in the tensile, flexural, and impact properties and also thermal stability. The biodegradation process was enhanced with the addition of seaweed. According to the authors, the main indicator for biodegradation by moisture and microorganism is the weight loss of material. The incorporation of seaweed waste increased this indicator from 53% to 75% and 70% to 89%, after two and four weeks of soil burial, respectively, implying that these additives were efficient to accelerate the biodegradation process. The authors explained that these results can be associated with the more hydrophilic behavior of seaweed, which increased the hygroscopicity of materials that stimulates the growth of microorganisms during degradation, accelerating biodegradation [18, 21, 25, 50, 62, 74, 106, 112].

Prapruddivongs et al. studied the biodegradation behavior of PLA and chemically cross-linked PLA with and without incorporation of silica (SiO_2). The addition of SiO_2 ended in more water absorption with wide surface erosion, which led to faster biodegradation. The higher content of loaded SiO_2 improved the water absorption, accelerating this process. Moreover, commercial silica showed to be more effective at accelerating biodegradation than rice husks silica for both biopolymers (cross-linked or not) due to its greater hydrophilicity [27, 43, 110].

Bootklad and Kaewtatip developed composites with TPS incorporating chicken eggshell powder (EP) as a filler and studied their biodegradation process. The authors compared the effect of incorporating eggshell powder (95% of calcium carbonate [CC] and 5% of organic materials) and the commercial CC on the properties of developed thermoplastic [33, 58]. They noted that the eggshell powder created a strong adhesion with TPS (especially because of organic compounds) and improved the water resistance and thermal stability of this thermoplastic. Through soil burial test and weight loss, it was observed that the composites incorporated with EP had a faster degradation process than composites with CC, which can be associated with the presence of organic compounds in eggshell powder facilitating the microorganisms attack. However, both EP and CC delayed the biodegradation rate of TPS (without additives), which showed higher weight loss. Both fillers hindered the water and microorganisms' penetration into the matrix and increased its hydrophobicity, slowing degradation rate with pure TPS. Overall, comparing both additives, the authors suggested that CC can be replaced by eggshell powder to improve the properties of TPS composites [1, 2, 25, 78, 113].

Biodegradable materials as TPS have some restrictions as hydrophilic characteristics, poor mechanical properties, and water resistance, besides poor long-term stability [48, 113]. Because of that, many researchers have been trying to improve these properties by applying additives in biodegradable matrices as fibers [114], clay [115], and carbon nanotubes [83]. Some fillers used to improve biopolymer properties are nonbiodegradable materials, but, currently, the focus of most research is the development of bio-fillers because it can improve compatibility between matrix and additive, besides can make biodegradation by microbial pathway easier [113], for example, lignocellulosic material as cellulose microfibers or nanofibers from sugar cane [114].

Silva et al. studied the effect of lignin as accelerator of biodegradation of PLA/lignin composites [2, 59]. The authors highlighted that the incorporation of organic compounds from renewable sources can lead to a decrease in biodegradation time. In general, PLA is degraded from 6 to 24 months according to the conditions, but the use of organic additives is an alternative to reduce this. The results showed that lignin was able to accelerate the biodegradation of PLA and the composite with 10% of lignin was the best condition because it demonstrated similarity to thermal and mechanical properties neat PLA but with highest biodegradation rate comparing to all contents (0– 15% lignin) [1, 29].

Biocomposites based on cassava starch were studied by Maran et al., and they were considered appropriate for being disposed of in landfills, for example, because of their biodegradation capacity [78]. The hydrophilic character of materials more intensively promotes water absorption generating suitable conditions for the access of soil microorganisms [78]. Also, it can use these materials as a source of energy/carbon to grow. This loss of matrix components could be associated with a decrease in weight and mechanical property of materials during biodegradation. Plasticizers as glycerol and biopolymers as starch intensify the water absorption due to their strong hydrophilic characteristic, which leads to an increase in solubility and instability of intermolecular interactions, besides higher water availability within the material, promoting microorganisms colonization, with consequent disintegration. Thus, loss in weight and mechanical properties can be treated as indicators of degradation, becoming important criteria to evaluate biodegradation. Fractures, breaches, cavities, and holes also offer good information about it [4, 89].

In summary, the use of biopolymers has been increasing, but their limitations have demanded the development of composites with the addition of reinforcement fillers to improve mechanical properties, such as nanoclays, carbon nanotubes, and nanosilica, and addition of plasticizers and blends with other polymers. However, the biodegradation process of these materials is a concern to be assessed during the development of composites [17, 111]. Because of that, the scientific research field has been going deeper into studies with composites applying biocompatible fillers that can favor and promote biodegradation by microorganisms [113]. Moreover, active agents such as natural extracts, essential oils, and nanostructures have also been studied to improve mechanical and barrier characteristics, besides achieve active properties [2, 66, 80, 99, 108, 109].

13.5 Considerations

Biodegradable materials generally have shown many advantages. Biodegradable polymers have been researched, but polymers based on renewable sources are most desirable. The main advantage of biodegradable polymers is that they are composted with organic waste and intended to enrich the soil [37, 53]. There are a large number of tests that are used to determine the extent of degradation of polymers either alone or in blended forms. Many are respirometric, determining the amount of carbon dioxide released on exposure to fungi, bacteria, activated sludge (aerobically or anaerobically), compost, or soil. Some tests use the loss of weight or change in physical properties such as tensile strength and comparison of spectroscopic [10, 15].

The use of biodegradable polymers in many fields of industry, instead of synthetic materials, can significantly help to protect the natural environment [41, 75]. They can be used in packaging for multiple applications. Technology is now available for the production, processing, converting, and forming of a variety of biodegradable packaging. Several biopolymers and their blends have a broad and versatile range of physical properties and other advantageous characteristics and biodegradation rates and can be employed in a wide range of applications [13, 32, 116]. In case foods are currently purchased through different means, including e-commerce, in addition to preserving the environment, the packaging material must ensure nutritional and microbiological safety, as well as protection, integrity, consumer information, and increased shelf life of food.

The introduction of bionanocomposites in films, as well as the use of several methodologies, such as layer-by-layer, deposition, or spin coating, enabled the emergence of active and intelligent packaging, making products more practical and safe, maintaining the nutritional and sensory value of the product to be consumed. However, important points also need to be researched and analyzed especially when referring to bionanocomposites, both for obtaining films and for nanostructured coatings. The evaluation of the desired and unwanted migration and the study of the cytotoxicity of these compounds are important to ensure safety for the consumer and meet the regulatory standards of each country.

References

1 Rhim, J. and Ng, P.K.W. (2007). Natural biopolymer-based nanocomposite films for packaging. *Crit. Rev. Food Sci. Nutr.* 8398: 411–433.

2 Scaffaro, R., Maio, A., Sutera, F. et al. (2019). Degradation and recycling of films based on biodegradable polymers: a short review. *Polymers* 71 (651): 1–20.

3 Sanyang, M.L., Sapuan, S.M., Jawaid, M. et al. (2015). Effect of plasticizer type and concentration on tensile, thermal and barrier properties of biodegradable films based on sugar palm (*Arenga pinnata*) starch. *Polymers* 7: 1106–1124.

4 Maran, J.P., Sivakumar, V., Thirugnanasambandham, K., and Sridhar, R. (2014). Degradation behavior of biocomposites based on cassava starch buried under indoor soil conditions. *Carbohydr. Polym.* 101: 20–28. https://doi.org/10.1016/j.carbpol.2013.08.080.

5 Rubino, C., Bonet Aracil, M., Liuzzi, S. et al. (2021). Wool waste used as sustainable nonwoven for building applications. *J. Cleaner Prod.* 278: 123905.

6 Balaguer, M.P., Aliaga, C., Fito, C., and Hortal, M. (2016). Compostability assessment of nano-reinforced poly(lactic acid) films. *Waste Manage.* 48 (February): 143–155. https://doi.org/10.1016/j.wasman.2015.10.030.

7 Rhim, J., Park, H., and Ha, C. (2013). Progress in polymer science bio-nanocomposites for food packaging applications. *Prog. Polym. Sci.* 38 (10–11): 1629–1652. https://doi.org/10.1016/j.progpolymsci.2013.05.008.

8 Leja, K. and Lewandowicz, G. (2010). Polymer biodegradation and biodegradable polymers – a review. *Pol. J. Environ. Stud.* 19 (2): 255–266.

9 Vroman, I. and Tighzert, L. (2009). Biodegradable polymers. *Materials (Basel)* 2: 307–344.

10 Lucas, N., Bienaime, C., Belloy, C. et al. (2008). Chemosphere polymer biodegradation: mechanisms and estimation techniques. *Chemosphere* 73: 429–442.

11 Alizadeh-sani, M., Mohammadian, E., and Mcclements, D.J. (2020). Eco-friendly active packaging consisting of nanostructured biopolymer matrix reinforced with TiO_2 and essential oil: application for preservation of refrigerated meat. *Food Chem.*: 126782. https://doi.org/10.1016/j.foodchem.2020.126782.

12 Maisanaba, S., Pichardo, S., Jordá-beneyto, M. et al. (2014). Cytotoxicity and mutagenicity studies on migration extracts from nanocomposites with potential use in food packaging. *Food Chem. Toxicol.* 66: 366–372. https://doi.org/10.1016/j.fct.2014.02.011.

13 Zhong, Y., Godwin, P., Jin, Y., and Xiao, H. (2020). Biodegradable polymers and green-based antimicrobial packaging materials: a mini-review. *Adv. Ind. Eng. Polym. Res.* 3 (1): 27–35. https://doi.org/10.1016/j.aiepr.2019.11.002.

14 Dimkpa, C.O., Mclean, J.E., Britt, D.W., and Anderson, A.J. (2013). Antifungal activity of ZnO nanoparticles and their interactive effect with a biocontrol bacterium on growth antagonism of the plant pathogen Fusarium graminearum. *Biometals*: 1–39.

15 Verma, D. and Fortunati, E. (2019). Biopolymer processing and its composites: an introduction. In: *Biomass, Biopolymer-Based Materials, and Bioenergy* (eds. D. Verma, E. Fortunati, S. Jain and X. Zhang), 1–24. Elsevier Ltd https://doi.org/10.1016/B978-0-08-102426-3.00001-1.

16 Hall, A.R. and Geoghegan, M. (2018). Polymers and biopolymers at interfaces. *Rep. Prog. Phys.* 81: 036601.

17 Galiano, F., Briceño, K., Marino, T. et al. (2018). Advances in biopolymer-based membrane preparation and applications. *J. Membr. Sci* 564 (May): 562–586. https://doi.org/10.1016/j.memsci.2018.07.059.

18 Mowafi, S. and Taleb, M.A. (2018). Utilization of proteinic biopolymers: current status and future prospects. *J. Text. Color. Polym. Sci.* 15 (1): 15–31.

19 Priyadarshi, R. and Rhim, J. (2020). Chitosan-based biodegradable functional films for food packaging applications. *Innovative Food Sci. Emerg. Technol.*: 102346. https://doi.org/10.1016/j.ifset.2020.102346.

20 Plackett, B.D.V., Holm, V.K., Johansen, P. et al. (2006). Characterization of L-Polylactide and L-Polylactide – polycaprolactone co-polymer films for use in cheese-packaging applications and science. *Packag. Technol. Sci.* 19: 1–24.

21 Alizadeh-sani, M., Khezerlou, A., and Ehsani, A. (2018). Fabrication and characterization of the bionanocomposite film based on whey protein biopolymer loaded with TiO_2 nanoparticles, cellulose nano fibers and rosemary essential oil. *Ind. Crop Prod.* 124 (March): 300–315. https://doi.org/10.1016/j.indcrop.2018.08.001.

22 Arroyo, B.J., Bezerra, A.C., Oliveira, L.L. et al. (2020). Antimicrobial active edible coating of alginate and chitosan add ZnO nanoparticles applied in guavas

(*Psidium guajava* L.). *Food Chem.* 309 (August 2018): 125566. https://doi.org/10.1016/j.foodchem.2019.125566.

23 Alizadeh-sani, M., Rhim, J., and Azizi-lalabadi, M. (2019). Preparation and characterization of functional sodium caseinate/guar gum/TiO_2/cumin essential oil composite film. *Int. J. Biol. Macromol.* https://doi.org/10.1016/j.ijbiomac.2019.11.004.

24 Bocz, K., Szolnoki, B., Marosi, A. et al. (2013). Flax fibre reinforced PLA/TPS biocomposites flame retarded with multifunctional additive system. *Polym. Degrad. Stab.*: 1–11. https://doi.org/10.1016/j.polymdegradstab.2013.10.025.

25 Gurgel, M., Vieira, A., Altenhofen, M. et al. (2011). Natural-based plasticizers and biopolymer films: a review. *Eur. Polym. J.* 47 (3): 254–263. https://doi.org/10.1016/j.eurpolymj.2010.12.011.

26 Lomelí Ramírez, M.G., Satyanarayana, K.G., Iwakiri, S. et al. (2011). Study of the properties of biocomposites. Part I. Cassava starch-green coir fibers from Brazil. *Carbohydr. Polym.* 86 (4): 1712–1722.

27 Weber, C.J., Haugaard, V., Festersen, R., and Bertelsen, G. (2002). Production and applications of biobased packaging materials for the food industry. *Food Addit. Contam.* 19: 172–177.

28 Belgacem, M.N. and Gandini, A. (2008). Major sources, properties and applications as thermoplastic materials. In: *Monomers, Polymers and Composites* (eds. M.N. Belgacem and A. Gandini), 321–342. Elsevier.

29 da Silva, T.F., Menezes, F., Montagna, L.S., and Ana Paula Lemes, F.R.P. (2019). Effect of lignin as accelerator of the biodegradation process of poly(lactic acid)/lignin composites. *Mater. Sci. Eng., B* 251: 114441.

30 Decaen, P., Rolland-sabaté, A., Colomines, G. et al. (2019). Influence of ionic plasticizers on the processing and viscosity of starch melts. *Carbohydr. Polym.*: 115591. https://doi.org/10.1016/j.carbpol.2019.115591.

31 Isotton, F.S., Bernardo, G.L., Baldasso, C. et al. (2015). The plasticizer effect on preparation and properties of etherified corn starchs films. *Ind. Crops Prod.* 76: 717–724. https://doi.org/10.1016/j.indcrop.2015.04.005.

32 Abdorreza, M.N., Cheng, L.H., and Karim, A.A. (2011). Effects of plasticizers on thermal properties and heat sealability of sago starch films. *Food Hydrocolloids* 25 (1): 56–60. https://doi.org/10.1016/j.foodhyd.2010.05.005.

33 Stepto, R.F.T. (2006). Understanding the processing of thermoplastic starch. *Macromol. Symp.* 245–246 (1): 571–577.

34 Duffy, L.L., Osmond-mcleod, M.J., Judy, J., and King, T. (2018). Investigation into the antibacterial activity of silver, zinc oxide and copper oxide nanoparticles against poultry-relevant isolates of *Salmonella* and *Campylobacter*. *Food Control* 92: 293–300. https://doi.org/10.1016/j.foodcont.2018.05.008.

35 Kodsangma, A., Homsaard, N., Nadon, S. et al. (2020). Effect of sodium benzoate and chlorhexidine gluconate on a bio-thermoplastic elastomer made from thermoplastic starch-chitosan blended with epoxidized natural rubber. *Carbohydr. Polym.*: 116421. https://doi.org/10.1016/j.carbpol.2020.116421.

36 Luckachan, G.E. and Pillai, C.K.S. (2011). Biodegradable polymers - a review on recent trends and emerging perspectives. *J. Polym. Environ.* 19 (3): 637–676.

37 Chocyk, D., Gładyszewska, B., Ciupak, A. et al. (2015). Influence of water addition on mechanical properties of thermoplastic starch foils. *Int. Agrophys.* 29: 267–273.

38 Mahieu, A., Terrié, C., and Youssef, B. (2014). Thermoplastic starch films and thermoplastic starch/polycaprolactone blends with oxygen scavenging properties: influence of water content. *Ind. Crops Prod.* https://doi.org/10.1016/j.indcrop.2014.11.037.

39 Sahari, J., Sapuan, S.M., Zainudin, E.S., and Maleque, M.A. (2013). Thermo-mechanical behaviors of thermoplastic starch derived from sugar palm tree (*Arenga pinnata*). *Carbohydr. Polym.* 92 (2): 1711–1716. https://doi.org/10.1016/j.carbpol.2012.11.031.

40 Lendvai, L., Apostolov, A., and Karger-kocsis, J. (2017). Characterization of layered silicate-reinforced blends of thermoplastic starch (TPS) and poly(butylene adipate-*co*-terephthalate). *Carbohydr. Polym.* https://doi.org/10.1016/j.carbpol.2017.05.100.

41 Ferri, J.M., Fenollar, O., and Balart, R. (2016). The effect of maleinized linseed oil (MLO) on mechanical performance of poly(lactic acid)-thermoplastic starch (PLA-TPS) blends. *Carbohydr. Polym.* 147: 60–68. https://doi.org/10.1016/j.carbpol.2016.03.082.

42 Schmitt, H., Guidez, A., Prashantha, K. et al. (2015). Studies on the effect of storage time and plasticizers on the structural variations in thermoplastic starch. *Carbohydr. Polym.* 115: 364–372. https://doi.org/10.1016/j.carbpol.2014.09.004.

43 Taylor, P., Zhang, Y., Rempel, C., and Liu, Q. (2014). Thermoplastic starch processing and characteristics—a review. *Crit. Rev. Food Sci. Nutr.* 54 (August 2014): 1353–1370.

44 Rhim, J.W. (2011). Effect of clay contents on mechanical and water vapor barrier properties of agar-based nanocomposite films. *Carbohydr. Polym.* 86 (2): 691–699. https://doi.org/10.1016/j.carbpol.2011.05.010.

45 Mościcki, L., Mitrus, M., Wójtowicz, A. et al. (2012). Application of extrusion-cooking for processing of thermoplastic starch (TPS). *Food Res. Int. J.* 47: 291–299.

46 Dean, K., Yu, L., Bateman, S., and Wu, D.Y. (2007). Gelatinized starch/biodegradable polyester blends: processing, morphology, and properties. *J. Appl. Polym. Sci.* 103: 802–811.

47 Chow, W.S., Leu, Y.Y., Arifin, Z., and Ishak, M. (2016). Mechanical, thermal and morphological properties of injection molded poly(lactic acid)/calcium carbonate nanocomposites. *Period. Polytech. Mech. Eng.* 60 (1): 15–20.

48 Magalhães, N.F. and Andrade, C.T. (2010). Calcium bentonite as reinforcing nanofiller for thermoplastic starch. *J. Braz. Chem. Soc.* 21 (2): 202–208.

49 Cheng, H., Yang, Y., Li, S. et al. (2015). Modification and extrusion coating of polylactic acid films. *J. Appl. Polym. Sci.* 42472: 1–8.

50 Vinod, A., Sanjay, M.R., Suchart, S., and Jyotishkumar, P. (2020). Renewable and sustainable biobased materials: an assessment on biofibers, biofilms,

biopolymers and biocomposites. *J. Cleaner Prod.* 258: 120978. https://doi.org/10.1016/j.jclepro.2020.120978.

51 Ngai, K.S., Ramesh, S., Ramesh, K., and Juan, J.C. (2017). Electrical, dielectric and electrochemical characterization of novel poly(acrylic acid)-based polymer electrolytes complexed with lithium tetrafluoroborate. *Chem. Phys. Lett.* https://doi.org/10.1016/j.cplett.2017.11.042.

52 Kumar, S., Shukla, A., Pratim, P. et al. (2018). Biodegradable hybrid nanocomposites of chitosan/gelatin and silver nanoparticles for active food packaging applications. *Food Packag. Shelf Life* 16 (March): 178–184. https://doi.org/10.1016/j.fpsl.2018.03.008.

53 Sadasivuni, K.K., Saha, P., Adhikari, J. et al. (2019). Recent advances in mechanical properties of biopolymer composites: a review. *Polym. Compos.*: 1–28.

54 Palai, B., Biswal, M., Mohanty, S., and Nayak, S.K. (2019). In situ reactive compatibilization of polylactic acid (PLA) and thermoplastic starch (TPS) blends ; synthesis and evaluation of extrusion blown films thereof. *Ind. Crops Prod.* 141 (August): 111748. https://doi.org/10.1016/j.indcrop.2019.111748.

55 Ayana, B., Suin, S., and Khatua, B.B. (2014). Highly exfoliated eco-friendly thermoplastic starch (TPS)/poly(lactic acid)(PLA)/clay nanocomposites using unmodified nanoclay. *Carbohydr. Polym.* 110: 430–439. https://doi.org/10.1016/j.carbpol.2014.04.024.

56 Tan, I.K.P., Foong, C.P., Tan, H.T. et al. (2020). Polyhydroxyalkanoate (PHA) synthase genes and PHA-associated gene clusters in *Pseudomonas* spp. and *Janthinobacterium* spp. isolated from Antarctica. *J. Biotechnol.* https://doi.org/10.1016/j.jbiotec.2020.03.006.

57 Ross, G., Ross, S., and Tighe, B.J. (2017). Bioplastics: new routes, new products, Chapter 23. In: *Brydson's Plastics Materials*, 8e (ed. M. Gilbert), 631–652. Butterworth-Heinemann.

58 Gomes, M., Azevedo, H., Malafaya, P. et al. (2013). Natural Polymers in tissue engineering applications, Chapter 6. In: *Plastics Design Library, Handbook of Biopolymers and Biodegradable Plastics* (eds. C. van Blitterswijk, P. Thomsen, A. Lindahl, et al.), 385–425. William Andrew Publishing.

59 George, A., Sanjay, M.R., Sriusk, R. et al. (2020). A comprehensive review on chemical properties and applications of biopolymers and their composites. *Int. J. Biol. Macromol.* https://doi.org/10.1016/j.ijbiomac.2020.03.120.

60 Kurdikar, D., Gruys, K.J., Gerngross, T.U., and Slater, S.C. (2001). greenhouse gas profile of a plastic material derived from a genetically modified plant. *J. Ind. Ecol.* 4 (3): 107–122.

61 Mooney, B.P. (2009). The second green revolution? Production of plant-based biodegradable plastics. *Biochem. J.* 418: 219–232.

62 Edna, R. (2008). Environmental impact of compostable polymer materials, Chapter 10. In: *Compostable Polymer Materials*, 2e, 182–199. Elsevier.

63 Sabapathy, P.C., Devaraj, S., Meixner, K. et al. (2020). Recent developments in Polyhydroxyalkanoates (PHAs) production – a review. *Bioresour. Technol.* 306 (February): 123132. https://doi.org/10.1016/j.biortech.2020.123132.

64 Yates, M.R. and Barlow, C.Y. (2013). Life cycle assessments of biodegradable, commercial biopolymers - a critical review. *Resour. Conserv. Recycl.* 78: 54–66. https://doi.org/10.1016/j.resconrec.2013.06.010.

65 Rojas-Graü, M.A., Avena-Bustillos, R.J., Olsen, C. et al. (2007). Effects of plant essential oils and oil compounds on mechanical, barrier and antimicrobial properties of alginate-apple puree edible films. *J. Food Eng.* 81 (3): 634–641.

66 Kan, J., Liu, J., Yong, H. et al. (2019). Development of active packaging based on chitosan-gelatin blend films functionalized with Chinese hawthorn (*Crataegus pinnatifida*) fruit extract. *Int. J. Biol. Macromol.* 140: 384–392. https://doi.org/10.1016/j.ijbiomac.2019.08.155.

67 Paixão, L.C., Lopes, I.A., Kardec, A., and Barros, D. (2019). Alginate bio films plasticized with hydrophilic and hydrophobic plasticizers for application in food packaging. *J. Appl. Polym. Sci.* 48263: 1–11.

68 Özeren, H.D., Olsson, R.T., Nilsson, F., and Hedenqvist, M.S. (2020). Prediction of plasticization in a real biopolymer system (starch) using molecular dynamics simulations. *Mater. Des.* 187: 108387. https://doi.org/10.1016/j.matdes.2019.108387.

69 Li, X., Zhang, H., He, L. et al. (2018). Flexible nano fibers-reinforced silk fibroin films plasticized by glycerol. *Composites Part B* 152 (June): 305–310.

70 Basiak, E., Lenart, A., and Debeaufort, F. (2018). How glycerol and water contents affect the structural and functional properties of starch-based. *Polymers (Basel)* 10 (418): 1–18.

71 Heydari, A. and Vossoughi, M. (2013). Functional properties of biodegradable corn starch nanocomposites for food packaging applications. *Mater. Des.* 50: 954–961. https://doi.org/10.1016/j.matdes.2013.03.084.

72 Gao, W., Liu, P., Li, X. et al. (2019). The co-plasticization effects of glycerol and small molecular sugars on starch-based nanocomposite fi lms prepared by extrusion blowing. *Int. J. Biol. Macromol.* 133: 1175–1181. https://doi.org/10.1016/j.ijbiomac.2019.04.193.

73 Esmaeili, M., Pircheraghi, G., and Bagheri, R. (2020). Optimizing the mechanical and physical properties of thermoplastic starch via tuning the molecular microstructure through co-plasticization by sorbitol and glycerol. *Polym. Int.* 66: 809–819.

74 Sanyang, M.L. and Sapuan, S.M. (2015). Development of expert system for biobased polymer material selection: food packaging application. *J. Food Sci. Technol.* 52 (October): 6445–6454.

75 Ma, X.F., Yu, J.G., and Wan, J.J. (2006). Urea and ethanolamine as a mixed plasticizer for thermoplastic starch. *Carbohydr. Polym.* 64: 267–273.

76 Khan, B., Bilal, M., Niazi, K. et al. (2019). Effect of ultra-violet cross-linking on the properties of boric acid and glycerol co-plasticized thermoplastic starch films. *Food Packag. Shelf Life* 19: 184–192.

77 Gattin, R., Copinet, A., Bertrand, C., and Couturier, Y. (2002). Biodegradation study of a starch and poly(lactic acid) co-extruded material in liquid, composting and inert mineral media. *Int. Biodeterior. Biodegrad.* 50 (1): 25–31.

78 Dehnad, D., Emam-djomeh, Z., and Mirzaei, H. (2014). Optimization of physical and mechanical properties for chitosan – nanocellulose biocomposites. *Carbohydr. Polym.* 105: 222–228. https://doi.org/10.1016/j.carbpol.2014.01.094.

79 Müller, C.M.O., Laurindo, J.B., and Yamashita, F. (2011). Effect of nanoclay incorporation method on mechanical and water vapor barrier properties of starch-based films. *Ind. Crops Prod.* 33 (3): 605–610.

80 Yeddes, W., Djebali, K., Aidi, W. et al. (2020). Gelatin-chitosan-pectin fi lms incorporated with rosemary essential oil: Optimized formulation using mixture design and response surface methodology. *Int. J. Biol. Macromol.* 154: 92–103.

81 Hajji, S., Younes, I., Affes, S. et al. (2018). Optimization of the formulation of chitosan edible coatings supplemented with carotenoproteins and their use for extending strawberries postharvest life. *Food Hydrocolloids* 83: 375–392.

82 Portugal Zegarra, M.d.C.C., Santos, A.M.P., Silva, A.M.A.D., and Melo, E.d.A. (2018). Chitosan films incorporated with antioxidant extract of acerola agroindustrial residue applied in chicken thigh. *J. Food Process. Preserv.* 42: 1–12.

83 Famá, L.M., Pettarin, V., Goyanes, S.N., and Bernal, C.R. (2011). Starch/multi-walled carbon nanotubes composites with improved mechanical properties. *Carbohydr. Polym.* 83 (3): 1226–1231.

84 Shalumon, K.T., Anulekha, K.H., Nair, S.V. et al. (2011). Sodium alginate/poly(vinyl alcohol)/nano ZnO composite nanofibers for antibacterial wound dressings. *Int. J. Biol. Macromol.* 49 (3): 247–254. https://doi.org/10.1016/j.ijbiomac.2011.04.005.

85 Nayantara, K.P. (2018). Biosynthesis of nanoparticles using eco-friendly factories and their role in plant pathogenicity: a review. *Biotechnol. Res. Innovation* 2 (1): 63–73. https://doi.org/10.1016/j.biori.2018.09.003.

86 Losurdo, M. (2004). Relationships among surface processing at the nanometer scale, nanostructure and optical properties of thin oxide films. *Thin Solid Films* 456: 301–312.

87 Rojas-Graü, M.A., Soliva-Fortuny, R., and Martín-Belloso, O. (2010). Edible coatings: past, present and future. *Stewart Postharvest Rev.* 6 (3): 1–5.

88 Acevedo-Fani, A., Soliva-Fortuny, R., and Martín-Belloso, O. (2017). Nanoemulsions as edible coatings. *Curr. Opin. Food Sci.* 15: 43–49.

89 Acevedo-Fani, A., Salvia-Trujillo, L., Rojas-Graü, M.A., and Martín-Belloso, O. (2015). Edible films from essential-oil-loaded nanoemulsions: physicochemical characterization and antimicrobial properties. *Food Hydrocolloids* 47: 168–177. https://doi.org/10.1016/j.foodhyd.2015.01.032.

90 Carbone, M., Tommasa, D., and Sabbatella, G. (2016). Silver nanoparticles in polymeric matrices for fresh food packaging. *J. King Saud Univ. Eng. Sci.* 28 (4): 273–279. https://doi.org/10.1016/j.jksus.2016.05.004.

91 Akbar, A. and Kumar, A. (2014). Zinc oxide nanoparticles loaded active packaging, a challenge study against *Salmonella typhimurium* and *Staphylococcus aureus* in ready- to-eat poultry meat. *Food Control* 38: 88–95. https://doi.org/10.1016/j.foodcont.2013.09.065.

92 Giannakas, A., Patsaoura, A., Barkoula, N.M., and Ladavos, A. (2017). A novel solution blending method for using olive oil and corn oil as plasticizers in chitosan based organoclay nanocomposites. *Carbohydr. Polym.* 157: 550–557. https://doi.org/10.1016/j.carbpol.2016.10.020.

93 Guerreiro, A.C., Gago, C.M.L., Faleiro, M.L. et al. (2015). The effect of alginate-based edible coatings enriched with essential oils constituents on *Arbutus unedo* L. fresh fruit storage. *Postharvest Biol. Technol.* 100: 226–233. https://doi.org/10.1016/j.postharvbio.2014.09.002.

94 Seydim, A.C. and Sarikus, G. (2006). Antimicrobial activity of whey protein based edible films incorporated with oregano, rosemary and garlic essential oils. *Food Res. Int.* 39 (5): 639–644.

95 Abdollahi, M. and Rezaei, M.F.G. (2012). A novel active bionanocomposite film incorporating rosemary essential oil and nanoclay into chitosan. *J. Food Eng.* 111: 343–350.

96 Poveda, J.M., Loarce, L., Alarcón, M. et al. (2018). Revalorization of winery by-products as source of natural preservatives obtained by means of green extraction techniques. *Ind. Crops Prod.* 112 (December 2017): 617–625. https://doi.org/10.1016/j.indcrop.2017.12.063.

97 Musso, Y.S., Salgado, P.R., and Mauri, A.N. (2019). Smart gelatin films prepared using red cabbage (*Brassica oleracea* L.) extracts as solvent. *Food Hydrocolloids* 89 (October 2018): 674–681. https://doi.org/10.1016/j.foodhyd.2018.11.036.

98 Osdaghi, E., Mohsen, S., Fazliarab, A. et al. (2015). Characterization, geographic distribution and host range of *Curtobacterium flaccumfaciens*: an emerging bacterial pathogen in Iran. *Crop Prot.* 78: 185–192. https://doi.org/10.1016/j.cropro.2015.09.015.

99 Liu, B., Xu, H., Zhao, H. et al. (2017). Preparation and characterization of intelligent starch/PVA films for simultaneous colorimetric indication and antimicrobial activity for food packaging applications. *Carbohydr. Polym.* 157: 842–849. https://doi.org/10.1016/j.carbpol.2016.10.067.

100 Adhikari, B., Chaudhary, D.S., and Clerfeuille, E. (2010). Effect of plasticizers on the moisture migration behavior of low-amylose starch films during drying. *Drying Technol.* 28: 468–480.

101 Gomes, V., Souza, L., and Fernando, A.L. (2016). Nanoparticles in food packaging: biodegradability and potential migration to food — a review. *Food Packag. Shelf Life* 8: 63–70. https://doi.org/10.1016/j.fpsl.2016.04.001.

102 García, V., Sendón, R., Bustos, J. et al. (2019). Estimates of dietary exposure of Spanish population to packaging contaminants from cereal based foods contained in plastic materials. *Food Chem. Toxicol.* 128 (April): 180–192. https://doi.org/10.1016/j.fct.2019.04.003.

103 Silvestre, C., Duraccio, D., and Cimmino, S. (2011). Food packaging based on polymer nanomaterials. *Prog. Polym. Sci.* 36 (12): 1766–1782. https://doi.org/10.1016/j.progpolymsci.2011.02.003.

104 Dimitrijevic, M., Karabasil, N., Boskovic, M., and Teodorovic, V. (2015). Safety aspects of nanotechnology applications in food packaging. *Procedia Food Sci.* 5: 57–60. https://doi.org/10.1016/j.profoo.2015.09.015.

105 Metak, A.M., Nabhani, F., and Connolly, S.N. (2015). Migration of engineered nanoparticles from packaging into food products. *LWT Food Sci. Technol.* 64 (2): S156–S162. https://doi.org/10.1016/j.lwt.2015.06.001.

106 Jumaidin, R., Sapuan, S.M., Jawaid, M. et al. (2017). Effect of seaweed on mechanical, thermal, and biodegradation properties of thermoplastic sugar palm starch/agar composites. *Int. J. Biol. Macromol.* 99: 265–273. https://doi.org/10.1016/j.ijbiomac.2017.02.092.

107 Martínez, M., León de Pinto, G., Bozo de González, M. et al. (2008). New structural features of *Spondias purpurea* gum exudate. *Food Hydrocolloids* 22 (7): 1310–1314.

108 Amjadi, S., Emaminia, S., Heyat, S., and Pourmohammad, S. (2019). Preparation and characterization of gelatin-based nanocomposite containing chitosan nano fiber and ZnO nanoparticles. *Carbohydr. Polym.* 216 (December 2018): 376–384. https://doi.org/10.1016/j.carbpol.2019.03.062.

109 Chen, S., Wu, M., Lu, P. et al. (2020). Development of pH indicator and antimicrobial cellulose nanofibre packaging film based on purple sweet potato anthocyanin and oregano essential oil. *Int. J. Biol. Macromol.* 149: 271–280. https://doi.org/10.1016/j.ijbiomac.2020.01.231.

110 Prapruddivongs, C., Apichartsitporn, M., and Wongpreedee, T. (2018). Effect of silica resources on the biodegradation behavior of poly(lactic acid) and chemical crosslinked poly(lactic acid) composites. *Polym. Test.* 71: 87–94. https://doi.org/10.1016/j.polymertesting.2018.08.026.

111 Pinto, A.M., Gonçalves, C., Gonçalves, I.C., and Magalhães, F.D. (2016). Effect of biodegradation on thermo-mechanical properties and biocompatibility of poly(lactic acid)/graphene nanoplatelets composites. *Eur. Polym. J.* 85: 431–444. https://doi.org/10.1016/j.eurpolymj.2016.10.046.

112 Jumaidin, R., Sapuan, S.M., Jawaid, M. et al. (2017). Thermal, mechanical, and physical properties of seaweed/sugar palm fibre reinforced thermoplastic sugar palm starch/agar hybrid composites. *Int. J. Biol. Macromol.* 97: 606–615. https://doi.org/10.1016/j.ijbiomac.2017.01.079.

113 Bootklad, M. and Kaewtatip, K. (2013). Biodegradation of thermoplastic starch/eggshell powder composites. *Carbohydr. Polym.* 97 (2): 315–320. https://doi.org/10.1016/j.carbpol.2013.05.030.

114 Gilfillan, W.N., Nguyen, D.M.T., Sopade, P.A., and Doherty, W.O.S. (2012). Preparation and characterisation of composites from starch and sugar cane fibre. *Ind. Crops Prod.* 40: 45–54. https://doi.org/10.1016/j.indcrop.2012.02.036.

115 Kaewtatip, K. and Tanrattanakul, V. (2012). Structure and properties of pregelatinized cassava starch/kaolin composites. *J. Mater.* 37: 423–428. https://doi.org/10.1016/j.matdes.2011.12.039.

116 Mittal, V., Akhtar, T., and Luckachan, G. (2015). PLA, TPS and PCL binary and ternary blends: structural characterization and time-dependent morphological changes. *Colloids Polym. Sci.*: 573–585.

14

Nanobiotechnology in Food Preservation and Molecular Perspective

S. Agriopoulou, E. Stamatelopoulou, V. Skiada, and T. Varzakas

University of the Peloponnese, Department of Food Science and Technology, Kalamata, Messinia 24100, Greece

14.1 Introduction

Nanotechnology or "nanoscience" uses materials and structures, in the nanoscale range, usually 100 nm or less in at least one dimension, with a nanometer being 10^{-9} m [1, 2]. The term "nanotechnology" was used for the first time in Tokyo in 1974 [3]. The term "nano" has its roots in the Greek word for "dwarf" [4]. Any material that has at least one dimension in nanoscale (1–100 nm) is called nanomaterial (NM) [5]. Nano-sized materials change their optical, magnetic, electrical, and other properties, and for this reason they are widely used in many scientific fields [3]. Moreover, particles at the scale of nanometer cause changes in the surface area and solubility [4]. Many nanomaterials are naturally produced, and some are artificially synthesized [3]. The competence of nanotechnology is proven in many technologies, like electronics, energy, and catalysts [4, 6] with agri-food production being the major application area [5]. As food quality assurance is an important and top priority in the food sciences sector, there is a continuing effort by the scientific community to find innovative and up-to-date technologies in this direction [6]. Food properties as well as food preservation can be significantly improved with the help of nanotechnology [6]. The production of "nano-foods" occurs with the application of nanotechnology and takes place in any productive phase starting from the field and ending with the final product [5].

Biotechnology uses principles of biology and through various techniques develops useful products at the genetic, molecular, and cellular level in various fields including the agri-food sector [7]. Nanotechnology and biotechnology have combined to provide a new and dynamic branch of "nanobiotechnology," which is applied through the study of biological sciences [7]. Current researches have documented the production of nanoparticles (NPs) through metabolic processes from microorganisms, fungi, and plant extracts [8–10]. The development of appropriate nano-preservatives could help preserve food both through appropriate food packaging and through postharvest management techniques [11].

Nanotechnology-Enhanced Food Packaging, First Edition. Edited by Jyotishkumar Parameswaranpillai, Radhakrishnan Edayileveettil Krishnankutty, Aswathy Jayakumar, Sanjay Mavinkere Rangappa, and Suchart Siengchin.

Preservation and protection from unfavorable conditions, throughout the supply chain, are the dominant goals of food packaging [12]. The packaging materials are intended to reduce or delay the deterioration of food quality, preventing the transportation of gases, such as oxygen from the external environment to foods, for a specific period of time until it is used by the consumer [13, 14]. Research on food packaging materials aims to produce materials that will primarily protect the food as well as extend the shelf life by keeping away pathogenic microorganisms [15]. The common packaging materials are paper, plastic, glass, and metals. Conventional plastic like polyethylene terephthalate (PET), poly(vinyl chloride) (PVC), polyethylene (PE), polypropylene (PP), polystyrene (PS), and polyamide (PA) are the common plastic food packaging materials, which are derived from petroleum [16]. Although all these materials have strong mechanical and physical characteristics, they are nonbiodegradable materials with many waste management issues [17].

Food packaging systems have evolved significantly in recent years due to advances in nanotechnology. Research areas related to barrier properties and to antimicrobial properties have made significant progress. Water, gas, and light are usually the main factors to consider in the barrier properties of a food container. Moreover, the development of sensors capable of detecting gases and microorganisms has been important, and properties related to biodegradability and biocompatibility have been improved [12]. A package can prevent the exchange of gases with the environment, thus extending the life of the products, something that happens in carbonated drinks packaging [18]. Concerning the acceptance of nanotechnology by the public in the food sector, from studies related to consumer behavior, there is a greater acceptance of nanotechnology in the external environment such as food packaging than inside the food [16].

This chapter summarizes recent advances in the development of effective nano-preservatives, multifunctional devices, and systems that could be used both to maintain food quality and to extend shelf life through food packaging and postharvest management.

14.2 Nanobiotechnology Aspects in Food Preservation and Food Packaging

In order to ensure the safety and quality of food products and to avoid contamination from pathogens and spoilage microorganisms, various food preservations technologies and food packaging systems are applied. Nanobiotechnology has developed a wide range of proper nanomaterials which have been shown to enhance the effectiveness of food preservation and food packaging by prolonging the shelf life of food products. Advances in nanotechnology in the field of food preservation have been achieved through the development of antimicrobial agents, of nanoencapsulation, of enhancement in physical properties, and of protection against chemical deteriorates [19].

Food packaging is mainly intended to protect food products as they are likely to be degraded by a variety of physical, chemical, or biological agents, losing part or all of their primary quality [13]. In addition to detection and avoidance of foodborne pathogens, the incorporation of food packaging and nanobiotechnology aims to act against heavy metal reduction, allergens, pesticides, additives, and drugs and to inhibit the biofilm formation [19]. Smart, intelligent, active, and antimicrobial packaging are the main categories of food packaging, which nanobiotechnology has significantly contributed with promising nanomaterials and nanoscale substances such as bacteriocins [20, 21].

14.3 Classification of Nanomaterials and Molecular Basis of Application

According to the recommendation of European Commission, "Nanomaterial means a natural, incidental or manufactured material containing particles, in an unbound state or as an aggregate or as an agglomerate and where, for 50% or more of the particles in the number size distribution, one or more external dimensions is in the size range 1 nm–100 nm" [22]. A conventional light microscope as well as the human naked eye cannot see the nanomaterials [4]. Although these materials have the same chemical composition with the conventional ones, they have different physical and chemical properties as they have very small size [23]. In fact, these nanoscale materials exhibit a high surface-area-to-volume ratio and surface activity [15]. Nanomaterials (NMs), are classified based on the number of dimensions. So there are zero-dimensional (0D), one-dimensional (1D), two-dimensional (2D), and three-dimensional (3D) nanomaterials [24]. Nanoparticles (NPs), nanocomposites (NCs), nanoclays (NCs), nanoemulsions (NEs), nanosensors (NSs), nanostructures (NSTs) (including nanoliposomes, nanospheres, nanocapsules, nanofibers, and nanotubes), and nanobarcodes are some of the important nanomaterials that have been used in food preservation (Figure 14.1) [6, 25].

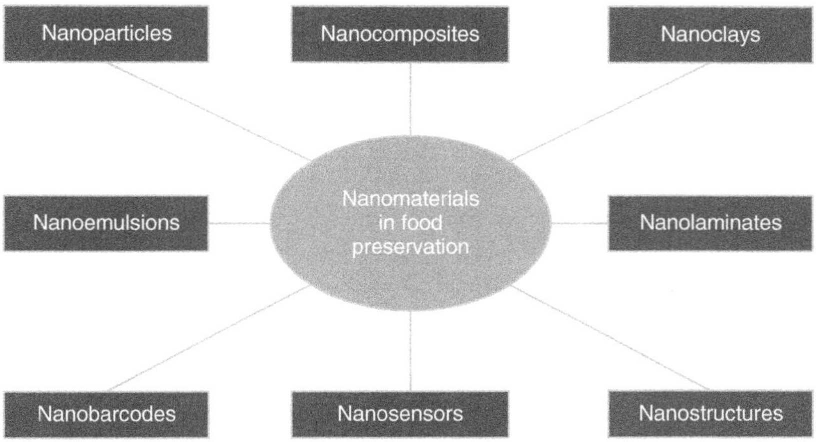

Figure 14.1 Nanomaterials in food preservation.

14.3.1 Nanoparticles

Nanoparticle is defined as tinny material having all three dimensions less than 100 nm [5]. NPs can be separated into metallic NPs, polymeric NPs, magnetic NPs, lipid-based NPs, carbon-based NPs, semiconductor NPs, and ceramic NPs, based on physical and chemical features [26]. The hydrophobic or hydrophilic character of the nanoparticles is the determining factor for their applications. Inorganic nanoparticles have been used in both food packaging and food processing. Recently, studies on NPs of metallic and metallic oxides have increased because of their optical, electronic, catalytic, and antibacterial properties [27]. Metal oxide nanoparticles (MONPs), due to their high stability and durability and their low toxicity, have been extensively studied in science and technology. Among them, zinc oxide (ZnO), copper oxide (CuO), and silver (AgNPs) have been used for their antimicrobial properties, and NPs such as Ag, Au, iron oxide (Fe_3O_4), ZnO, titanium dioxide (TiO_2), magnesium oxide (MgO), and CuO act antimicrobially by inhibiting biofilm formation. Also, they have UV exclusion properties in food packaging and biomedical applications ([27–29]. Metallic nanoparticles, like gold (AuNPs), AgNPs, and platinum (PtNPs), are easily synthesized, are stable, and have been used for food preservation and food packaging applications [30, 31]. Carbon-based nanoparticles are of great interest due to their excellent physicochemical properties and relatively higher biosecurity [32]. The fullerenes (0D), the carbon nanotubes (1D), the graphene (2D), and the graphene coil (multidimensional) belong to the category of carbon-based nanoparticles [33]. Nanoparticles have also been used to bind and remove mycotoxins in food [34].

14.3.1.1 Silver-Based Nanoparticles

AgNPs have been shown to be effective against many microbes, although it is not completely understood, which mechanism provides the main antibacterial effect. AgNPs release silver ions (Ag +) and form free radicals, capable of destroying the membrane. Bacterial cell adhesion and penetration, as well as DNA degradation and overgrowth of reactive oxygen species (ROS), leading to early apoptosis, are some possible mechanisms associated with the bactericidal effects of AgNPs [35]. AgNPs have been used as bacteriostatic agents since antiquity because of their antimicrobial activity. In nanoscale, silver-based particles include AgNPs, silver nanocluster (AgNC), and silver-based alloy materials [28]. AgNPs have been used in many fields, medical, healthcare, and food industry, as they have exhibited exceptional physical and chemical properties. Their antimicrobial activity against microorganisms makes them particularly important in various applications [36].

The fruit extract of *Vitis vinifera* (black grapes) has the ability to form silver nanoparticles that are capable of producing antibacterial substances. The shelf life of lemons and strawberries was extended when nanofiber Ag/PVA was applied. It is an effective method of extending the storage duration of fruits and a new consideration to their implementation in food preservation [37].

AgNPs were recently prepared and characterized from Khalil et al. [38]; AgNPs showed remarkable antifungal activity and inhibited mycotoxin production. In particular, the results showed a decrease in aflatoxin and ochratoxin from *Aspergillus flavus* and *Aspergillus ochraceus*, respectively, at noncytotoxic doses. However, their use in food packaging should be further studied [38].

14.3.1.2 Titanium Dioxide (TiO_2) Nanoparticles

Titanium dioxide (TiO_2) nanoparticles also have bactericidal properties, with the main mechanisms of toxicity of these nanoparticles being ROS generation. They are also capable of causing cell death through lipid peroxidation and oxidative stress [39]. White color in dairy products such as milk and cheese can be improved by using titanium dioxide (TiO_2) [40]. In addition, significant reduction in the growth of *Escherchia coli*, *Listeria monocytogenes*, and *Staphylococcus aureus* in lamb meat was observed when it was coated with biodegradable nanocomposite TiO_2 nanoparticles and rosemary essential oil (EO), extending its shelf life by 12–15 days [41].

14.3.1.3 Zinc Oxide (ZnO) Nanoparticles

Zinc oxide nanoparticles (ZnONPs) belong to multifunctional inorganic nanoparticles with proven antimicrobial activity. Zinc oxide has been characterized as Generally Recognized as Safe (GRAS) by the US Food and Drug Administration (FDA), and the incorporation with nanostructures can improve their properties [42]. ZnONPs were used as nanofillers to make packaging materials.

Edible coatings combined with nanoparticles have been used to delay the tanning response of fresh produce. In particular, nanosized ZnO active packaging coating increased the shelf life of fresh Fuji apple slices [42]. Materials made from phosphorescent nanosized ZnO can detect lactic and acetic acids, alcohols, and amines, substances produced from spoiled foods [40]. Pediocin (a bacteriocin) associated to ZnO nanoparticles has been combined to develop nanocomposite films with antimicrobial effect against *S. aureus* and *L. monocytogenes* [43].

14.3.2 Nanocomposites

A nanocomposite is a multiphase solid hybrid material, with at least one phase and one dimension smaller than 100 nm, and it is usually enhanced by up to 5% by weight of nanoparticles. Nanocomposites are applicable to a variety of materials suitable for food packaging [5].

14.3.2.1 Cellulose-Based Nanocomposites

Cellulose, compared with all polysaccharides (carbohydrates) in nature, is the most abundant [44]. It can be obtained both from herbal and bacterial sources [45]; it is renewable and biodegradable [46], but it cannot be used as a coating because it is insoluble in water [44]. Carboxymethylcellulose, methylcellulose, hydroxypropyl cellulose, hydroxypropyl methylcellulose, and cellulose acetate are water-soluble and are some of the most commonly known cellulose derivatives for applications in food packaging materials such as cellulose-based films [16, 46]. Cellulose is combined with other materials as it is a strengthening agent [17]. Nanocellulose, a bio-based nanomaterial, is defined as cellulose having at least one dimension in the nanoscale [12]. Nanofibrillated cellulose and cellulose nanocrystals are the dominant cellulose-based nanofillers, which are used with the purpose to improve mechanical, thermal, and barrier properties of food packaging films [12].

14.3.2.2 Chitosan-Based Nanocomposites

Chitosan (CH) is derived from chitin through alkaline deacetylation or enzymatic hydrolysis and belongs to the category of polysaccharides. Crustacean, fish, and seafood are the main sources of chitosan and chitin, which are produced as by-products [15]. Reduction and avoidance of chemicals in the food industry also occurs with the use of chitosan. Chitosan acts as a secondary antioxidant because it can prevent the initiation of lipid oxidation by binding metals. In addition, the oxidation rate of foods is reduced due to the low oxygen permeability of chitosan. When chitosan is combined with other natural substances such as glucose, its antioxidant effect increases, but its antimicrobial effect against *S. aureus*, *Bacillus subtilis*, *E. coli*, and *Pseudomonas* remains the same. In addition, when chitosan is applied to the surface of fruits as well as eggs, the rate of respiration and transpiration is reduced by forming a protective barrier. However, when combined with xylan, its antimicrobial and antioxidant properties are enhanced, making it a promising tool for extending the storage period of many foodstuffs. Food quality characteristics can be improved with the use of chitosan, while at the same time a reduction of the microbiological load can be achieved [47].

The ability of chitosan to make films in combination with its antimicrobial properties makes it an interesting biopolymer that is applicable to food packaging and food preservation by spraying, dipping, coating, or wrapping. The combination of CH with nanometals like zinc oxide (ZnO), magnesium oxide (MgO), TiO_2, copper oxide (CuO), iron oxide (Fe_3O_4), and silicon dioxide (SiO_2) has been found in various applications in food packaging, improving both quality shelf life parameters and many packaging material properties [15].

Kumar et al. [48] reviewed the recent advances in chitosan and chitosan-based nanocomposites, which have been found in various food packaging applications to prolong the shelf life of fruits, vegetables, meat, and fish products. The use of chitosan against various microorganisms such as Gram-positive and Gram-negative bacteria, filamentous fungi, and yeast is well documented [49–51].

The ability of chitin, chitosan, and their derivatives to scavenge free radicals and the chelation of metal ions allow biopolymers to exert their antioxidant activity [52–54]. Additionally significant improvements have been reported in antioxidant activities, as the degree of deacetylation of chitin increases [55].

Biopolymers such as polysaccharides, proteins, and lipids are mixed with chitosan to produce films and coatings with improved functional properties for food packaging applications [56]. Polysaccharide mixtures are advantageous compared with mixtures with proteins and lipids as they are cheaper, there are many sources for the films, and they are characterized for stability and good heat sealability and water solubility [57].

14.3.2.3 Protein-Derived Bionanocomposites

The mechanical, thermal, and barrier properties of biopolymers can be significantly improved with the contraction of bionanocomposites. Bionanocomposites consist of two or more phases, a continuous and a discontinuous phase. Proteins, lipids, polysaccharides, and nucleic acids compose the continuous phase, whereas the discontinuous phase comprises of silicates, carbon nanotubes, Ag, and TiO_2. These materials are biocompatible and biodegradable with excellent physical, thermal, barrier, and optical properties.

The formation of the bionanocomposite creates the transfer of tension between the biopolymer matrix and the nanofiller [58–61].

The nature of nanoparticles contributes significantly to the improvement of the properties of materials. The high surface of the nanofiller is responsible for differences observed in the composition of the material, the crystallinity, and the mobility [62].

The dispersion of nanofillers within the polymer matrix has been improved by both chemical and physical approaches, with surface modification of nanofiller to be one of these approaches.

Surface modification of the nanofiller can improve the compatibility between NP and polymer matrix. Long chain alkylammonium ions or organosilanes are used for surface grafting of the clay [63, 64].

There are organic and inorganic nanofillers. Nanocellulose (nanocrystals or fibers), zein nanocrystals, chitosan, and starch nanocrystals are organic nanofillers, whereas layered nanoclays, ZnO [65], TiO_2, silver ions (Ag^+) [60, 66], and carbon nanotubes are examples of inorganic nanofillers.

Nanoclays are one of the most popular nanofillers, which improve the properties of protein-based materials [67, 68].

Dry and wet processes can prepare protein-derived bionanocomposites and can be divided into three phases. During the first phase the low-energy interactions between polypeptide chains are broken down. The polypeptide chain is then structurally arranged and oriented. Finally, during the last phase, all solvents and disruptive agents are removed, and a new 3D polymer network is formed [58]; the stabilization of which is achieved with new bonds and interactions [69].

In order to obtain protein-based bionanocomposites films for food packaging applications, milk, soy, corn gelatin, and wheat gluten proteins are widely employed, although fish myofibrillar protein-based bionanocomposites films could also be prepared [70].

Zubair and Ullah [71] reported on recent advances in proteins-derived bionanocomposites used in food packaging applications. Protein-based biomaterials are sustainable, renewable, biodegradable, and have a low carbon footprint. Protein-based materials cannot be used in food packaging applications since they have low mechanical strength, low thermal, barrier, and inferior physicochemical properties and can be replaced by bionanocomposites. Soy protein isolates (SPI) and gelatin proteins are the most reported products used in food packaging. Silicate nanofillers have been widely used as nanofiller, with montmorillonites (MMT) to be one of the best [71].

14.3.2.4 Polylactic Acid Nanobiocomposites

Biopolymer-based composites such as polylactic acid (PLA) are at the present time paying increased interest to the food packaging applications as there are many consumers and manufacturers, which are increasingly interested in environmentally friendly products. PLA applies to food packaging because it is involved with features related to renewable sources, bioavailability, biocompatibility, and compostability.

Biopolymer-based composites films of PLA with 5% ZnO when compared with plain PLA showed lower permeability to O_2 and CO_2 and antimicrobial activity in *E. coli* [68, 72].

14.3.3 Nanoclays

Nanoclays, also known as phyllosilicates, are found in nature as aluminum silicates and have a leaflike structure [6, 12]. They act as barrier against oxygen and carbon dioxide [73]. Resistance to moisture and gases increases when biopolymers are combined with nanoclays, making these materials ideal for packaging applications [18]. Montmorillonite, saponite, and ectorite are the most common silicates to be used as reinforcements with polymers. Microstructure of polycaprolactone/clay nanocomposite film was used in food preservation with good mechanical properties [74]. The barrier properties of nanoclay nanocomposite are improved as there is a problematic and long distance that permeants have to travel through nanosynthesis [11].

14.3.4 Nanoemulsions

Nanoemulsions have been used to carry antimicrobial, antioxidant, anti-brown, aromatic, and coloring agents in food products, causing positive effects on food quality and affecting consumer acceptance related to nutritional and health benefits [75]. In nanoemulsions two immiscible phases are dispersed into each other, and stabilization is achieved using an amphiphilic surfactant [76]. The size of nanoemulsions' droplet is very small ($r < 100$ nm) [77] and in order to form energy is required. Low-energy and high-energy methods are the main methods for generation of nanoemulsions [78]. Gentle or no stirring is required in the first method in order to produce small droplets, while microfluidizer, or high-pressure homogenizer, is required in the second method [78, 79]. Phase inversion and self-emulsification are also included in low-energy methods. In a very recent work, Jiang et al. [80] presented the latest developments for encapsulation of curcumin in nanoemulsions, with the preparation technologies including the phase reversal temperature, phase reversal synthesis, ultrasound, high-pressure homogenization, and microfluidization [80]. Nanoemulsions combine with natural antimicrobial agents such as essential oils (EOs) and cause the death of microorganisms by primarily destroying their membranes [41]. Clove bud and oregano essential oils nanoemulsion caused reduction in yeasts and molds during 15 days storage in sliced bread, when it was applied as edible film [81]. Application of chitosan nanoemulsion containing *Cuminum cyminum* (0.05%w/v) essential oil as a delivery system caused reduced microbial growth during postharvest *Agaricus bisporus* maintenance, causing their shelf life extension [82].

14.3.5 Nanosensors

Nanosensors have either their size, or its sensitivity, or the sensor distance from the object to the nanoscale, and are capable of detecting contaminants, mycotoxins, gases, pesticides, and microorganisms in foods [1, 28, 83]. The use of nanotechnology in the food industry with the use of nanosensors is a new concept, while its application in pharmaceuticals is well-known [40]. A possible use of nanotechnology with the development of nanosensors in combination with polymers can be a useful tool for detecting infectious and pathogenic chemicals in the food system during storage.

Nanosensors incorporated in food packaging materials can monitor any physical, chemical or biological modifications at any stage of food processing. In addition, smart packaging with specialized nanosensors and nanotechnologies are capable of detecting toxins, pathogenic bacteria, and chemicals [84, 85]. Nanosensors provide quick and efficient detection of microorganisms, toxins, and parasites compared with existing traditional sensors. Degradable vitamins and antioxidants can also be detected with the help of nanoparticles. Using time–temperature and oxygen indices, food freshness can be monitored. Nanosensors have unique properties that improve food safety [85]. The design of appropriate biosensors has helped to detect mycotoxins in food. Aflatoxins produced by *A. flavus* and *Aspergillus parasiticus* contaminating food have been detected using a magnetic nano-gold immunosensor. In addition, in smart packaging, various nanobiosensors are used to the detection of microorganisms, e.g. *L. monocytogenes*, *E. coli*, and *Salmonella* spp. [84]. A nanobiosensor was employed to carry out selective determination of lactate in milk and human serum samples [86].

Moreover, a biosensor has recently been developed, which alters the resonance frequency and, in concert with the increasing mass in a probe array, detects the growth of *E. coli*. For the detection of multiple toxins and microorganisms, probes with different molecular identification elements can be used with an array mounted on a single chip [85].

Nanosensors can help detect modifications in the color of food products as well as the gases produced by deterioration. Gold, platinum, and palladium are the main primary materials for gas sensors. Gold-based nanoparticles can detect the aflatoxin B1 in milk samples. In addition, certain nanosensors can detect pesticides on the surface of vegetables and fruits. Sensitivity of the sensors can be improved by packaging consisting of single-walled DNA and carbon nanotubes [87]. Table 14.1 presents some examples of applications of nanosensors in food preservation.

Nanosensors can be classified into the following:

(a) **Nanosensors that detect atmospheric effects (external conditions)**
Adding nanosensors to the food packaging allows to monitor all atmospheric effects on products from the factory to their sale. Vendors will be able to read the data to control the temperature and humidity of fresh fruit and to decide how to store them, even whether or not they will receive the goods. These sensors require further investigation because, although the size of the labels used to transmit data is greater than 1000 nm, these labels are small and invisible.

(b) **Nanosensors for the detection of microorganisms and chemicals in the packaging (internal state)**
Chemical sensors and biosensors operate similarly by reacting to specific chemicals or bacteria causing changes. These changes are detected by the nanosensors

Table 14.1 Applications of nanosensors in food preservation.

Food	Targeting element	Nanomaterial used for the construction of nanosensors	Nanosensors developed for analysis	Detection method	Limit of detection	References
Chicken white meat	Histamine	AuNPs	Gold nanoparticles (GNPs) with particle size 11–19 nm function	Colorimetry (LSPR)	0.6 μM	[88]
Tuna fish	Histamine	Magnetic nanoparticles	Chitosan molecularly imprinted polymers (CHI/MIPs) coated on Fe_3O_4	Spectrophotometry	0.013 μM (1.5 ng/ml)	[89]
Red wine	Histamine	Graphene nanoribbons and AgNPs	Neurotransmitter (voltammetric sensor)	Square wave voltammetry	0,049 μμ	[90]
Tap water and seawater samples	Nitrite	Hyperbranched polyethyleneimine scaffolds–AgNPs	hPEI-Ag NC-based nanosensor	Fluorescence quenching	100 nM	[91]
Milk, apple juice	*Escherichia coli Salmonella, Listeria, E. coli O157*	Subsequent aggregation of plasmonic gold nanoparticles	Signal amplification via cysteine-loaded liposomes	Liposome-amplified plasmonic liposomes immunoassay	6.7 attomolar	[92]
Dairy products	*Brucella* spp.	Fluorescent–silica nanoparticle	Fluorescent, micellar silica nanosensors	Fluorescence	50 μl	[93]

[83]. Chemical or biological sensors can be effectively integrated into food packaging for fast and precise "in situ" analysis, although these devices must be applied for mass production, in a reliable, easy-to-use, environmentally friendly, and safe food contact [83].

Many biosensors have been designed but few are traded. The Toxin Guard, for example, is based on antibody–antigen reactions and is used in polymer packaging film indicating the occurrence of pathogenic bacteria (*Salmonella*, *Campylobacter*, *E. coli O157*, and *Listeria*), parasites (*Cyclospora*), insecticides, and marker modifiers. Particular modifications in the shape or color of the polymeric membranes indicate the presence of pathogens [83].

Another commercial device is Food Sentinel Systems, a combination of immunochemical method with a uniquely designed Universal Product Code (UPC) system. A barcode is applied to a membrane containing a specific pathogenic antibody, which when it identifies contaminants such as *Salmonella* spp., *E. coli 0157: H7*, or *L. monocytogenes*, a dark-line membrane is formed [83].

Histidine is a biogenic amine formed in foods such as fish with bacterial decarboxylation of the amino acid histidine[89, 94]. As histidine causes food poisoning, its quantification is necessary. The US FDA has set 50 ppm (50 mg/kg) of histamine as a chemical indicator for fish deterioration. Detection of histidine in food products based on nanomaterials, with the detection based on optical (fluorescence, surface-enhanced Raman spectroscopy [SERS], localized surface plasmon resonance) and electrochemical methods (impedimetric, voltammetry, potentiometric) has been used as it is very sensitive and easy to operate [94].

14.3.6 Nanostructures

NSTs are suitable carrier vehicles for the delivery of antimicrobial substances used to control spoilage and pathogenic microorganisms during food production and storage [25]. Nanoliposomes, nanospheres, nanocapsules, nanotubes, and nanofibers are some common NSTs.

The main characteristic of nanoliposomes is the occurrence of phospholipid bilayer with an aqueous core [25]. In addition to their delivery of vitamins, nutraceuticals, and nutrients, nanoliposomes can also deliver antibiotic compounds into food materials, thus extending their shelf life [95]. The lipid composition of the bacterial membrane seems to be related to the degree of fluid in the fusion of liposomes bacteria. In particular, the high content of phosphatidylethanolamine in Gram-negative bacteria improves fusion with fluid liposomes [96].

Nanospheres have the three dimensions in the nanoscale, while nanotubes have the two dimensions in the nanoscale [16]. Nanospheres are solid colloidal particles with the active ingredients embedded in a polymeric matrix [97]. Carbon nanotubes belong to carbon-based NMs, are very advantageous materials, and they can be combined with the food packaging matrix in order to ameliorate their mechanical strength and elasticity, flexibility, and gas barrier or to adapt their electrical conductivity [4, 98, 99].

Nanocapsules control the release of the core with the desired component [100] and entrap the odor and unwanted components, contributing in the preservation of the food [6]. Nanoencapsulation retains their functional properties. During nanoencapsulation, combinations of bioactive solids, liquids, or gases are incorporated, absorbed, or scattered into small vesicles with a nanometer scale diameter [30]. Bioactive molecules and nanoparticles can also be used for prevention of food oxidation and degradation. Selenium and cellulose NPs can be combined with food packaging in order to delay or inhibit ROS responsible for deterioration of food quality. Nanoencapsulation of phenols can also avoid food degradation, especially of fatty foods [18]. Encapsulated functional ingredients in the form of a droplet can also slow down the chemical degradation, something that happened in the case of nanoemulsion encapsulating curcumin, where the overall antioxidant activity of curcumin nanoemulsion decreased [101].

Nanofibers are defined as fibers having diameters in a nanometer range [102] and are used as a thickener [73]. Essential oils can be incorporated into nanofibers to prolong the shelf life of fresh products [18]. Electrospinning technique in combination with phase separation, synthesis, and drawing is the main option for the production of nanosized fibers. This technique is constantly gaining interest as it is characterized by high efficiency and simplicity [103]. Electrospinning nanofibers characterized by unique physical, chemical, and mechanical properties including porosity, high surface-to-volume ratio, and encapsulation of active ingredients [104]. Electrospinning nanofibers have also been used for the reduction of risk contamination of food-borne pathogenic bacteria such as *E. coli* [105].

Polysaccharide from *Dendrobium officinale* (DOP) is a useful natural substance for removing free radicals, preventing fatty liver, and lowering blood sugar [106, 107] also having a wide antibacterial ability against a number of bacteria [108, 109]. Zhu et al. [110] studied the fabrication and characterizations of polyethylene oxide (PEO)/*Dendrobium officinale* polysaccharide (DOP) antibacterial nanofibers by electrospinning technique. The addition of adipic acid in PEO/DOP nanofibers demonstrated better antibacterial activity against *E. coli* on pork samples, without effect on pork quality.

14.4 Nanomaterials and Active and Intelligent Food Packaging Applications

14.4.1 Active Packaging

Novel nanomaterials or nanostructure materials (NSMs) have been synthesized, and their properties have been investigated [111]. The extended shelf life and microbial safety of food products are guaranteed by incorporation of NSMs into packaging systems [23]. Active packaging is a system when the food and the inside of the package interact, so that the quality of the food can be maintained or even improved [112]. Comparing the active with the conventional packaging, the active packaging contributes significantly to the preservation of the food through both the antioxidant and antimicrobial compounds it releases inside the packaging as well as through the capture of oxygen and water vapor, which are degrading agents of quality [19].

A subcategory of active packaging is antimicrobial packaging with various substances to incorporate into the packaging, preventing, or inhibiting the growth of microorganisms (spoiling agents and the pathogens). Besides, preventing the growth of pathogenic microorganisms is the most serious challenge in the development of innovative antimicrobial packaging materials [113]. Inorganic compounds at the nanoscale exhibit strong antibacterial activities as they are characterized by high surface area-to-volume ratio [30]. Metal oxide nanoparticles, such as silver oxide (Ag_2O), titanium dioxide, (TiO_2), silicon (Si), copper oxide (CuO), zinc oxide (ZnO), calcium oxide (CaO), and magnesium oxide (MgO), have showed antibacterial activity and have been extensively used as antimicrobial films for food packaging [30]. The mechanistic action by which these NPs inhibit bacterial growth has not been fully elucidated. Some studies have suggested that these particles release metal ions and form ROS and reactive nitrogen species (RNS), destroying the bacterial cell wall or membrane [28, 113]. Also the intrinsic antimicrobial activity of nanoparticles has recently been attributed to the direct destruction and decay of cellular proteins, amino acids, DNA, or lipids. In addition, cationic nanoparticles interact with the negatively charged surface of cell membranes, interrupting their continuity [114]. It is recommended to avoid direct contact of these materials with food as migration issues may arise. The migration of nanoparticles from container into the food can be affected by various factors, including the chemical composition of the food, the pH, the radius of the nanoparticles and the temperature [83].

Another subcategory of active food packaging is the oxygen scavenger [16]. The occurrence of oxygen in a food package could have a significant effect on the food quality. The damage that can occur concerns both the appearance through the presence of the brown color and its deterioration through the development of aerobic microorganisms [97]. Moreover some other qualitative characteristics could be affected by the presence of oxygen such as texture, flavor, and nutritive value [21]. Many types of oxygen scavengers could incorporate with packaging substrate in order to prevent the deterioration of the quality due to oxidation reactions [97]. NPs of TiO_2, ZnO, and zero valent iron particles were the raw material for the production of oxygen scavenger materials, intended for food preservation [73, 115]. Rovera et al. recently presented an overview of the nano-inspired oxygen barrier coatings suitable for applications in food packaging technology [116]. As all these nanomaterials work as nanofillers, the explanation that was given for the improved barrier properties concerns the longest route that a gas is required to travel compared to a conventional packaging material [11]. Some recent studies of the use of nanomaterials in active food packaging applications are presented in Table 14.2.

14.4.2 Intelligent ("Smart") Packaging

The smart food packaging checks the status of the packaging and informs the consumer of the quality and safety of the packaged product as well as of the integrity

Table 14.2 Recent studies of the use of nanomaterials in active food packaging applications.

Food matrix	Category	Nanomaterials	Effect	References
Grapes	Active food packaging	Nanocomposites of chitosan/gelatin and silver nanoparticles	Extend the shelf life for two weeks	[117]
Fresh-cut apple	Active food packaging	Polylactic acid film incorporated with ZnO nanoparticle	Maintain the quality for two weeks	[118]
Peach	Biodegradable packaging films	Chitosan-rice starch/nano-ZnO nanocomposites	Control the microbial growth and spoilage	[119]
Lamb meat	Active films	Cellulose nanofiber/whey protein/containing TiO_2 nanoparticles and rosemary essential oil	Reduced microbial growth, extend the shelf life for 6–15 d	[120]
Fruits and vegetables	Active food packaging	Silver nanoparticles	Absorption and decomposition of ethylene emitted	[121]
Chicken meat	Active food packaging	Polylactide films and bimetallic Ag–Cu nanoparticles and essential oil	Maximum antibacterial action during 21 d at 4 °C	[122]
Salmon	Active films	Chitosan/montmorillonite/α-tocopherol film	Inhibition of oxidation reactions	[123]
Bread	Active food packaging	Carboxymethyl cellulose–chitosan–ZnO NPs nanocomposite	Extend the shelf life by 15 d	[124]
Meat and poultry products	Active food packaging	Essential oils and nanoparticles in pullulan films	Control foodborne pathogens	[125]
Lamb meat	Active food packaging	Nanocomposite films containing TiO_2 nanoparticles and rosemary essential oil	Extend the shelf life by 12–15 d	[41]

of the container and any breach. A smart packaging could provide accurate information on the remaining shelf life of the particular food, and nutritional information, associated with the air intake in a vacuum package, with the development of microorganisms, with the leaking gases from the packaging, or with the existence chemical compounds. Gases produced by spoilage can be detected by smart packaging through nanosensors at surprisingly low levels. Hydrogen, hydrogen sulfide, nitrogen oxides, sulfur dioxide, and ammonia are the most common ones [73, 126].

14.5 Nanomaterials and Postharvest Quality Parameters

In the postharvest management of fresh fruits and vegetables, nanotechnology contributes to their better preservation by antifungal-containing food packaging materials that are capable of inhibiting the development of postharvest diseases as well as by monitoring quality parameters with nanosensors. Additionally innovative food packaging films can be used to prevent the effects of gases and dangerous rays [127].

14.5.1 Edible Coatings and Films in Food Packaging

Recently, interest in the use of edible materials in food packaging has increased [128]. An edible packaging should consist of materials that are edible as it can be consumed with the packaged foods. Edible coatings and films are the main types of edible packaging [16]. All ingredients used to form edible packaging should be categorized as GRAS [129].

Edible coatings and films are applied differently to foods, and although they often appear as synonyms, a distinction is necessary [130]. Edible coating systems are thin layers of edible substance applied on the outer surface of foods, by dipping or washing them, are considered part of food packaging, and are an alternative solution to preservation for many categories of food, like perishable foods, especially fresh fruits and vegetables [131, 132]. Edible coatings are used in food packaging to decrease weight loss and prevent microbiological deterioration by controlling the rate of gases exchange and deliver antimicrobial compounds [128]. The use of edible coatings in the postharvest management of fruits and vegetables is an emerging technology that significantly contributes to extend the preservation time. Especially in the postharvest storage of fresh-cut fruits, prolonging shelf life through edible coatings is achieved by decreasing moisture, respiration, and gas exchange as well as with the reduction or even suppression of physiological disorders [133, 134]. The edible films are thin and solid layers or sheets and wrap or cover foodstuffs as independent constructions [135]. The thickness of an edible film cannot exceed 254 μm and is applied as an external-separate material [16]. There are differences in the application of edible coatings and films to foods since there are different types of edible packaging. Coatings are applied in a liquid form as the food is immersed in the appropriate solution, whereas films are applied as a solid wrap to the food product [135].

Edible coatings have been applied in fresh-cut fruits and vegetables, nuts, seeds, and cheeses using various nanosystems, including polymeric NPs, NEs, and NCs, among others, in order to control the release of EOs, polyphenols, and fat-soluble vitamins. EOs of oregano and thyme employ the active agents of carvacrol and thymol, respectively, which exert a good antimicrobial effect. Regarding antioxidant capacity it reduces the rate of reactions and enzymatic activity of polyphenol oxidases, peroxidases, and lipoxygenases. Nanosystems require the exploration of ways to control the release of components into the environment and act as a barrier along with the polymer matrix in which they are integrated [42, 100, 136, 137]. Moreover edible coatings can act as carriers of bioactive molecules, such as antioxidants and antimicrobials agents like organic acids, polypeptides, and nutrients and flavors [129].

The main changes observed are related to mechanical and optical properties, antioxidant and antimicrobial effects, and the possibility to achieve controlled release especially during cold storage of minimally processed foods [138]; Arfat et al. [139]. All these properties are extremely useful in solving problems related to the microbial safety of food during the storage period.

The particle size on a nanometer scale characterizes the materials, which in turn acquire distinct and improved properties. During storage, hydrophilic and lipophilic substances with antimicrobial and antioxidant properties can be released targeting the increase in shelf life of various products. Edible coatings are prepared with the use of natural polymers and characterized as nontoxic, cheap, and readily available. Biodegradable synthetic polymers and liquid and solid lipids at room temperature are used to prepare nanosystems. Zambrano-Zaragosa et al. [42] reported on the use of nanosystems such as nanoparticles, nanotubes, nanocomposites, and nanoemulsions in edible coatings for food preservation [42]. In the case of cheese, edible nanocoatings and films can be used as an alternative to nonedible coatings and films as they help to reduce the weight loss and to control respiration [128].

Edible films and coatings with alginate, pectin, and gellan gum, when applied to fresh-cut apples, can help in the maintenance of the desired quality characteristics. The coated apples retain their color and their stability throughout refrigerated storage. Also, regarding the microbiological results, there was a decrease in mesophilic and psychophilic measurements during all storage period. Moreover, coating the polysaccharides maintained the antioxidant activity of the cut apples [140]. Table 14.3 presents some recent studies of the use of nanotechnological edible coatings and films.

14.5.2 Nanomaterials and the Potential against Postharvest Disease and Ethylene Production

Postharvest diseases of fruits and vegetables are the source of significant economic losses throughout the supply chain as fruits and vegetables come into contact, and they are inevitably attacked by hundreds of species of pathogenic microorganisms during their postharvest life. Especially after the harvest, the plant organs gradually lose their resistance, so the pathogens grow rapidly and cause serious damage with the growth of postharvest diseases. The main contributors to the development of these diseases are plant pathogenic microorganisms such as fungi and bacteria

Table 14.3 Recent studies of the use of nanotechnological edible coatings and films in postharvest preservation of fruits and vegetables.

Food matrix	Category	Nanomaterials	Effect	References
Refrigerated strawberries	Edible coatings	Thymol nanoemulsion/quinoa protein/chitosan	Lower fungal and yeast load 4 d extension in shelf life	[141]
Fresh-cut "Fuji" apples	Edible coatings	Lemongrass essential oil nanoemulsion	Completely inhibition of natural microflora inactivation of *Escherichia coli*	[134]
Cherry tomatoes	Edible films	Chitosan and alginate nanostructure	Reported to show good antimicrobial activity	[110]
Fresh-cut pineapple	Edible coatings	Citral nanoemulsion	Prolong the shelf life for 12 d	[142]
Cucumber	Edible coatings	Chitosan nanoparticles with *Zataria multiflora* essential oil	Enhance antioxidant activity and prolong shelf life	[143]
Grapes	Edible coatings	Lemongrass oil nanoemulsion	Inhibition of *Salmonella typhimurium* and *E. coli O157:H7*	[144]
Pomegranate arils	Edible coatings	Nano-ZnO/carboxymethyl cellulose	Decreased total yeasts, molds, and weight loss	[145]
Strawberries	Edible coatings	Nano-ZnO	Inhibition of microbial growth, delayed weight loss, preserving functional nutrients of strawberries	[146]
Fresh-cut kiwi fruit	Edible coatings	Nano-ZnO	Lower ethylene content and water loss, maintain the texture in comparison with control treatment	[147]
Mangoes	Edible coatings	Chitosan/nano-TiO_2 composite	Maintain the nutrient composition	[148]
Fresh-cut "Formosa" papaya	Edible coatings	Whitemouth croaker (*Micropogonias furnieri*) protein isolate and organo-clay nanocomposite	Lower microbial growth, lower mass loss, during 12 d of storage	[133]

Table 14.3 (Continued)

Food matrix	Category	Nanomaterials	Effect	References
Lemon	Edible coating	Chitosan–clay nanocomposite	Extend the shelf life for 63 d, lower mass loss	[149]
Grapes	Edible coating	Polymeric nanocapsules containing *Thymus vulgaris* L. essential oil	Extend the shelf life, lower incidence of gray mold in grapes stored for 6 mo	[150]
Fresh-cut “Gala” apples	Edible coating	Chitosan-based nanoparticles	Higher antimicrobial action and reduced total microbial contamination levels	[151]
Fresh-cut “Red Delicious” apples	Edible coating	Tocopherol/nopal mucilage (*Opuntia ficus indica*) nanoemulsion	Control the browning index and texture	[118]

[127]. Among pathogenic fungi, *Aspergillus niger*, *Alternaria alternata*, *Botrytis cinerea*, *Penicillium digitatum*, *Penicillium expansum*, *Colletotrichum musae*, and *Colletotrichum gloeosporioides* can attack many agricultural products, while among bacteria, *E. coli*, *S. aureus*, *L. monocytogenes*, *Pseudomonas aeruginosa*, and *Salmonella* spp. can also cause serious losses during storage period and transportation [127]. Nanomaterials could be used instead of synthetic chemical fungicides that are traditionally used to control postharvest diseases, as consumers are concerned about the use of chemical fungicides with great demand for fungicide-free products [128]. The unique functions of nanomaterials make them a promising and helpful option for sustainable horticulture as well as in the prevention of postharvest diseases [152].

Silver nanoparticles showed antifungal activity against the harmful citrus pathogenic fungi, *A. alternata*, *P. digitatum*, and *Alternaria citri*, when they were tested *in vitro* trials. Antifungal activity of silver nanoparticles has been attributed to the potential destruction of DNA and proteins, causing death in fungal spore cells [153]. Zinc oxide nanoparticles showed antifungal activity by inhibiting the growth of both *B. cinerea*, causing deformation in fungal hyphae, and *P. expansum*, preventing the growth of conidiophores and conidia, *in vitro* experiments [154]. In another *in vitro* investigation, soybean protein isolate/cinnamaldehyde/zinc oxide bionanocomposites were examined for antifungal properties against *A. niger*. The antifungal performance of bionanocomposite coating was 1.25-fold stronger than that of soybean protein isolate/cinnamaldehyde treatment [155].

The shelf life of fruits and vegetables is greatly influenced by the presence of ethylene, which is widely known as ripening hormone. Postharvest techniques, for fresh fruits and vegetables preservation, such as controlled atmosphere (CA), modified atmosphere (MA), and modified active packaging (MAP) regulate ethylene production, resulting in a longer shelf life of fruits and vegetables. Treatments with Ag+ and polyvinyl pyrrolidone-coated AgNPs of the model plant *Arabidopsis thaliana* provided information on the molecular mechanisms of the plant response, associated with reduced ethylene production [156]. Expand of shelf life of Chinese jujube with AgNPs nano-packing material, effecting the preservation quality by absorbing and decomposing ethylene, was also reported [157]. Moreover the exact mechanism of interaction of nanobiotechnology and plant physiology is not clear, and more studies are needed to explore the way that the biosynthesis of ethylene is inhibited [11].

14.6 Regulations and Safety Aspects

Regulatory systems are necessary in order to manage the potential risks posed by the use of nanotechnology in food sector [158]. Application of specific legislative framework of nanotechnology in food could cause the widespread use of nanomaterials in food packaging. Lack of regulations on consumer safety in the United States increases the challenge of implementing nanotechnology regulations [1]. In the United States, Regulation 258/97 defines that if an nanomaterial is used as a primary ingredient, then it needs to be considered as a "Novel Food" [159]. When the term "food additive" is employed according to the European Commission Regulation No.1333/2008, even if its composition is approved, it does not mean that it can be applied in nanoform, and the nanosystem should be considered as a different additive; therefore, it must be considered as a new, safe material before it can be marketed [160]. According to European Commission Regulation No. 2015/2283, the European Food Safety Authority (EFSA) must verify that for the participation of mechanically manufactured nanoparticles in food production, their evaluation for their safety assessment was done with the latest generation of analytical techniques [161]. Guidance on risk assessment of the application of nanosciences and nanotechnologies in the food chain and feed has been recently published by EFSA [162]. Although it provides an organized pathway to conduct a food safety assessment of nanomaterial in food/feed, it also emphasizes that additional research is needed to assess the safety of nanomaterials, to characterize them, and to measure their reactivity.

The toxicological risk posed by the use of nanomaterials in food packaging and in particular in the case of edible nano-packaging is significantly linked to the migration phenomenon as well as the occurrence of toxic effects on the exposed human body [116]. The size, the shape, the solubility, the physical and chemical properties, and the chemical composition significantly define the toxicity of nanomaterials [163]. Little information about the potential toxicity, behavior, bioavailability, biodistribution, and bioaccumulation of NMs has not yet allowed international regulations to be adopted [113]. Long-term exposure to nanoparticles can cause oxidative stress in human cells, damage to the kidneys and liver, and damage to DNA [164]. Inhalation and penetration through the skin tissue into people who work in nanotechnology production units is considered extremely common [165]. The distribution of food in Europe requires Declaration of Compliance for the safety of food contact materials even with approval of other regulatory organizations, such as FDA. Replacing traditional packaging materials with NMs requires further research to guarantee safe use by consumers as well as more toxicological studies about the migration of NMs into packaged food [126]. Besides, during the study of migration, the functions of nanomaterials may be differentiated, due to both the different interactions with the various foods and their ingredients and the various thermodynamic conditions [121].

The way in which nanoparticles are transported into the food system is characterized as a diffusion process similar to that characterized by Fick's second law. Therefore, the development of innovative packaging materials that will contain nanomaterials should be supplemented by corresponding migration surveys, which will identify elements that will concern both global and special migration [166]. Moreover, the placement of an additional layer of biopolymer material could possibly prevent the release of NMs, improving both the permeability of the package and providing additional mechanical support [68].

14.7 Conclusions and Outlook

Nanotechnology is recognized as an emerging technology that finds application in all branches of industry including food science and technology. The importance of nanotechnology, especially in food preservation, is great, since a variety of materials find many applications, prolonging the shelf life of foods and increasing preservation time. Food preservation in all cases should be related to materials that are completely safe and compatible with food systems. In recent years, consumers have shown a tendency for greater consumption for natural products and mainly for products of vegetable origin. Especially for this food category, nanotechnology can offer solutions in their preservation, through active and smart packaging, through edible coatings, and through the development of a wide range of capable nanosystems as important tools and efficient options for the control, limit, and improvement of safety and qualitative parameters of foodstuffs, which are highly desirable in food technology.

For the most NMs used in food preservation, the mechanism of action against pathogenic microorganisms is not completely understood. Nanobiotechnology undoubtedly contributes to the preservation of food both by providing promising and innovative packaging NMs with excellent characteristics and by development of efficient nanopreservatives, aiming at preserving quality and minimizing food spoilage for almost all categories of foods. Innovative NMs even achieve real-time food quality monitoring, providing an efficient option in implementation in food preservation. The incorporation of these materials in food preservation is an emerging trend that raises significant inconveniences, since some of them, through ingestion, inhalation, and exposure end up in the human body. Because the characteristics of these materials are completely unlike from those of the originals, it is imperative that more studies on *in vitro* and *in vivo* levels be carried out, both to address in detail all the safety issues that may arise from human exposure and consumption and to understand the mechanistic actions between NMs and food systems. At the same time, the synthesis of suitable alternatives NMs to replace conventional ones should be accompanied by low-cost and eco-friendly processes, which will combine optimal preservation conditions with high food quality.

References

1 Adeyeye, S.A.O. and Fayemi, O.E. (2019). Nanotechnology and food processing: between innovations and consumer safety. *J. Culin. Sci. Technol.* 17: 435–452.

2 Agriopoulou, S. (2016a). Active packaging for food applications. *EC Nutr.* 62: 86–87.

3 Prakash, J., Vignesh, K., Anusuya, T. et al. (2019). Application of nanoparticles in food preservation and food processing. *J. Food Hyg. Saf.* 34: 317–324.
4 Jafarizadeh-Malmiri, H., Sayyar, Z., Anarjian, N., and Berenjian, A. (2019). Challenges for nanobiotechnology. In: *Nanobiotechnology in Food: Concepts, Applications and Perspectives* (eds. H. Jafarizadeh-Malmiri, Z. Sayyar, N. Anarjan and A. Berenjian), 1–17. Cham: Springer.
5 Sekhon, B.S. (2014). Nanotechnology in agri-food production: an overview. *Nanotechnol. Sci. Appl.* 7: 31–53.
6 Pradhan, N., Singh, S., Ojha, N. et al. (2015). Facets of nanotechnology as seen in food processing, packaging, and preservation industry. *Biomed Res. Int.* 2015: 1–47.
7 Qamar, S.A., Asgher, M., Khalid, N., and Sadaf, M. (2019). Nanobiotechnology in health sciences: current applications and future perspectives. *Biocatal. Agric. Biotechnol.* 22: 101388.
8 Geraldes, A.N., da Silva, A.A., Leal, J. et al. (2016). Green nanotechnology from plant extracts: synthesis and characterization of gold nanoparticles. *Adv. Nanopart.* 5: 176–185.
9 Shafiq, S.A., Al-Shammari, R.H., and Majeed, H.Z. (2016). Study of biosynthesis silver nanoparticles by *Fusarium graminearum* and test their antimicrobial activity. *Int. J. Innov. Appl. Stud.* 15: 43–50.
10 Shafiq, M., Anjum, S., Hano, C. et al. (2020). An overview of the applications of nanomaterials and nanodevices in the food industry. *Foods* 9: 148.
11 Bhuyan, D., Greene, G.W., and Das, R.K. (2019). Prospects and application of nanobiotechnology in food preservation: molecular perspectives. *Crit. Rev. Biotechnol.* 39: 759–778.
12 Mei, L. and Wang, Q. (2020). Advances in using nanotechnology structuring approaches for improving food packaging. *Annu. Rev. Food Sci. Technol.* 11: 339–364.
13 Dos Santos, C.A., Ingle, A.P., and Rai, M. (2020). The emerging role of metallic nanoparticles in food. *Appl. Microbiol. Biotechnol.* 104: 2373–2383.
14 Wang, L., Chen, C., Wang, J. et al. (2020). Cellulose nanofibrils versus cellulose nanocrystals: comparison of performance in flexible multilayer films for packaging applications. *Food Packag. Shelf Life* 23: 100464.
15 Shahbazi, Y. and Shavisi, N. (2019). Current advancements in applications of chitosan based nano-metal oxides as food preservative materials. *Nanomed. Res. J.* 4: 122–129.
16 Ramos, O.L., Pereira, R.N., Cerqueira, M.A. et al. (2018). Bio-based nanocomposites for food packaging and their effect in food quality and safety. In: *Food Packaging and Preservation* (eds. A.M. Grumezescu and A.M. Holban), 271–306. Elsevier Inc.
17 Din, M.I., Ghaffar, T., Najeeb, J. et al. (2020). Potential perspectives of biodegradable plastics for food packaging application-review of properties and recent developments. *Food Addit. Contam. Part A* https://doi.org/10.1080/19440049.2020.1718219.

18 Eleftheriadou, M., Pyrgiotakis, G., and Demokritou, P. (2017). Nanotechnology to the rescue: using nano-enabled approaches in microbiological food safety and quality. *Curr. Opin. Biotechnol.* 44: 87–93.

19 Bajpai, V.K., Kamle, M., Shukla, S. et al. (2018). Prospects of using nanotechnology for food preservation, safety, and security. *J. Food Drug Anal.* 26: 1201–1214.

20 Agriopoulou, S., Stamatelopoulou, E., Sachadyn-Król, M., and Varzakas, T. (2020c). Lactic acid bacteria as antibacterial agents to extend the shelf life of fresh and minimally processed fruits and vegetables: quality and safety aspects. *Microorganisms* 8: 952.

21 Topuz, F. and Uyar, T. (2020). Antioxidant, antibacterial and antifungal electrospun nanofibers for food packaging applications. *Food Res. Int.* 130: 108927.

22 J. Potočnik (2011). Commission recommendation of 18 October 2011 on the definition of nanomaterial (2011/696/EU). *Off. J. Eur. Union*, L 257/38: 38–40.

23 Patel, A., Patra, F., Shah, N., and Khedkar, C. (2018). Application of nanotechnology in the food industry: present status and future prospects. In: *Impact of Nanoscience in the Food Industry* (eds. A.M. Grumezescu and A.M. Holban), 1–27. Elsevier Inc.

24 Cerqueira, M.A., Vicente, A.A., and Pastrana, L.M. (2018). Nanotechnology in food packaging: opportunities and challenges. In: *Micro and Nano Technologies, Nanomaterials for Food Packaging* (eds. M.Â.P.R. Cerqueira, J.M. Lagaron, L.M.P. Castro and A.A.M., O.S. de Vicente), 1–11. Elsevier.

25 Lopes, N.A. and Brandelli, A. (2018). Nanostructures for delivery of natural antimicrobials in food. *Crit. Rev. Food Sci. Nutr.* 58: 2202–2212.

26 Khan, I., Saeed, K., and Khan, I. (2019). Nanoparticles: properties, applications and toxicities. *Arab. J. Chem.* 12: 908–931.

27 Oun, A.A., Shankar, S., and Rhim, J.W. (2020). Multifunctional nanocellulose/metal and metal oxide nanoparticle hybrid nanomaterials. *Crit. Rev. Food Sci. Nutr.* 60: 435–460.

28 Basavegowda, N., Mandal, T.K., and Baek, K.-H. (2020). Bimetallic and trimetallic nanoparticles for active food packaging applications: a review. *Food Bioprocess Technol.* 13: 30–44.

29 Tarazona, A., Gómez, J.V., Mateo, E.M. et al. (2019). Antifungal effect of engineered silver nanoparticles on phytopathogenic and toxigenic *Fusarium* spp. and their impact on mycotoxin accumulation. *Int. J. Food Microbiol.* 306: 108259.

30 Jagadish, K., Shiralgi, Y., Chandrashekar, B.N. et al. (2018). Ecofriendly synthesis of metal/metal oxide nanoparticles and their application in food packaging and food preservation. In: *Impact of Nanoscience in the Food Industry* (eds. A.M. Grumezescu and A.M. Holban), 197–216. Elsevier Inc.

31 Vinci, G. and Rapa, M. (2019). Noble metal nanoparticles applications: recent trends in food control. *Bioengineering* 6: 10.

32 Xin, Q., Shah, H., Nawaz, A. et al. (2019). Antibacterial carbon-based nanomaterials. *Adv. Mater.* 31: 1–15.

33 Pan, M., Yin, Z., Liu, K. et al. (2019). Carbon-based nanomaterials in sensors for food safety. *Nanomaterials* 9: 1–23.

34 Agriopoulou, S., Stamatelopoulou, E., and Varzakas, T. (2020a). Advances in occurrence, importance, and mycotoxin control strategies: prevention and detoxification in foods. *Foods* 9: 137.

35 Murphy, M., Ting, K., Zhang, X. et al. (2015). Current development of silver nanoparticle preparation, investigation, and application in the field of medicine. *J. Nanomater.* 2015: 1–12.

36 Zhang, X.-F., Liu, Z.-G., Shen, W., and Gurunathan, S. (2016). Silver nanoparticles: synthesis, characterization, properties, applications, and therapeutic approaches. *Int. J. Mol. Sci.* 17: 1534.

37 Kowsalya, E., Mosa Christas, K., Balashanmugam, P. et al. (2019). Biocompatible silver nanoparticles/poly(vinyl alcohol) electrospun nanofibers for potential antimicrobial food packaging applications. *Food Packag. Shelf Life* 21: 100379.

38 Khalil, N.M., Abd El-Ghany, M.N., and Rodríguez-Couto, S. (2019). Antifungal and anti-mycotoxin efficacy of biogenic silver nanoparticles produced by *Fusarium chlamydosporum* and *Penicillium chrysogenum* at non-cytotoxic doses. *Chemosphere* 218: 477–486.

39 Pathakoti, K., Morrow, S., Han, C. et al. (2013). Photoinactivation of *Escherichia coli* by sulfur-doped and nitrogen-fluorine-codoped TiO_2 nanoparticles under solar simulated light and visible light irradiation. *Environ. Sci. and Technol.* 47: 9988–9996.

40 D'Souza, A.A., Kumari, D., and Banerjee, R. (2017). Nanocomposite biosensors for point-of-care. Evaluation of food quality and safety. In: *Nanosensors* (ed. A.M. Grumezescu), 629–676. Elsevier Inc.

41 Alizadeh Sani, M., Ehsani, A., and Hashemi, M. (2017). Whey protein isolate/cellulose manofibre/TiO_2 nanoparticle/rosemary essential oil nanocomposite film: its effect on microbial and sensory quality of lamb meat and growth of common foodborne pathogenic bacteria during refrigeration. *Int. J. Food Microbiol.* 251: 8–14.

42 Zambrano-Zaragoza, M.L., González-Reza, R., Mendoza-Muñoz, N. et al. (2018). Nanosystems in edible coatings: a novel strategy for food preservation. *Int. J. Mol. Sci.* 19: 705.

43 Espitia, P.J., Soares, N., F., Teófilo, R.F. et al. (2013). Physical-mechanical and antimicrobial properties of nanocomposite films with pediocin and ZnO nanoparticles. *Carbohydr. Polym.* 94: 199–208.

44 Mahajan, B.V.C., Tandon, R., Swati, K., and Sidhu, M.K. (2018). Natural coatings for shelf-life enhancement and quality maintenance of fresh fruits and vegetables – a review. *J. Postharvest Technol.* 6: 12–26.

45 Alavi, M. and Nokhodchi, A. (2020). An overview on antimicrobial and wound healing properties of ZnO nanobiofilms, hydrogels, and bionanocomposites based on cellulose, chitosan, and alginate polymers. *Carbohydr. Polym.* 227: 115349.

46 Roy, S. and Rhim, J.-W. (2020). Carboxymethyl cellulose-based antioxidant and antimicrobial active packaging film incorporated with curcumin and zinc oxide. *Int. J. Biol. Macromol.* 148: 666–676.

47 Riaz Rajoka, M.S., Mehwish, H.M., Wu, Y. et al. (2020). Chitin/chitosan derivatives and their interactions with microorganisms: a comprehensive review and future perspectives. *Crit. Rev. Biotechnol.* 40: 365–379.

48 Kumar, S., Mukherjee, A., and Dutta, J. (2020). Chitosan based nanocomposite films and coatings: emerging antimicrobial food packaging alternatives. *Trends Food Sci. Technol.* 97: 196–209.

49 Hosseinnejad, M. and Jafari, S.M. (2016). Evaluation of different factors affecting antimicrobial properties of chitosan. *Int. J. Biol. Macromol.* 85: 467–475.

50 Pan, Y., Huang, X., Shi, X. et al. (2015). Antimicrobial application of nanofibrous mats self-assembled with quaternized chitosan and soy protein isolate. *Carbohydr. Polym.* 133: 229–235.

51 Raafat, D. and Sahl, H.G. (2009). Chitosan and its antimicrobial potential – acritical literature survey. *Microb. Biotechnol.* 2: 186–201.

52 Anraku, M., Gebicki, J.M., Iohara, D. et al. (2018). Antioxidant activities of chitosans and its derivatives in in vitro and in vivo studies. *Carbohydr. Polym.* 199: 141–149.

53 Xie, W., Xu, P., and Liu, Q. (2001). Antioxidant activity of water-soluble chitosan derivatives. *Bioorg. Med. Chem. Lett.* 11: 1699–1701.

54 Yen, M.-T., Yang, J.-H., and Mau, J.-L. (2008). Antioxidant properties of chitosan from crab shells. *Carbohydr. Polym.* 74: 840–844.

55 Sabaghi, M., Maghsoudlou, Y., and Habibi, P. (2015). Enhancing structural properties and antioxidant activity of kefiran films by chitosan addition. *Food Struct.* 5: 66–71.

56 Zhu, Y., Cui, H., Li, C., and Lin, L. (2019). A novel polyethylene oxide/*Dendrobium officinale* nanofiber: preparation, characterization and application in pork packaging. *Food Packag. Shelf Life* 21: 100329.

57 Li, J., Sun, Q., Sun, Y. et al. (2019). Improvement of banana postharvest quality using a novel soybean protein isolate/cinnamaldehyde/zinc oxide bionanocomposite coating strategy. *Sci. Hortic.* 258: 108786.

58 Arfat, Y.A., Benjakul, S., Prodpran, T. et al. (2014). Properties and antimicrobial activity of fish protein isolate/fish skin gelatin film containing basil leaf essential oil and zinc oxide nanoparticles. *Food Hydrocolloids* 41: 265–273.

59 Azeredo, H., Mattoso, L., and McHugh, T. (2011). Nanocomposites in food packaging – a review. In: *Advances in Diverse Industrial Applications of Nanocomposites* (ed. B. Reddy), 57–78. InTech.

60 Kanmani, P. and Rhim, J.W. (2014a). Physical, mechanical and antimicrobial properties of gelatin based active nanocomposite films containing AgNPs and nanoclay. *Food Hydrocolloids* 35: 644–652.

61 Trovatti, E., Fernandes, S.C.M., Rubatat, L. et al. (2012). Sustainable nanocomposite films based on bacterial cellulose and pullulan. *Cellulose* 19: 729–737.

62 Müller, K., Bugnicourt, E., Latorre, M. et al. (2017). Review on the processing and properties of polymer nanocomposites and nanocoatings and their applications in the packaging, automotive and solar energy fields. *Nanomaterials* 7: 47.

63 Bertuoli, P.T., Piazza, D., Scienza, L.C., and Zattera, A.J. (2014). Preparation and characterization of montmorillonite modified with 3-aminopropyltriethoxysilane. *Appl. Clay Sci.* 87: 46–51.

64 Huskić, M., Žigon, M., and Ivanković, M. (2013). Comparison of the properties of clay polymer nanocomposites prepared by montmorillonite modified by silane and by quaternary ammonium salts. *Appl. Clay Sci.* 85: 109–115.

65 Shankar, S., Teng, X., Li, G., and Rhim, J.-W. (2015). Preparation, characterization, and antimicrobial activity of gelatin/ZnO nanocomposite films. *Food Hydrocolloids* 45: 264–271.

66 Kanmani, P. and Rhim, J.W. (2014b). Physicochemical properties of gelatin/silver nanoparticle antimicrobial composite films. *Food Chem.* 148: 162–169.

67 Jawaid, M. and Swain, S.K. (2017). *Bionanocomposites for Packaging Applications.* New York, NY: Springer.

68 Rouf, T.B. and Kokini, J.L. (2018). Natural biopolymer-based nanocomposite films for packaging applications. In: *Bionanocomposites for Packaging Applications* (eds. M. Jawaid and S. Swain), 149–177. New York, NY: Springer.

69 Cuq, B., Gontard, N., and Guilbert, S. (1998). Proteins as agricultural polymers for packaging production. *Cereal Chem.* 75: 1–9.

70 Zink, J., Wyrobnik, T., Prinz, T., and Schmid, M. (2016). Physical, chemical and biochemical modifications of protein-based films and coatings: an extensive review. *Int. J. Mol. Sci.* 17: 1376.

71 Zubair, M. and Ullah, A. (2019). Recent advances in protein derived bionanocomposites for food packaging applications. *Crit. Rev. Food Sci. Nutr.* 60: 406–434.

72 Marra, A., Silvestre, C., Duraccio, D., and Cimmino, S. (2016). Polylactic acid/zinc oxide biocomposite films for food packaging application. *Int. J. Biol. Macromol.* 88: 254–262.

73 Vilena De Francisco, E. and García-Estepa, R.M. (2018). Nanotechnology in the agrofood industry. *J. Food Eng.* 238: 1–11.

74 Ludueña, L.N., Alvarez, V.A., and Vazquez, A. (2007). Processing and microstructure of PCL/clay nanocomposites. *Mater. Sci. Eng. A* 460-461: 121–129.

75 Chaudhary, S., Kumar, S., Kumar, V., and Sharma, R. (2020). Chitosan nanoemulsions as advanced edible coatings for fruits and vegetables: composition, fabrication and developments in last decade. *Int. J. Biol. Macromol.* 152: 154–170.

76 Singh, Y., Meher, J.G., Raval, K. et al. (2017). Nanoemulsion: concepts, development and applications in drug delivery. *J. Control. Release* 252: 28–49.

77 Salvia-Trujillo, L., Martín-Belloso, O., and McClements, D.J. (2016). Excipient nanoemulsions for improving oral bioavailability of bioactives. *Nanomaterials* 6: 1–16.

78 Yang, Y., Marshall-Breton, C., Leser, M.E. et al. (2012). Fabrication of ultrafine edible emulsions: comparison of high-energy and low-energy homogenization methods. *Food Hydrocolloids* 29: 398–406.

79 Ren, G., Sun, Z., Wang, Z. et al. (2019). Nanoemulsion formation by the phase inversion temperature method using polyoxypropylene surfactants. *J. Colloid Interface Sci.* 540: 177–184.

80 Jiang, T., Liao, W., and Charcosset, C. (2020). Recent advances in encapsulation of curcumin in nanoemulsions: a review of encapsulation technologies, bioaccessibility and applications. *Food Res. Int.* 132: 109035.

81 Otoni, C.G., de Moura, M.R., Aouada, F.A. et al. (2014). Antimicrobial and physical-mechanical properties of pectin/papaya puree/cinnamaldehyde nanoemulsion edible composite films. *Food Hydrocolloids* 41: 188–194.

82 Karimirad, R., Behnamian, M., and Dezhsetan, S. (2019). Application of chitosan nanoparticles containing *Cuminum cyminum* oil as a delivery system for shelf life extension of *Agaricus bisporus*. *LWT* 106: 218–228.

83 Caon, T., Martelli, S.M., and Fakhouri, F.M. (2017). New trends in the food industry: application of nanosensors in food packaging. In: *Nanobiosensors* (ed. A.M. Grumezescu), 773–804. Elsevier Inc.

84 Berekaa, M.M. (2015). Nanotechnology in food industry; advances in food processing, packaging and food safety. *Int. J. Curr. Microbiol. Appl. Sci.* 4: 345–357.

85 Pathakoti, K., Manubolu, M., and Hwang, H.M. (2017). Nanostructures: current uses and future applications in food science. *J. Food Drug Anal.* 25: 245–253.

86 Economou, α., Karapetis, S.K., Nikoleli, G.-P. et al. (2017). Enzyme-based sensors. In: *Advances in Food Diagnostics* (eds. F. Toldrá and L.M.L. Nollet), 231–250. Wiley.

87 Hamad, A.F., Han, J.-H., Kim, B.-C., and Rather, I.A. (2018). The intertwine of nanotechnology with the food industry. *Saudi J. Biol. Sci.* 25: 27–30.

88 El-Nour, K.M.A., Salam, E.T.A., Soliman, H.M., and Orabi, A.S. (2017). Gold nanoparticles as a direct and rapid sensor for sensitive analytical detection of biogenic amines. *Nanoscale Res. Lett.* 12: 231.

89 Hashemi, M., Nazari, Z., and Noshirvani, N. (2017). Synthesis of chitosan based magnetic molecularly imprinted polymers for selective separation and spectrophotometric determination of histamine in tuna fish. *Carbohydr. Polym.* 177: 306–314.

90 Kumar, N. and Goyal, R.N. (2018). Silver nanoparticles decorated graphene nanoribbon modified pyrolytic graphite sensor for determination of histamine. *Sens. Actuators, B Chem.* 268: 383–391.

91 Chen, C., Yuan, Z., Chang, H.T. et al. (2016). Silver nanoclusters as fluorescent nanosensors for selective and sensitive nitrite detection. *Anal. Methods* 8: 2628–2633.

92 Bui, M.-P.N., Ahmed, S., and Abbas, A. (2015). Single-digit pathogen and attomolar detection with the naked eye using liposome-amplified plasmonic immunoassay. *Nano Lett.* 15: 6239–6246.

93 Vyas, S.S., Jadhav, S.V., Majee, S.B. et al. (2015). Development of immunochromatographic strip test using fluorescent, micellar silica nanosensors for rapid detection of *B. abortus* antibodies in milk samples. *Biosens. Bioelectron.* 70: 254–260.

94 Yadav, S., Nair, S.S., Sai, V.V.R., and Satija, J. (2019). Nanomaterials based optical and electrochemical sensing of histamine: progress and perspectives. *Food Res. Int.* 119: 99–109.

95 Zou, L., Peng, S., Liu, W. et al. (2015). A novel delivery system dextran sulfate coated amphiphilic chitosan derivatives-based nanoliposome: capacity to improve *in vitro* digestion stability of (-)-epigallocatechin gallate. *Food Res. Int.* 69: 114–120.

96 Ma, Y., Wang, Z., Zhao, W. et al. (2013). Enhanced bactericidal potency of nanoliposomes by modification of the fusion activity between liposomes and bacterium. *Int. J. Nanomed.* 8: 2351–2360.

97 Ranjan, S., Dasgupta, N., Chakraborty, A.R. et al. (2014). Nanoscience and nanotechnologies in food industries: opportunities and research trends. *J. Nanopart. Res.* 16: 2464.

98 Agriopoulou, S., Stamatelopoulou, E., and Varzakas, T. (2020b). Advances in analysis and detection of major mycotoxins in foods. *Foods* 9: 518.

99 Lin, L., Gu, Y., and Cui, H. (2019). Moringa oil/chitosan nanoparticles embedded gelatin nanofibers for food packaging against *Listeria monocytogenes* and *Staphylococcus aureus* on cheese. *Food Packag. Shelf Life* 19: 86–93.

100 Zambrano-Zaragoza, M.L., Quintanar-Guerrero, D., Real, A.D. et al. (2017). The release kinetics of β-carotene nanocapsules/xanthan gum coating and quality changes in fresh-cut melon (cantaloupe). *Carbohydr. Polym.* 157: 1874–1882.

101 Sari, P., Mann, B., Kumar, R. et al. (2015). Preparation and characterization of nanoemulsion encapsulating curcumin. *Food Hydrocolloids* 43: 540–546.

102 Varzakas, T. and Kafetzopoulos, D. (2019). Strategy reformulation for development of healthier food products: emerging technologies and novel ingredients. In: *Reformulation as a Strategy for Developing Healthier Food Products* (eds. V. Raikos and V. Ranawana), 199–217. Springer.

103 Aydogdu, A., Sumnu, G., and Sahin, S. (2018). A novel electrospun hydroxypropyl methylcellulose/polyethylene oxide blend nanofibers: morphology and physicochemical properties. *Carbohydr. Polym.* 181: 234–246.

104 Mendes, A.C., Stephansen, K., and Chronakis, I.S. (2017). Electrospinning of food proteins and polysaccharides. *Food Hydrocolloids* 68: 53–68.

105 Cui, H., Wu, J., Li, C., and Lin, L. (2017). Improving anti-listeria activity of cheese packaging via nanofiber containing nisin-loaded nanoparticles. *LWT-Food Sci. Technol.* 81: 233–242.

106 Fan, Y., He, X., Zhou, S. et al. (2009). Composition analysis and antioxidant activity of polysaccharide from *Dendrobium Denneanum*. *J. Food Meas. Charact.* 45: 169–173.

107 Luo, A., He, X., Zhou, S. et al. (2010). Purification, composition analysis and antioxidant activity of the polysaccharides from *Dendrobium nobile* Lindl. *Carbohydr. Polym.* 79: 1014–1019.

108 Fan, H., Meng, Q., Xiao, T., and Zhang, L. (2018). Partial characterization and antioxidant activities of polysaccharides sequentially extracted from *Dendrobium officinale. Int. J. Biol. Macromol.* 12: 1054–1164.

109 Teixeira da Silva, J.A. and Ng, T.B. (2017). The medicinal and pharmaceutical importance of *Dendrobium* species. *Appl. Microbiol. Biotechnol.* 101: 2227–2239.

110 Zhu, J., Wu, H., and Sun, Q. (2019). Preparation of crosslinked active bilayer film based on chitosan and alginate for regulating ascorbate-glutathione cycle of postharvest cherry tomato (*Lycopersicon esculentum*). *Int. J. Biol. Macromol.* 130: 584–594.

111 Nikoleli, G.-P., Nikolelis, D.P., Siontorou, C.G. et al. (2018). Novel biosensors for the rapid detection of toxicants in foods. In: *Advances in Food and Nutrition Research* (ed. F. Toldrá), 57–102. Elsevier Inc.

112 Agriopoulou, S. (2016b). Nanotechnology in food packaging. *EC Nutr.* 42: 1137–1141.

113 Huang, Y., Mei, L., Chen, X., and Wang, Q. (2018). Recent developments in food packaging based on nanomaterials. *Nanomaterials* 8: 1–29.

114 Jamil, B., Bokhari, H., and Imran, M. (2017). Mechanism of action: how nano-antimicrobials act? *Curr. Drug Targets* 18: 363–373.

115 Foltynowicz, Z., Bardenshtein, A., Sängerlaub, S. et al. (2017). Nanoscale, zero valent iron particles for application as oxygen scavenger in food packaging. *Food Packag. Shelf Life* 11: 74–83.

116 Rovera, C., Ghaani, M., and Farris, S. (2020). Nano-inspired oxygen barrier coatings for food packaging applications: an overview. *Trends Food Sci. Technol.* 97: 210–220.

117 Kumar, S., Shukla, A., Baul, P.P. et al. (2018). Biodegradable hybrid manocomposites of chitosan/gelatin and silver nanoparticles for active food packaging applications. *Food Packag. Shelf Life* 16: 178–184.

118 Li, W., Li, L., Cao, Y. et al. (2017). Effects of PLA film incorporated with ZnO nanoparticle on the quality attributes of fresh-cut apple. *Nanomaterials* 7: 207.

119 Kaur, M., Kalia, A., and Thakur, A. (2017). Effect of biodegradable chitosan–rice-starch nanocomposite films on post-harvest quality of stored peach fruit. *Starch/Staerke* 69: 1600208.

120 Alizadeh Sani, M., Mohammadian, E., and Mcclements, D.J. (2020). Eco-friendly active packaging consisting of nanostructured biopolymer matrix reinforced with TiO_2 and essential oil: application for preservation of refrigerated meat. *Food Chem.*: 126782.

121 Huang, J.-Y., Li, X., and Zhou, W. (2015). Safety assessment of nanocomposite for food packaging application. *Trends Food Sci. Technol.* 45: 187–199.

122 Ahmed, J., Arfat, Y.A., Bher, A. et al. (2018). Active chicken meat packaging based on polylactide films and bimetallic Ag–Cu nanoparticles and essential oil. *J. Food Sci.* 83: 1299–1310.

123 Dias, M.V., Azevedo, V.M., Santos, T.A. et al. (2019). Effect of active films incorporated with montmorillonite clay and α-tocopherol: potential of nanoparticle migration and reduction of lipid oxidation in salmon. *Packag. Technol. Sci.* 32: 39–47.

124 Noshirvani, N., Ghanbarzadeh, B., Mokarram, R.R., and Hashemi, M. (2017). Novel active packaging based on carboxymethyl cellulose-chitosan-ZnO NPs nanocomposite for increasing the shelf life of bread. *Food Packag. Shelf Life* 11: 106–114.

125 Morsy, M.K., Khalaf, H.H., Sharoba, A.M. et al. (2014). Incorporation of essential oils and nanoparticles in pullulan films to control foodborne pathogens on meat and poultry products. *J. Food Sci.* 79: M675–M684.

126 Al-Tayyar, N.A., Youssef, A.M., and Al-hindi, R. (2020). Antimicrobial food packaging based on sustainable bio-based materials for reducing foodborne pathogens: a review. *Food Chem.* 310: 125915.

127 Ruffo Roberto, S., Youssef, K., Hashim, A.F., and Ippolito, A. (2019). Nanomaterials as alternative control means against postharvest diseases in fruit crops. *Nanomaterials* 9: 1–20.

128 Costa, M.J., Maciel, L.C., Teixeira, J.A. et al. (2018). Use of edible films and coatings in cheese preservation: opportunities and challenges. *Food Res. Int.* 107: 84–92.

129 Flores-López, M.L., Cerqueira, M.A., de Rodríguez, D.J., and Vicente, A.A. (2016). Perspectives on utilization of edible coatings and nano-laminate coatings for extension of postharvest storage of fruits and negetables. *Food Eng. Rev.* 8: 292–305.

130 Otoni, C.G., Avena-Bustillos, R.J., Azeredo, H.M.C. et al. (2017). Recent advances on edible films based on fruits and vegetables—a review. *Compr. Rev. Food Sci. Food Saf.* 16: 1151–1169.

131 Hasan, S.M.K., Ferrentino, G., and Scampicchio, M. (2019). Nanoemulsion as advanced edible coatings to preserve the quality of fresh-cut fruits and vegetables: a review. *Int. J. Food Sci. Technol.* 55: 1–10.

132 Tumbarski, Y., Nikolova, R., Petkova, N. et al. (2019). Biopreservation of fresh strawberries by carboxymethyl cellulose edible coatings enriched with a bacteriocin from *Bacillus methylotrophicus* BM47. *Food Technol. Biotechnol.* 57: 230–237.

133 Cortez-Vega, W.R., Pizato, S., De Souza, J.T.A., and Prentice, C. (2014). Using edible coatings from whitemouth croaker (*Micropogonias furnieri*) proteini solate and organo-clay nanocomposite for improve the conservation properties of fresh-cut 'Formosa' papaya. *Innov. Food Sci. Emerg. Technol.* 22: 197–202.

134 Salvia-Trujillo, L., Rojas-Graü, M.A., Soliva-Fortuny, R., and Martín-Belloso, O. (2015). Use of antimicrobial nanoemulsions as edible coatings: impact on safety and quality attributes of fresh-cut Fuji apples. *Postharvest Biol. Technol.* 105: 8–16.

135 Yousuf, B., Qadri, O.S., and Srivastava, A.K. (2018). Recent developments in shelf-life extension of fresh-cut fruits and vegetables by application of different edible coatings: a review. *LWT-Food Sci. Technol.* 89: 198–209.

136 Aytac, Z., Ipek, S., Durgun, E. et al. (2017). Antibacterial electrospun zein nanofibrous web encapsulating thymol/cyclo- dextrin-inclusion complex for food packaging. *Food Chem.* 233: 117–124.

137 Dang, X., Cao, X., Ke, L. et al. (2017). Combination of cellulose nanofibers and chain-end-functionalized polyethylene and their applications in nanocomposites. *J. Appl. Polym. Sci.* 134: 1–11.

138 Otoni, C.G., Pontes, S.F., Medeiros, E.A., and de Soares, N., F.F. (2014). Edible films from methylcellulose and nanoemulsions of clove bud (*Syzygium Aromaticum*) and oregano (*Origanum Vulgare*) essential oils as shelf life extenders for sliced bread. *J. Agric. Food Chem.* 62: 5214–5219.

139 Arfat, Y.A., Ahmed, J., Hiremath, N. et al. (2017). Thermo-mechanical, rheological, structural and antimicrobial properties of bionanocomposite films based on fish skin gelatin and silver-copper nanoparticles. *Food Hydrocolloids* 62: 191–202.

140 Moreira, M.R., Cassani, L., Martín-Belloso, O., and Soliva-Fortuny, R. (2015). Effects of polysaccharide-based edible coatings enriched with dietary fiber on quality attributes of fresh-cut apples. *J. Food Sci. Technol.* 52: 7795–7805.

141 Robledo, N., López, L., Bunger, A. et al. (2018). Effects of antimicrobial edible coating of thymol nanoemulsion/quinoa protein/chitosan on the safety, sensorial properties, and quality of refrigerated strawberries (Fragaria × Ananassa) under commercial storage environment. *Food Bioprocess Technol.* 11: 1566–1574.

142 Prakash, A., Baskaran, R., and Vadivel, V. (2020). Citral nanoemulsion incorporated edible coating to extend the shelf life of fresh cut pineapples. *LWT* 118: 108851.

143 Mohammadi, A., Hashemi, M., and Hosseini, S.M. (2016). Postharvest treatment of nanochitosan-based coating loaded with zataria multiflora essential oil improves antioxidant activity and extends shelf-life of cucumber. *Innov. Food Sci. Emerg. Technol.* 33: 580–588.

144 Cui, H., Abdel-Samie, M.A.S., and Lin, L. (2019). Novel packaging systems in grape storage – A review. *J. Food Process Eng.* 42 (6): e13162.

145 Koushesh Saba, M. and Amini, R. (2017). Nano-ZnO/carboxymethyl cellulose-based active coating impact on ready-to-use pomegranate during cold storage. *Food Chem.* 232: 721–726.

146 Sogvar, O.B., Saba, M.K., Emamifar, A., and Hallaj, R. (2016). Influence of nano-ZnO on microbial growth, bioactive content and postharvest quality of strawberries during storage. *Innov. Food Sci. Emerg. Technol.* 35: 168–176.

147 Meng, X., Zhang, M., and Adhikari, B. (2014). The effects of ultrasound treatment and nano-zinc oxide coating on the physiological activities of fresh-cut kiwifruit. *Food Bioprocess Technol.* 7: 126–132.

148 Xing, Y., Yang, H., Guo, X. et al. (2020). Effect of chitosan/nano-TiO_2 composite coatings on the postharvest quality and physicochemical characteristics of mango fruits. *Sci. Hortic.* 2020: 109135.

149 Taghinezhad, E. and Ebadollahi, A. (2017). Potential application of chitosan-clay coating on some quality properties of lemon during storage. *Agric. Eng. Int. CIGR J* 19: 189–194.

150 Pina-Barrera, A.M., Alvarez-Roman, R., Baez-Gonzalez, J.G. et al. (2019). Application of a multisystem coating based on polymeric nanocapsules containing essential oil of *Thymus vulgaris* L. to increase the shelf life of table grapes (*Vitis vinifera* L.). *IEEE Trans. Nanobiosci.* 18: 549–557.

151 Pilon, L., Spricigo, P.C., Miranda, M. et al. (2015). Chitosan nanoparticle coatings reduce microbial growth on fresh-cut apples while not affecting quality attributes. *Int. J. Food Sci. Technol.* 50: 440–448.

152 Feregrino-Perez, A.A., Magaña-López, E., Guzmán, C., and Esquivel, K. (2018). A general overview of the benefits and possible negative effects of the nanotechnology in horticulture. *Sci. Hortic.* 238: 126–137.

153 Abdelmalek, G.A.M. and Salaheldin, T.A. (2016). Silver nanoparticles as a potent fungicide for citrus phytopathogenic fungi. *J. Nanomed. Res.* 3: 00065.

154 He, L., Liu, Y., Mustapha, A., and Lin, M. (2011). Antifungal activity of zinc oxide nanoparticles against *Botrytis cinerea* and *Penicillium expansum*. *Microbiol. Res.* 166: 207–215.

155 Li, K., Zhu, J., Guan, G., and Wu, H. (2019). Preparation of chitosan-sodium alginate films through layer-by-layer assembly and ferulic acid crosslinking: film properties, characterization, and formation mechanism. *Int. J. Biol. Macromol.* 122: 485–492.

156 Kaveh, R., Li, Y.S., Ranjbar, S. et al. (2013). Changes in *Arabidopsis thaliana* gene expression in response to silver nanoparticles and silver ions. *Environ. Sci. Technol.* 47: 10637–10644.

157 Li, H., Li, F., Wang, L. et al. (2009). Effect of nano-packing on preservation quality of chinese jujube (*Ziziphus jujuba Mill.* var. *inermis (Bunge) Rehd*). *Food Chem.* 114: 547–552.

158 Singh, P. (2018). Nanotechnology in food preservation. *Food Sci. Res. J.* 9: 435–441.

159 Gallocchio, F., Belluco, S., and Ricci, A. (2015). Nanotechnology and food: brief overview of the current scenario. *Procedia Food Sci.* 5: 85–88.

160 Regulation (EC) No 1333/2008 of the European Parliament and of the Council of 16 December 2998 on Food Additives. *Off. J. Eur. Union,* 2008, L354: pp. 16–33.

161 Regulation (EU) 2015/2283 of the European parliament and of the Council of 25 November 2015 on novel foods, Amending Regulation (EU) No 1169/2011 of the European Parliament and of the Council and repealing Regulation (EC) No 258/97 of the European Parliament and of the Council and Commission Regulation (EC) No 1852/2001, *Off. J. Eur. Union L*, 2015, 327, 1–22.

162 EFSA (European Food Safety Authority), Scientific Committee, Hardy, A., Benford, D. et al. (2018). Guidance on risk assessment of the application of nanoscience and nanotechnologies in the food and feed chain: Part 1, human and animal health. *EFSA J.* 16: 5327.

163 Acevedo-Fani, A., Soliva-Fortuny, R., and Martín-Belloso, O. (2017). Nanostructured emulsions and nanolaminates for delivery of active ingredients: improving food safety and functionality. *Trends Food Sci. Technol.* 60: 12–22.

164 Naseer, B., Srivastava, G., Qadri, O.S. et al. (2018). Importance and health hazards of nanoparticles used in the food industry. *Nanotechnol. Rev.* 7 (6): 623–641.

165 Youssef, A.M. (2013). Polymer nanocomposites as a new trend for packaging applications. *Polym. Plast. Technol. Eng.* 52: 635–660.

166 Souza, V.G.L. and Fernando, A.L. (2016). Nanoparticles in food packaging: biodegradability and potential migration to food—a review. *Food Packag. Shelf Life* 8: 63–70.

15

Environmental and Toxicological Aspects of Nanostructures in Food Packaging

Sabarish Radoor[1], Jasila Karayil[2], Jyothi M. Shivanna[3], Aswathy Jayakumar[4], Sandhya A. Varghese[1], Radhakrishnan E. Krishnankutty[4], Jyotishkumar Parameswaranpillai[1], and Suchart Siengchin[1]

[1] *King Mongkut's University of Technology North Bangkok, The Sirindhorn International Thai-German Graduate School of Engineering (TGGS), Department of Mechanical and Process Engineering, 1518 Wongsawang Road, Bangsue, Bangkok 10800, Thailand*
[2] *Directorate of Technical Education, Government Women's Polytechnic College, Calicut, Kerala 673009, India*
[3] *Jain University, Jain Global Campus, Center for Nano and Material Sciences, Bangalore 562112, Karnataka, India*
[4] *Mahatma Gandhi University, School of Biosciences, P.D. Hills (P.O.), Kottayam, Kerala 686560, India*

15.1 Introduction

Among urban population, there is high demand for the packed and processed food items such as meat, fish, dairy products, etc. It is no doubt that in the coming years, the packed food will replace the conventional food storage system. Food packing sector witnessed a high demand in the last few decades and is escalating in a tremendous phase. Reports show that the global food packaging market size will reach US$ 456.6 billion by 2027. An ideal food packing should be bio- and eco-friendly. The following are the preferable properties for an ideal food packing [1, 2]:

1. It should be free from toxic materials.
2. It should be economical.
3. It should not alter the properties of food such as flavor, texture, aroma, nutritional value, etc.
4. It should protect food from moisture and light.
5. It should be stable toward temperature.
6. It should possess good mechanical properties.

The diverse functionality of nanoparticle (NP) makes it one of the promising packaging candidates for food packaging. Nano-based food packaging materials are classified as improved, active, and smart (or intelligent) food packaging. The improved food packaging material is fabricated by incorporating NP into polymer matrix. This type of food packing material displayed good mechanical and barrier properties. It also exhibits barrier properties against oxygen and moisture. Incorporation of NP also extends the shelf life of food materials. In active food packaging, the NP guards

Nanotechnology-Enhanced Food Packaging, First Edition. Edited by Jyotishkumar Parameswaranpillai, Radhakrishnan Edayileveettil Krishnankutty, Aswathy Jayakumar, Sanjay Mavinkere Rangappa, and Suchart Siengchin.

the food by interacting directly with it. Here the inherent property of NP such as antibacterial property, biodegradation, etc. will help in the protection of food. Intelligent or smart packaging is another category of food packaging material, which received considerable attention from researchers. It has the ability to sense oxygen level, specific smell, high temperature, pathogens, etc. [3–6].

15.2 Nanoparticles in Food Packaging

15.2.1 Nanoclay

Nanoclay is one of the highly explored nanostructured materials for food packaging application. Owing to its eco-friendly and cost-effective nature, the global consumption of nanoclay is increasing steeply, and by 2023 the global nanoclay market is expected to cross US$ 3.3 billion. The attractive features of nanoclay include unique platelet structure, high surface area (750 mg/g), and large aspect ratio (200–1000) [7, 8]. Studies show that nanoclay-based food packaging is superior in terms of barrier properties, chemical resistance, and thermal stability. It also shows improved resistance against bacteria, thereby extending the shelf life of food materials. The food packages developed using nanoclay are transparent and flexible, which make the product more convenient to use. Montmorillonite (MMT) is the most promising nanoclay for fabrication of food packaging materials, as it possesses fairly high aspect ratio, good compatibility with polymer matrix, and high surface area [9, 10]. Makwana et al. [11] developed a novel bionanocomposite film from agar, carboxymethyl cellulose (CMC), and Ag–MMT. The film has good mechanical, thermal, and antibacterial properties. The authors suggested it as potential material to extend the shelf life of packed food items. Nanoclay has been used to develop beer bottles and soft drinks and juice container. Nanoclay-based food packing is commercialized in many countries in different trade names such as Durethan, NanoTuff, Imperm, etc. [12].

15.2.2 Nanosilver

Nanosilver has been used to develop several active food packaging materials. The antibacterial, antifungal, antiviral, anti-mold properties of nanosilver enable it to safeguard packed food items such as vegetable, fruits, meat, etc. [13, 14]. Several mechanisms have been proposed to explain the antibacterial property of silver NPs. The high affinity of silver NPs for phosphorus and sulfur enables it to interact with the bacterial protein. Thus nanosilver can inhibit the replication of bacterial DNA and leads to cell damage [15]. There are ample reports on nanosilver-based polymer composite for food packaging application. Previous reports show that low-density polyethylene (LDPE)-Ag package is very effective to enhance the shelf life of food. For instance, Emamifar et al. [16] reported that LDPE/Ag and ZnONPs nanocomposite could improve the storage period of orange juice. Meanwhile, a good preservation of stored barberries was achieved using Ag-LDPE [17]. Ethylene vinyl alcohol (EVOH)/AgNP is reported to be a promising candidate to preserve chicken, pork, and cheese [18]. The composite protects the food by inhibiting the growth of *Escherichia coli* and *Staphylococcus aureus*, which is commonly seen in contaminated food [13]. Tavakoli et al. [19] in their work demonstrated that nanosilver can extend the shelf life of hazelnut, almond, pistachio, and walnut. Their studies show that nanosilver inhibits the growth of pathogens and decreases the peroxide content of the sample. Another unique feature of nanosilver is its ability to inhibit the production of ethylene. As a result, it can delay fruit ripening and thus extend the life of packed fruits. Bags Fresher Longer, Fresh Box Nanosilver Food Container, and Nanosilver Food are trade names of some of the commercially available nanosilver used in food packaging [20].

15.2.3 Zinc Oxide (ZnO) NP

According to the US Food and Drug Administration, ZnONP is one of the safest metal oxides, which could be used in food packaging without compromising its color and flavor. The attractive feature of ZnO is its low cost and good antibacterial activity. ZnO on entering the body releases Zn^{2+}, which is responsible for inducing antibacterial property. Food packaging incorporated with ZnO has excellent barrier property, UV blocking power, and good mechanical stability [21, 22]. Vasile et al. [23] developed a promising food packaging material by incorporating ZnO:Cu/Ag NPs into polylactic acid (PLA). The composites exhibit excellent barrier and biocidal activity. The migration of NPs into the food was also found to be low, which is quite advantageous for food packing application. Kumar and coworkers reported that agar/ZnO NP extends the shelf life of green grapes [24].

15.2.4 Titanium Dioxide (TiO_2)

TiO_2 is another NP that is employed for fabricating active food packaging. Owing to its high photostability, TiO_2 can minimize the UV degradation of packed food items. It also possesses remarkable antibacterial activity against *E. coli*, *S. aureus*, etc. TiO_2 is also reported to be a good oxygen scavenger. Hence TiO_2-incorporated food packaging is beneficial for enhancing the longevity of food [25]. Othman et al. [26] induces antibacterial property to LDPE film by coating it with TiO_2 NPs. The authors also observed that with increase in concentration of TiO_2, the antibacterial property also increases. Li et al. [27] proposed **nano-TiO_2/LDPE** (low-density polyethylene) as effective material to enhance the life of packed strawberry. Gumiero et al. [28] reported that **TiO_2 NP**-incorporated **HDPE** film provides structural and microbiological stability for cheese.

15.2.5 Silicon Dioxide (SiO_2)

On account of its nontoxic and bio-friendly nature, SiO_2NPs have been utilized to design novel food packing materials. The addition of nano-SiO_2 significantly improves the barrier, thermal, and mechanical property of food packaging [29, 30].

Table 15.1 The food packaging developed from copper, zinc, and carbon nanotube.

Nanomaterial	Food items	Observations	References
Cu/cellulose	Melon and pineapple juice	Excellent antifungal activity and reduced spoilage-related yeasts	[33]
Cu/polyethylene	Peda/Indian sweet diary product	Superior antimicrobial property against *E. coli* and *S. aureus*	[34]
Cu/HDPE		Good mechanical, oxygen barrier, and antibacterial activity	[35]
PHBV/g-CNTs	Beverage	Good water vapor barrier, mechanical, and thermal property	[36]
PLA/CNT/chitosan	Strawberry	Inhibit the growth of *E. coli* and *S. aureus*	[37]
ZnO/chitosan/Neem oil	Carrot	High tensile strength, barrier, and antibacterial property	[38]
PLA/ZnO		Good mechanical, barrier, and antibacterial property	[39]
ZnO/chitosan film		Thermal, mechanical, and antibacterial property	[40]
ZnO/chitosan/PE		Water affinity and antibacterial property against *E. coli, S. aureus,* and *S. enterica*	[41]

Wu et al. [31] reported an enhancement in the tensile property of polypropylene on incorporating SiO_2 NP. The incorporation of SiO_2 is also found to improve the water resistance property and oxygen barrier of polymer. This is beneficial for safeguarding food from spoilage and oxidation. Bi et al. [32] developed a novel antioxidant and antibacterial packaging using TPGS (D-α-tocopheryl polyethylene glycol 1000 succinate), chitosan (CS), and SiO_2NPs. The film exhibits superior mechanical, barrier, antioxidant, and antimicrobial property. The film is found to enhance the oxidative stability of soya bean oil.

The food packaging developed from copper, zinc, and carbon nanotube is presented in Table 15.1.

15.3 Toxicity Measurement of Nanoparticles Used in Food Industry

The properties of NPs such as size, charge, crystallinity, solubility, etc. are found to affect the toxic behavior of NP (Figure 15.1). The size of the NPs is inversely related to its toxicity. This is due to the correlation between size and surface area. Hence, small NP tends to interact strongly with biomolecules. Besides this, small NPs can

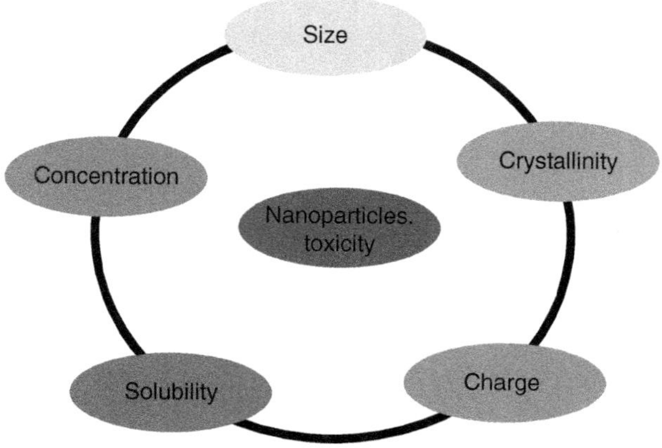

Figure 15.1 Different factors that affect the overall toxicity.

easily penetrate into the biological cell [42]. The size-dependent toxicity is reported for gold, silver, and zinc oxide NPs [43, 44]. Previous reports show that silver NPs with small size shows greater toxicity than large-sized counterpart. Meanwhile, the ZnO with low dimension has greater intestinal uptake. The toxicity of NP is concentration dependent. At high concentration, most of the NPs are toxic to humans. The toxicity of NPs could be tuned by varying its crystallinity. Reports show that crystalline NPs displayed greater toxicity than its amorphous analog. The influence of charge of NPs on the toxicological effect is also reported. Studies show that NPs with positive charge are more toxic than neutral and negative NPs. NPs with high positive charge have greater toxicity, which is attributed to the greater electrostatic interaction between positively charged NPs and negatively charged surface of mammalian cell wall. Solubility is another important property, which affects the toxicity of NPs. Highly soluble NPs are more toxic than insoluble one [45, 46].

15.4 Nanotoxicity

In recent years, there is rapid increase in the production of nano-based food packaging. However, there is risk factor associated with NPs as it can migrate from food packaging into food items. NP can enter human body through the skin, lungs, and gastrointestinal tract (Figure 15.2) [47].

The following are the toxic effects of NP used in food packaging:

15.4.1 Silver Nanoparticles (Ag NPs)

Silver NPs come into the environment through plastics, textile, cosmetic, medical, and food packaging industries. Although silver NP possesses good antibacterial activity, it does have some risk associated with it [13, 14]. Studies show that there is possibility of silver nanocomposite to migrate from food packaging and cause serious impact on humans [48]. Therefore, in the last decade researchers have

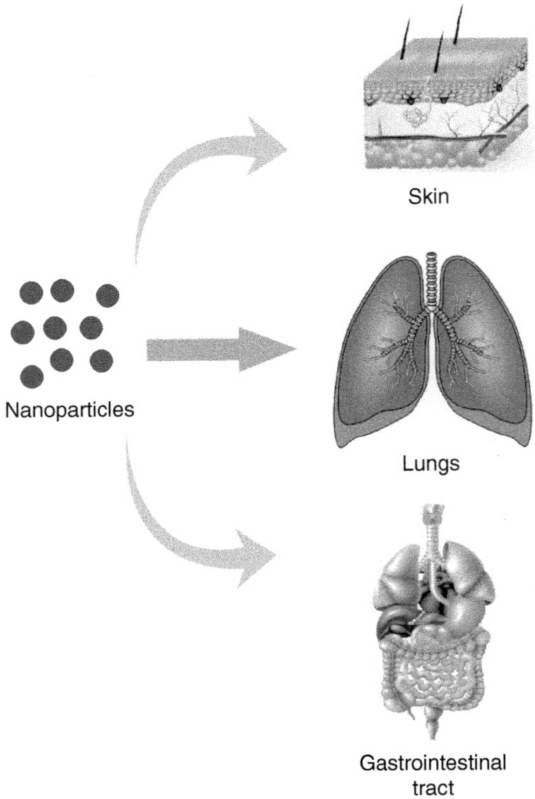

Figure 15.2 Possible route of nanoparticle entering into the body [47].

focused on the toxicological effect of nanosilver on humans as well as on animals. *In vitro* cytotoxicity studies revealed that nanosilver is toxic to rat and human fibroblasts [49–51]. The cytotoxic and genotoxic effect of silver NP on human cells is also reported. Asharani and coworker reported that nanosilver can induce the generation of reactive oxygen species (ROS). ROS can trigger apoptosis and is also responsible for damaging DNA and chromosome [52]. The toxicity of nanosilver is concentration dependent. At high concentration, silver NP causes genotoxic, hepatotoxic, and neurotoxic effect [53, 54]. Previous report shows that silver NP can migrate to the olfactory bulb and can concentrate on mitochondria, inducing apoptosis and cell death. The silver NP can also accumulate in several organs such as the liver, kidneys, testicle, and heart [55, 56]. Tang et al. [57] reported that silver NP causes irreversible neurologic toxicity in rats.

15.4.2 Titanium Nanoparticles (TiO_2 NPs)

TiO_2 NPs are the most used and furthermost manufactured nanomaterials with an annual production nearly 10 000 metric tons [58]. The extensive usage along with its abandoned disposal is indicating the great damage to ecosystem and living world [59]. The toxicity of TiO_2 is explained in three steps involving ROS production making electron–hole pairs under light irradiation of suitable energy, damage of cell walls and its lipid peroxidation, and finally outflow of cytoplasm, leading to attachment of TiO_2 to intercellular organelles [60]. In case of microorganisms, it is reported that just titanium with Ti^{4+} ion state could not contribute for its toxicity since it is insoluble in water [61]. But during metabolism aerobic cells produce ROS including superoxide anions, hydroxyl anion, and hydrogen peroxides, which are responsible for the cell damage as well as bacterial cell membranes [62]. Studies have reported the linear relation between concentration of OH ions and the cell viability. Hence it is understood that TiO_2 is toxic for microorganisms and that preconditions for ROS formation are TiO_2 in its positively charged form adsorbed on the surface of bacteria via electrostatic force [63]. The adsorbed titania endorse lipid peroxidation of cell membrane and ease the fluidity [64]. With an increase in ROS, lipid peroxidation increases and damages the cell membrane. As the cell membrane damages, titania will now enter into the cell and will be attached to cell organelles, leading to protein denaturation [65]. Surprisingly, electrostatic forces repel the titania, due to like polarity and also titania of around 50 nm particle size could cloud up and form several hundred nanometers, making the bigger size and creating difficulty in penetrating the cell [66]. The amount of light irradiated will also play an important role in the whole process. Overall, in case of microbes, titania could adsorb the light and promote ROS formation and intern promoting oxidative stress in bacteria [67].

In addition to oxidative stress, ROS also induce apoptosis. This oxidative stress damages lipids, proteins, carbohydrates, and DNAs. As discussed, lipid damage leads to cell membrane changes resulting in misfunction of key cellular functions, and DNA damage is due to formation of OH free radicals. Oxidative stress in continuation with titania exposure reduced the levels of hepatic malondialdehyde, leading to increase in the levels of glutamate, which is a reduced form of glutathione [68]. Excessive glutamate dysfunctions the neurons and results in neurodegenerative diseases. Studies have reported the activation of ionotropic *N*-methyl-D-aspartate (NMDA) receptors due to elevated levels of glutamate in hippocampal neurons [69]. These over-activated receptors open up Ca^{2+} channels, making the flow of Ca^{2+} to neurons. In the mean time, titania exposure depletes the glutathione of hepatic neurons and decreases Na^+K^+ – ATPase activity is the gills and intestines of *Danio rerio* as reported by Chen et al. [70]. Overall, the effect of this whole process leads to imbalance in ionic homeostasis of cells disrupting the growth, survival, and physiology of organisms. Exposure of titania has shown to induce apoptosis of microglia mouse brain and hippocampal neurons of rats, leading to gathering of Na^+ ions and decreased cytoplasmic K^+ ions [71]. However, the actual elucidation of titania toxicity needs further studies as it reaches environment in different forms and along with different NPs. The food chain effect of titania transmitting to offspring and reproduction species is yet to be explored along with its cross-time toxicity effects.

15.4.3 Silica Nanoparticle

Toxicological effect of food grade silica NPs on biological systems is less touched research topic compared with many other commercial NPs. The probable interactions between silica nanomaterials and food components or biological components are not usually focused [72]. Most of the work reported on *in vitro* and *in vivo* toxicity of silica NPs explained the dependence of toxicity on surface area and size. Amorphous silica has shown the higher toxicity than the submicron-sized one [73]. Si NPs less than 20 mg/ml showed their ineffectiveness toward cytotoxicity for RAW 264.7 macrophages [74]. The size-dependent studies were in contrast; one of the studies showed that the cells viability decreased with increased NPs cells and vice versa [75]. *In vitro* studies on silica NPs concluded multiple effects such as induction of endocytosis depending on ROS generation, resulting in damage of DNA, protein aggregates formation, endorsement of apoptotic signaling, cell cycle arrest, and membrane disruption [76]. Macrophages of the cells first capture the silica NPs upon exposure and are allowed to aggregate in vesicles and phagolysosomes [77]. Silica NPs also affect the gene expression-related activities. The phagocytosis is found to reduce by 50% upon exposure of Si NPs of 10 nm size [78]. Styrene oxide, a DNA-alkylating agent, is also found sensitive to silica [79]. Apart from the effects of physicochemical properties of silica NPs on biological components, the effect of zeta potentials is yet to be explored.

15.4.4 Clay Nanoparticle

Nanoclay for food packaging is emerging to be the most growing market and has reached a market of US$ 343.0 million in 2014 and predicted to reach high by 2022 [12]. The toxicity for nanoclay is found to be less than nanosilica and nanotitania. However, as in case of nano titania and silica, mitochondrial damage on exposure to nano clay and its derivatives is reported [80]. *In vitro* analysis of modified bentonite clay damages the human B lymphoblast cells [81]. Besides, the toxicity of nanoclay varied with modifying material, concentration, selected experimental models, and exposure time [82]. Clays modified with quaternary ammonium salts showed less toxicity. Nature of clay on toxicity was also investigated with an aid of HUVE cells and found to be Mt > bentonite = kaolinite > berionite [83]. Several biomarkers such as ROS generation, glutathione content, and also changes in organelles like mitochondria and lysosomes are explored to find the mechanism. The toxicity is assessed depending on the alteration in the metabolic rates, growth of cells, and transcription of genes. Likewise, in nanosilica, in case of nanoclay also, many contradictory results are observed. The response to oxidative stress varied with different cell lines exposed and the concentration [84]. DNA damage is another key factor for genotoxicity studies. EFSA has confirmed that MMTs underwent series of tests and found to be non-genotoxic [85]. Many reports established non-genotoxic effect or mutagenic potential of Mt [7, 85]. Reports on inflammation responses suggested the cytotoxicity against macrophage type cells along with hemolytic activity with respect to red blood cells; it must be noted that Mt. clay upholds neutrophils infection and makes them unavailable for phagocytosis of bacteria [86]. However, reports on *in vivo* studies are too less, and the mechanism still needs a scientific justification.

The toxic effect of zinc, copper, and carbon nanotube is displayed in Table 15.2.

Table 15.2 The toxic effect of zinc, copper, and carbon nanotube.

Nanoparticles	Toxic effects	References
Zinc oxide nanoparticles (ZnO NPs)	• It induces the production of reactive oxygen species (ROS) • High concentration of ZnO NPs causes apoptotic damage, oxidative stress, inflammation, and cell viability • ZnO NPs are reported to cause liver and kidney damage in mice • ZnO nanoparticles are toxic to aquatic organism • Fish exposed to ZnO nanoparticles has swollen epithelium cell	[87]
Copper nanoparticles	• Cytotoxic to human extra villous trophoblast cells • It can adversely affect the biosynthesis of amino acid and DNA • It induces the formation of reactive oxygen species (ROS) • It causes lipid peroxidation and protein oxidation • It is cytotoxic to mice and adversely affects reproductive system of mice • It causes hepatocytic necrosis in rat • High concentration of copper nanoparticles affects the kidney, liver, and spleen of mice • It adversely affects the mammalian and piscine cell lines	[88, 89]
Carbon nanotubes (CNTs)	• It can migrate from food packing and adversely affect the health of human beings • It can trigger the generation of free radial • It induces pulmonary toxicity, cytotoxicity, and genotoxicity • It causes detrimental effect on DNA • It induces the formation of ROS • It triggered the formation of granulomas and thus could cause chronic granulomatous diseases	[90–92]

15.5 Migration Issues of Nanoparticles

The use of NPs in food packaging likely involves the lower amounts of them packed into other polymers that sufficiently increase the performance of the material. These small amounts of NPs offer fire-resistant property and light and strong mechanical and thermal aspects, and they also inhibit the growth of microbes [93, 94]. Hence NPs in food packing also help in food quality preservation. Despite these potential applications, an important question arises regarding the movement of nanomaterials within the polymer composite matrix and also to the food materials. The migration of nanomaterials from polymer matrix to food material is due to pressure-driven phenomenon taking place under mechanical abrasion, UV exposure, or swelling interactions, and the migration is explained by Fickian diffusion process. In many cases this chemical release is due to weak bonding forces between polymer matrix and the NPs [95]. A semi-theoretical approach to explain the migration mechanism was first published by Simon et al. (2008); since then a large number of publications have come up. Maximum reports have considered migration of silver ions and were analyzed by ICP–MS and AAS. However, these studies fail to differentiate between ionic form of silver and nanoparticulate one. Migration of NPs was purely dependent on polymer material. One of the best examples is that polypropylene containers and polyethylene bags having silver NPs were tested for migration of silver ions. From containers, the migration was more and very less from bags, though bag had higher loading of silver than containers [96]. The containers are usually LDPE having a thick wall of millimeter range and will have lot of cut edge areas wherein maximum NPs are in direct contact with migration stimulants, whereas in thin, soft bags, the cut edge areas are less, and hence lesser amount of NPs are exposed to stimulants. This scientific theory was also supported by another two reports. Low dynamic viscosity of the polymer is another key factor, which concluded the migration of nanoclay from polyethylene. Along with that, sizes of nanomaterials also influenced the migration properties. Silver particles of 10 and 50 nm size were embedded in highly plasticized polyvinylchloride (PVC) films, and the migrations of both sizes to chicken breast were compared [97]. Change in sizes did not affect the migration much, and it was found to be around 1%. The competition between migration of two kinds of nanomaterials was studied by embedding silver and copper NPs into polyethylene film but failed to investigate which material migrated first to chicken breast.

Another important factor is the extent of migration – after migration of NPs from packaging materials to food. Once ingested the migration into bloodstream or adsorption or absorption by various organelles is found to be important. Inhalation, direct exposure of skin, and ingestion are the main probable ways for NPs to enter the body. Some of the studies reported migration of nanomaterials to the gastrointestinal tract, and also few will form bio-corona, nevertheless affecting the immunotoxicity and cell death [98]. However, very less research has been found on this aspect, giving opportunity for the young researchers to work upon. Even though numerous studies have been done on the migration part of nanomaterials, the results are not in agreement and are contradictory in few cases.

15.6 Environmental Impacts of Nanoparticles

The mass of NPs in the environment is guarded by two major factors, physicochemical properties of NPs and their interaction and competition with other pollutants in environment. Most of the food packaging materials reaching the domestic purpose are usually disposed into environment and their accumulation in different environment matrices like air, soil, and water. NPs exposed to air are more likely to further undergo sunlight and UV light exposure, leading to photochemical changes. These nanomaterials experience multiple transformations like size dissipation and condensation of volatile compounds [99]. In case of exposure to soil, NPs enter to soil pores and adhere to those particles because of their higher surface area. Silver NPs are found to aggregate and form Ag^+ ions soil-based sink depending on their concentration. The adsorbed NPs on soil may be taken up by plants, and the growth and germination will be affected [100]. The biodiversity of microorganism communities in the soil was also affected by titania and ZnO NPs [101]. Accumulation, degradation, diffusion, and interaction of NPs to aquatic environment for both flora and fauna are enormous.

Nevertheless, along with food packaging, many general positive aspects of NPs on environment involve reducing weights of aircraft and saving tons of fuel, increasing energy conversion efficiency, and producing clean energy, flexible rolls and solar panels, and many more. NPs are successfully implemented in purification of air, water, and soil [102, 103].

15.7 Conclusion

As NPs improve the mechanical, thermal, antibacterial, and barrier properties of food packaging system, there is a steep increase in the use of nanotechnology for the development of active and intelligent food packaging. However, recent studies show that NPs can migrate from food packaging, thus causing health issues in humans and animals. Hence in the past decades, researchers have monitored the fate of NP on entering the human body. The toxicity and environmental effect of NP is also studied by various researchers. The current report deals with toxic effect of different NPs used in food packaging. The toxicity of NPs could be tuned by varying its size, shape, concentration, charge, and solubility. Previous reports show that NPs can cross the cell barrier and enter the bloodstream. Hence it can interact with biomolecules and can adversely affect the cytokine production, cellular uptake, and immune system. NPs can also accumulate in several organs and can adversely affect its function. Studies show that NPs can induce apoptosis, genotoxicity, neurotoxicity, and oxidative stress. It is also responsible for low ATP production. NPs can accumulate in food chain and thus could adversely affect different organism. Thus, a proper risk and safety assessment of NPs is crucial.

Acknowledgments

The authors are thankful for the financial support by the King Mongkut's University of Technology North Bangkok (KMUTNB), Thailand, and grant funded by the Postdoctoral scholarship, Thailand (Grant No. KMUTNB-63-Post-03 to SR), and (Grant No. KMUTNB-BasicR-64-16).

References

1 Siegrist, M., Stampfli, N., Kastenholz, H., and Keller, C. (2008). Perceived risks and perceived benefits of different nanotechnology foods and nanotechnology food packaging. *Appetite* 51 (2): 283–290.

2 Mukhopadhyay, S.S. (2014). Nanotechnology in agriculture: prospects and constraints. *Nanotechnol. Sci. Appl.* 7: 63–71.

3 Sozer, N. and Kokini, J.L. (2009). Nanotechnology and its applications in the food sector. *Trends Biotechnol.* 27 (2): 82–89.
4 Bumbudsanpharoke, N. and Ko, S. (2015). Nano-food packaging: an overview of market, migration research, and safety regulations. *J. Food Sci.* 80 (5): R910–R923.
5 Rashidi, L. and Khosravi-Darani, K. (2011). The applications of nanotechnology in food industry. *Crit. Rev. Food Sci. Nutr.* 51 (8): 723–730.
6 Huang, Y., Mei, L., Chen, X., and Wang, Q. (2018). Recent developments in food packaging based on nanomaterials. *Nanomaterials* 8 (10): 1–29.
7 Kuorwel, K.K., Cran, M.J., Orbell, J.D. et al. (2015). Review of mechanical properties, migration, and potential applications in active food packaging systems containing nanoclays and nanosilver. *Compr. Rev. Food Sci. Food Saf.* 14 (4): 411–430.
8 Majeed, K., Jawaid, M., Hassan, A. et al. (2013). Potential materials for food packaging from nanoclay/natural fibres filled hybrid composites. *Mater. Design* 46: 391–410.
9 Khalaj, M.-J., Ahmadi, H., Lesankhosh, R., and Khalaj, G. (2016). Study of physical and mechanical properties of polypropylene nanocomposites for food packaging application: Nano-clay modified with iron nanoparticles. *Trends Food Sci. Technol.* 51: 41–48.
10 Rhim, J.-W., Park, H.-M., and Ha, C.-S. (2013). Bio-nanocomposites for food packaging applications. *Prog. Polym. Sci.* 38 (10–11): 1629–1652.
11 Makwana, D., Castaño, J., Somani, R.S., and Bajaj, H.C. (2020). Characterization of Agar-CMC/Ag-MMT nanocomposite and evaluation of antibacterial and mechanical properties for packaging applications. *Arabian J. Chem.* 13 (1): 3092–3099.
12 Bumbudsanpharoke, N. and Ko, S. (2019). Nanoclays in food and beverage packaging. *J. Nanomater.* 2019: 1–13.
13 Carbone, M., Donia, D.T., Sabbatella, G., and Antiochia, R. (2016). Silver nanoparticles in polymeric matrices for fresh food packaging. *J. King Saud Uni. Sci.* 28 (4): 273–279.
14 Echegoyen, Y. and Nerín, C. (2013). Nanoparticle release from nano-silver antimicrobial food containers. *Food Chem. Toxicol.* 62: 16–22.
15 Sen, S. and Pathak, Y. (2017). *Nanotechnology in Nutraceuticals: Production to Consumption*. Boca Raton: CRC Press, Taylor & Francis Group. xxi, 465 p.
16 Emamifar, A., Kadivar, M., Shahedi, M., and Soleimanian-Zad, S. (2010). Evaluation of nanocomposite packaging containing Ag and ZnO on shelf life of fresh orange juice. *Innovative Food Sci. Emerg. Technol.* 11 (4): 742–748.
17 Valipoor Motlagh, N., Hamed Mosavian, M.T., and Mortazavi, S.A. (2013). Effect of polyethylene packaging modified with silver particles on the microbial, sensory and appearance of dried barberry. *Packag. Technol. Sci.* 26 (1): 39–49.
18 Martínez-Abad, A., Lagaron, J.M., and Ocio, M.J. (2012). Development and characterization of silver-based antimicrobial ethylene–vinyl alcohol copolymer (EVOH) films for food-packaging applications. *J. Agric. Food. Chem.* 60 (21): 5350–5359.

19 Tavakoli, H., Rastegar, H., Taherian, M. et al. (2017). The effect of nano-silver packaging in increasing the shelf life of nuts: an in vitro model. *Italian Journal of Food Safety* 6 (4): 156–161.

20 Nakazato, G., RKT, K., Seabra, A.B., and Duran, N. (2017). Use of nanoparticles as a potential antimicrobial for food packaging. *Food Preserv.*: 413–447.

21 Espitia, P.J.P., Soares, N.F.F., Coimbra, J.S.R. et al. (2012). Zinc oxide nanoparticles: synthesis, antimicrobial activity and food packaging applications. *Food Bioprocess Technol.* 5 (5): 1447–1464.

22 Espitia, P.J.P., Otoni, C.G., and NFF, S. (2016). Zinc oxide nanoparticles for food packaging applications. *Antimicrob. Food Packag.* 8: 425–431.

23 Vasile, C., Rapa, M., Stefan, M. et al. (2017). New PLA/ZnO:Cu/Ag bionanocomposites for food packaging. *Exp. Polym. Lett.* 11 (7): 531–544.

24 Kumar, S., Boro, J.C., Ray, D. et al. (2019). Bionanocomposite films of agar incorporated with ZnO nanoparticles as an active packaging material for shelf life extension of green grape. *Heliyon* 5 (6): 1–8.

25 Verma, D.K., Goyal, M.R., and Suleria, H. (2020). *Nanotechnology and Nanomaterial Applications in Food, Health, and Biomedical Sciences*. Oakville, ON; Palm Bay, Florida: Apple Academic Press. xxx, 298 p.

26 Othman, S.H., Abd Salam, N.R., Zainal, N. et al. (2014). Antimicrobial activity of TiO_2 Nanoparticle-coated film for potential food packaging applications. *Int. J. Photoenergy* 2014: 1–6.

27 Li, D., Ye, Q., Jiang, L., and Luo, Z. (2017). Effects of nano-TiO_2-LDPE packaging on postharvest quality and antioxidant capacity of strawberry (*Fragaria ananassa* Duch.) stored at refrigeration temperature. *J. Sci. Food Agric.* 97 (4): 1116–1123.

28 Gumiero, M., Peressini, D., Pizzariello, A. et al. (2013). Effect of TiO_2 photocatalytic activity in a HDPE-based food packaging on the structural and microbiological stability of a short-ripened cheese. *Food Chem.* 138 (2–3): 1633–1640.

29 Ocak, B. (2019). Development of the mechanical and barrier properties of collagen hydrolysate/carboxymethyl cellulose films by using SiO_2 nanoparticles. *Pamukkale Univ. J. Eng. Sci.* 25 (3): 320–324.

30 Aji, A.I., Praseptiangga, D., Rochima, E. et al. (2018). Optical transparency and mechanical properties of semi-refined iota carrageenan film reinforced with SiO_2 as food packaging material. In: *AIP Conference Proceedings*, 030039-1–030039-6. AIP Publishing.

31 Wu, C., Zhang, M., Rong, M., and Friedrich, K. (2005). Silica nanoparticles filled polypropylene: effects of particle surface treatment, matrix ductility and particle species on mechanical performance of the composites. *Compos. Sci. Technol.* 65 (3–4): 635–645.

32 Bi, F., Zhang, X., Liu, J. et al. (2020). Development of antioxidant and antimicrobial packaging films based on chitosan, D-α-tocopheryl polyethylene glycol 1000 succinate and silicon dioxide nanoparticles. *Food Packag. Shelf Life* 24.

33 Llorens, A., Lloret, E., Picouet, P.A. et al. (2012). Metallic-based micro and nanocomposites in food contact materials and active food packaging. *Trends Food Sci. Technol.* 24 (1): 19–29.

34 Lomate, G.B., Dandi, B., and Mishra, S. (2018). Development of antimicrobial LDPE/Cu nanocomposite food packaging film for extended shelf life of peda. *Food Packag. Shelf Life* 16: 211–219.

35 Bikiaris, D.N. and Triantafyllidis, K.S. (2013). HDPE/Cu-nanofiber nanocomposites with enhanced antibacterial and oxygen barrier properties appropriate for food packaging applications. *Mater. Lett.* 93: 1–4.

36 Yu, H.-Y., Qin, Z.-Y., Sun, B. et al. (2014). Reinforcement of transparent poly(3-hydroxybutyrate-co-3-hydroxyvalerate) by incorporation of functionalized carbon nanotubes as a novel bionanocomposite for food packaging. *Compos. Sci. Technol.* 94: 96–104.

37 Liu, Y., Wang, S., Lan, W., and Qin, W. (2019). Fabrication of polylactic acid/carbon nanotubes/chitosan composite fibers by electrospinning for strawberry preservation. *Int. J. Biol. Macromol.* 121: 1329–1336.

38 Sanuja, S., Agalya, A., and Umapathy, M.J. (2015). Synthesis and characterization of zinc oxide–neem oil–chitosan bionanocomposite for food packaging application. *Int. J. Biol. Macromol.* 74: 76–84.

39 Marra, A., Silvestre, C., Duraccio, D., and Cimmino, S. (2016). Polylactic acid/zinc oxide biocomposite films for food packaging application. *Int. J. Biol. Macromol.* 88: 254–262.

40 Priyadarshi, R. and Negi, Y.S. (2016). Effect of varying filler concentration on zinc oxide nanoparticle embedded chitosan films as potential food packaging material. *J. Polym. Environ.* 25 (4): 1087–1098.

41 Al-Naamani, L., Dobretsov, S., and Dutta, J. (2016). Chitosan-zinc oxide nanoparticle composite coating for active food packaging applications. *Innovative Food Sci. Emerg. Technol.* 38: 231–237.

42 Duncan, T.V. (2011). Applications of nanotechnology in food packaging and food safety: barrier materials, antimicrobials and sensors. *J. Colloid Interface Sci.* 363 (1): 1–24.

43 Yu, K.O., Grabinski, C.M., Schrand, A.M. et al. (2008). Toxicity of amorphous silica nanoparticles in mouse keratinocytes. *J. Nanopart. Res.* 11 (1): 15–24.

44 Borm, P.J.A. and Kreyling, W. (2004). Toxicological hazards of inhaled nanoparticles—potential implications for drug delivery. *J. Nanosci. Nanotechnol.* 4 (5): 521–531.

45 Dhawan, A. and Sharma, V. (2010). Toxicity assessment of nanomaterials: methods and challenges. *Anal. Bioanal.Chem.* 398 (2): 589–605.

46 Aillon, K.L., Xie, Y., El-Gendy, N. et al. (2009). Effects of nanomaterial physicochemical properties on in vivo toxicity. *Adv. Drug Delivery Rev.* 61 (6): 457–466.

47 Naseer, B., Srivastava, G., Qadri, O.S. et al. (2018). Importance and health hazards of nanoparticles used in the food industry. *Nanotechnol. Rev.* 7 (6): 623–641.

48 Singh, Z., Randhawa, J.K., Wilkhu, S.S., and Sharma, S. (2018). Synthesis, applications and toxicological aspects of silver nanoparticles. *Int. J. Res. Studies Zool.* 4 (2): 17–20.
49 Braydich-Stolle, L., Hussain, S., Schlager, J.J., and Hofmann, M.-C. (2005). In vitro cytotoxicity of nanoparticles in mammalian germline stem cells. *Toxicol. Sci.* 88 (2): 412–419.
50 Hussain, S.M., Hess, K.L., Gearhart, J.M. et al. (2005). In vitro toxicity of nanoparticles in BRL 3A rat liver cells. *Toxicol. In Vitro* 19 (7): 975–983.
51 Wen, H.-C., Lin, Y.-N., Jian, S.-R. et al. (2007). Observation of growth of human fibroblasts on silver nanoparticles. *J. Phys. Conf. Ser.* 61: 445–449.
52 AshaRani, P.V., Low Kah Mun, G., Hande, M.P., and Valiyaveettil, S. (2008). Cytotoxicity and genotoxicity of silver nanoparticles in human cells. *ACS Nano* 3 (2): 279–290.
53 El Mahdy, M.M., Eldin, T.A.S., Aly, H.S. et al. (2015). Evaluation of hepatotoxic and genotoxic potential of silver nanoparticles in albino rats. *Exp. Toxicol. Pathol.* 67 (1): 21–29.
54 Gonzalez-Carter, D.A., Leo, B.F., Ruenraroengsak, P. et al. (2017). Silver nanoparticles reduce brain inflammation and related neurotoxicity through induction of H2S-synthesizing enzymes. *Scientific Reports* 7 (1): 1–14.
55 VanWinkle, B.A., De Mesy Bentley, K.L., Malecki, J.M. et al. (2009). Nanoparticle (NP) uptake by type I alveolar epithelial cells and their oxidant stress response. *Nanotoxicology* 3 (4): 307–318.
56 Takenaka, S., Karg, E., Roth, C. et al. (2001). Pulmonary and systemic distribution of inhaled ultrafine silver particles in rats. *Environ. Health Perspect.* 109 (suppl 4): 547–551.
57 Tang, S. and Zheng, J. (2018). Antibacterial activity of silver nanoparticles: structural effects. *Adv. Healthcare Mater.* 7 (13): 1–10.
58 Szymańska, R., Kołodziej, K., Ślesak, I. et al. (2016). Titanium dioxide nanoparticles (100–1000 mg/l) can affect vitamin E response in *Arabidopsis thaliana*. *Environ. Pollut.* 213: 957–965.
59 Khosravi-Katuli, K., Prato, E., Lofrano, G. et al. (2017). Effects of nanoparticles in species of aquaculture interest. *Environ. Sci. Pollut. Res.* 24 (21): 17326–17346.
60 Hou, J., Wang, L., Wang, C. et al. (2019). Toxicity and mechanisms of action of titanium dioxide nanoparticles in living organisms. *J. Environ. Sci.* 75: 40–53.
61 Tong, T., Binh, C.T.T., Kelly, J.J. et al. (2013). Cytotoxicity of commercial nano-TiO_2 to Escherichia coli assessed by high-throughput screening: effects of environmental factors. *Water Res.* 47 (7): 2352–2362.
62 Priyanka, K.P., Varghese, T., Sukirtha, T.H., and Balakrishna, K.M. (2016). Microbicidal activity of TiO_2 nanoparticles synthesised by sol–gel method. *IET Nanobiotechnol.* 10 (2): 81–86.
63 Kiser, M.A., Westerhoff, P., Benn, T. et al. (2009). Titanium nanomaterial removal and release from wastewater treatment plants. *Environ. Sci. Technol.* 43 (17): 6757–6763.

64 Yin, J.-J., Liu, J., Ehrenshaft, M. et al. (2012). Phototoxicity of nano titanium dioxides in HaCaT keratinocytes—generation of reactive oxygen species and cell damage. *Toxicol. Appl. Pharmacol.* 263 (1): 81–88.

65 Applerot, G., Lipovsky, A., Dror, R. et al. (2009). Enhanced antibacterial activity of nanocrystalline ZnO due to increased ROS-mediated cell injury. *Adv. Funct. Mater.* 19 (6): 842–852.

66 Du, J. and Gebicki, J.M. (2004). Proteins are major initial cell targets of hydroxyl free radicals. *Int. J. Biochem. Cell Biol.* 36 (11): 2334–2343.

67 Dasari, T.P., Pathakoti, K., and Hwang, H.-M. (2013). Determination of the mechanism of photoinduced toxicity of selected metal oxide nanoparticles (ZnO, CuO, Co_3O_4 and TiO_2) to E. coli bacteria. *J. Environ. Sci.* 25 (5): 882–888.

68 Rajapakse, K., Drobne, D., Valant, J. et al. (2012). Acclimation of Tetrahymena thermophila to bulk and nano-TiO_2 particles by changes in membrane fatty acids saturation. *J. Hazard. Mater.* 221–222: 199–205.

69 Hong, F., Sheng, L., Ze, Y. et al. (2015). Suppression of neurite outgrowth of primary cultured hippocampal neurons is involved in impairment of glutamate metabolism and NMDA receptor function caused by nanoparticulate TiO_2. *Biomaterials* 53: 76–85.

70 Chen, J., Dong, X., Xin, Y., and Zhao, M. (2011). Effects of titanium dioxide nano-particles on growth and some histological parameters of zebrafish (*Danio rerio*) after a long-term exposure. *Aquat. Toxicol.* 101 (3–4): 493–499.

71 Zhao, J., Xu, L., Zhang, T. et al. (2009). Influences of nanoparticle zinc oxide on acutely isolated rat hippocampal CA3 pyramidal neurons. *Neuro Toxicol.* 30 (2): 220–230.

72 An, S.S.A., Lee, J.-A., Kim, M.-K. et al. (2014). Tissue distribution and excretion kinetics of orally administered silica nanoparticles in rats. *International Journal of Nanomedicine* 9: 251–260.

73 Kim, H.W., Ahn, E.-K., Jee, B.K. et al. (2008). Nanoparticulate-induced toxicity and related mechanism in vitro and in vivo. *J. Nanopart. Res.* 11 (1): 55–65.

74 Choi, J., Zhang, Q., Reipa, V. et al. (2009). Comparison of cytotoxic and inflammatory responses of photoluminescent silicon nanoparticles with silicon micron-sized particles in RAW 264.7 macrophages. *J. Appl. Toxicol.* 29 (1): 52–60.

75 Yuan, Y., Liu, C., Lu, J. et al. (2011). In vitro cytotoxicity and induction of apoptosis by silica nanoparticles in human HepG2 hepatoma cells. *Int. J. Nanomed.*

76 Nabeshi, H., Yoshikawa, T., Matsuyama, K. et al. (2011). Amorphous nanosilica induce endocytosis-dependent ROS generation and DNA damage in human keratinocytes. *Part. Fibre Toxicol.* 8 (1).

77 Marzaioli, V., Groß, C., Weichenmeier, I. et al. (2017). Specific surface modifications of silica nanoparticles diminish inflammasome activation and in vivo expression of selected inflammatory genes. *Nanomaterials* 7 (11).

78 Tyner, K., Bancos, S., and Stevens, D. (2014). Effect of silica and gold nanoparticles on macrophage proliferation, activation markers, cytokine production, and phagocytosis in vitro. *Int. J. Nanomed.*

79 Dalzon, B., Aude-Garcia, C., Collin-Faure, V. et al. (2017). Differential proteomics highlights macrophage-specific responses to amorphous silica nanoparticles. *Nanoscale* 9 (27): 9641–9658.

80 Maisanaba, S., Gutiérrez-Praena, D., Pichardo, S. et al. (2014). Toxic effects of a modified montmorillonite clay on the human intestinal cell line Caco-2. *J. Appl. Toxicol.* 34 (6): 714–725.

81 Zhang, M., Lu, Y., Li, X. et al. (2010). Studying the cytotoxicity and oxidative stress induced by two kinds of bentonite particles on human B lymphoblast cells in vitro. *Chem. Biol. Interact.* 183 (3): 390–396.

82 Janer, G., Fernández-Rosas, E., Mas del Molino, E. et al. (2013). In vitrotoxicity of functionalised nanoclays is mainly driven by the presence of organic modifiers. *Nanotoxicology* 8 (3): 279–294.

83 Murphy, E.J., Roberts, E., and Horrocks, L.A. (1993). Aluminum silicate toxicity in cell cultures. *Neuroscience* 55 (2): 597–605.

84 Sharma, A.K., Mortensen, A., Schmidt, B. et al. (2014). In-vivo study of genotoxic and inflammatory effects of the organo-modified montmorillonite Cloisite® 30B. *Mutat. Res./Genet. Toxicol. Environ. Mutagen.* 770: 66–71.

85 Aquilina, G., Bach, A., Bampidis, V. et al. (2013). Scientific opinion on the safety and efficacy of a preparation of bentonite-and sepiolite (Toxfin®dry) as feed additive for all species. *EFSA J.* 11 (4).

86 Dougherty, S.H., Flegel, V.D., Nelson, R.D. et al. (1985). Effects of soil infection potentiating factors on neutrophils in vitro. *Am. J. Surg.* 150 (3): 306–311.

87 Liu, J., Kang, Y., Yin, S. et al. (2017). Zinc oxide nanoparticles induce toxic responses in human neuroblastoma SHSY5Y cells in a size-dependent manner. *Int. J. Nanomed.* 12: 8085–8099.

88 Song, L., Connolly, M., Fernández-Cruz, M.L. et al. (2013). Species-specific toxicity of copper nanoparticles among mammalian and piscine cell lines. *Nanotoxicology* 8 (4): 383–393.

89 Zhang, C.-H., Wang, Y., Sun, Q.-Q. et al. (2018). Copper nanoparticles show obvious in vitro and in vivo reproductive toxicity via ERK mediated signaling pathway in female mice. *Int. J. Biol. Sci.* 14 (13): 1834–1844.

90 Mohanta, D., Patnaik, S., Sood, S., and Das, N. (2019). Carbon nanotubes: evaluation of toxicity at biointerfaces. *J. Pharm. Anal.* 9 (5): 293–300.

91 Barna, B., Judson, M., and Thomassen, M. (2014). Carbon nanotubes and chronic granulomatous disease. *Nanomaterials* 4 (2): 508–521.

92 Helland, A., Wick, P., Koehler, A. et al. (2007). Reviewing the environmental and human health knowledge base of carbon nanotubes. *Environ. Health Perspect.* 115 (8): 1125–1131.

93 Kotsilkova, R., Petkova, V., and Pelovski, Y. (2001). *J. Therm. Anal. Calorim.* 64 (2): 591–598.

94 Xu, B., Zheng, Q., Song, Y., and Shangguan, Y. (2006). Calculating barrier properties of polymer/clay nanocomposites: effects of clay layers. *Polymer* 47 (8): 2904–2910.

95 Duncan, T.V. and Pillai, K. (2014). Release of engineered nanomaterials from polymer nanocomposites: diffusion, dissolution, and desorption. *ACS Appl. Mater. Interfaces* 7 (1): 2–19.

96 von Goetz, N., Fabricius, L., Glaus, R. et al. (2013). Migration of silver from commercial plastic food containers and implications for consumer exposure assessment. *Food Addit. Contam. Part A.* 30 (3): 612–620.

97 Cushen, M., Kerry, J., Morris, M. et al. (2012). Nanotechnologies in the food industry – recent developments, risks and regulation. *Trends Food Sci. Technol.* 24 (1): 30–46.

98 Lynch, I., Ahluwalia, A., Boraschi, D. et al. (2013). The bio-nano-interface in predicting nanoparticle fate and behaviour in living organisms: towards grouping and categorising nanomaterials and ensuring nanosafety by design. *BioNanoMaterials* 14 (3–4): 195–216.

99 Soni, D., Naoghare, P.K., Saravanadevi, S., and Pandey, R.A. (2015). Release, transport and toxicity of engineered nanoparticles. Reviews of environmental contamination and toxicology. *Rev. Environ. Contam. Toxicol.*: 1–47.

100 Khodakovskaya, M., Dervishi, E., Mahmood, M. et al. (2009). Carbon nanotubes are able to penetrate plant seed coat and dramatically affect seed germination and plant growth. *ACS Nano* 3 (10): 3221–3227.

101 Ge, Y., Schimel, J.P., and Holden, P.A. (2011). Evidence for negative effects of TiO_2 and ZnO nanoparticles on soil bacterial communities. *Environ. Sci. Technol.* 45 (4): 1659–1664.

102 Kausar, A., Rafique, I., and Muhammad, B. (2017). Aerospace application of polymer nanocomposite with carbon nanotube, graphite, graphene oxide, and nanoclay. *Polym. Plast. Technol. Eng.* 56 (13): 1438–1456.

103 Azmi, W.H., Sharif, M.Z., Yusof, T.M. et al. (2017). Potential of nanorefrigerant and nanolubricant on energy saving in refrigeration system – a review. *Renewable Sustainable Energy Rev.* 69: 415–428.

Index

Nanotechnology-Enhanced Food Packaging, First Edition. Edited by Jyotishkumar Parameswaranpillai, Radhakrishnan Edayileveettil Krishnankutty, Aswathy Jayakumar, Sanjay Mavinkere Rangappa, and Suchart Siengchin.

d